Optical Properties and Structure of Tetrapyrroles

Optical Properties and Structure of Tetrapyrroles

Proceedings of a Symposium
held at the University of Konstanz
West Germany, August 12-17, 1984

Editors
Gideon Blauer · Horst Sund

Walter de Gruyter · Berlin · New York 1985

7227-0457

Editors

Gideon Blauer, Ph. D.
Professor of Biophysical Chemistry
Institute of Life Sciences
The Hebrew University
Jerusalem, Israel

Horst Sund, Dr. rer. nat.
Professor of Biochemistry
Fakultät für Biologie
Universität Konstanz
D-7750 Konstanz, West Germany

CHEMISTRY

Library of Congress Cataloging in Publication Data

Main entry under title:
Optical properties and structure of tetrapyrroles.
Bibliography: p.
Includes indexes.
1. Tetrapyrroles--Congresses.   2. Tetrapyrroles--Optical properties--
Congresses.   I. Blauer, Gideon.   II. Sund, Horst, 1926–
QD441.067   1985   547'593   84-29791
ISBN 0-89925-009-2 (U.S.)

CIP-Kurztitelaufnahme der Deutschen Bibliothek

Optical properties and structure of tetrapyrroles : proceedings of a
symposium held at the Univ. of Konstanz, West Germany,
August 12 – 17, 1984 / ed. Gideon Blauer ; Horst Sund. –
Berlin ; New York : de Gruyter, 1985. (Berlin . . .)
ISBN 3-11-010054-1
NE: Blauer, Gideon [Hrsg.] ; Universität <Konstanz> ; Tetrapyrroles
ISBN 0-89925-009-2 (New York . . .) Gewebe

3 11 010054 1   Walter de Gruyter · Berlin · New York
0-89924-009-2 Walter de Gruyter, Inc., New York

PREFACE

The present volume contains the proceedings of an international symposium
on "Optical Properties and Structure of Tetrapyrroles" held at the University of Konstanz, West Germany, from August 12 to 17, 1984. The invited
lectures and records of the discussions are included in this publication.

Both cyclic and open-chain tetrapyrroles participate in very important life
processes such as oxygen transport, electron transport, photosynthesis and
plant development. The present Symposium is devoted to the presentation and
exchange of recent advances in this field. The topic was mainly confined to
optical properties and structural relationships and their interdependence,
because these are fundamental for an understanding of functional aspects of
tetrapyrroles. At the same time, the contributions also include advances
in the chemistry, photochemistry and photophysics of various tetrapyrroles
and their complexes with metals and proteins. Analogous considerations lso
apply to animal bile pigments such as biliverdin and bilirubin, the former
being closely related to the light-activated plant pigments of phytochrome,
phycocyanin or phycoerythrin. Despite the chemical and physical differences
between cyclic and open-chain tetrapyrroles, much can be learned from a
comparison of both. Moreover, a great deal of analogous advanced methodology has been used in recent years for the investigation of both cyclic and
open-chain tetrapyrroles. The present Symposium also demonstrates significant advances made recently in structural and functional aspects of biliproteins of algae, cyanobacteria, etc.

The interdisciplinary character of this meeting is apparent from the contents of the various contributions, ranging from physics and chemistry to
plant physiology and medicine. In this respect it resembles previous meetings held in Konstanz in 1974 (Protein-Ligand Interactions) and in 1978
(Transport by Proteins), although the present topic is more specific. It
was one of the aims of both the present and previous meetings to promote
such an interdisciplinary exchange of results and ideas. For the same reason, emphasis was laid on free discussion following each of lectures pre-

sented. These discussions were in most cases immediately put into writing
by the participants involved and are included in this volume; they were
facilitated by circulation of preprints of the invited papers among all
participants. No refereeing of the papers contributed was carried out;
their contents are the sole responsibility of the authors,and the papers
have been reproduced as submitted by them.

Obviously, not all aspects related to the topic of the Symposium could be
included in the invited lectures, and many outstanding scientists working
in this area of research were unable to take part. The present publication
should enable all workers in this or related fields who could not participate in the meeting to become familiar with the results and ideas presented,
which may stimulate further research.

We would like to acknowledge the kind and effective cooperation of the participants in this Symposium. The generous financial support of this Symposium by the Stiftung Volkswagenwerk and by the Universität Konstanz is
gratefully acknowledged. Support extended by the Gesellschaft der Freunde
und Förderer der Universität Konstanz is greatly appreciated. Thanks are
due to the staff of de Gruyter Publishers for their continued help in the
preparation of the present volume. We are particularly grateful to the
secretary of the Symposium, Mrs. H. Allen, for her very efficient and
devoted assistance in the organization of this meeting and the preparation
of this publication.

Konstanz, October 1984   G. Blauer

                                                          H. Sund

CONTENTS

List of Contributors XI

SECTION I. GENERAL
            Chairman: S. Beychok

An Historical Survey of Porphyrins, their Related Spectra, and
    their Interpretation *by A.D. Adler and M. Gouterman*   3
    Discussion   16

Routes to Functional Porphyrin Assemblies (Molecular Complexes,
    Covalent Redox Pairs, and Reactive Domains in Vesicles) *by*
    J.-H. Fuhrhop and T. Lehmann....................................   19
    Discussion   40

Organism Dependent Pathways of Tetrapyrrole Biosynthesis *by*
    D. Dörnemann and H. Senger   43
    Discussion   60

SECTION II. PORPHYRINS AND METALLOPORPHYRINS
             Chairmen: H. Rapoport, P.-S. Song and H. Zuber

Transient Absorption Spectra and Excited State Kinetics of
    Transition Metal Porphyrins *by D. Holten and M. Gouterman*   63
    Discussion   89

Metal Complexes of Tetrapyrrole Ligands. XXXVII. Structure and
    Optical Spectra of Sandwich-Like Lanthanoid Porphyrins *by*
    J.W. Buchler and M. Knoff   91
    Discussion 104

Excited State Electron Transfer of Carbonyl[mesotetra(p-tolyl)
    porphyrinato osmium(II)]Complexes. One and Two Electron Photo-
    redox Processes *by A. Vogler, J. Kisslinger and J.W. Buchler* 107
    Discussion 119

Photophysical Properties of Hematoporphyrin and its Derivatives
    and their Use in Tumor Phototherapy *by A. Andreoni and R. Cubeddu* 121
    Discussion 133

Optical Properties of Ferriheme-Quinidine Complexes *by*
    G. Blauer 137
    Discussion 152

Resonance Raman Studies of Haemoglobin *by K. Nagai* 157
    Discussion 173

Optical Properties of Heme Bound to the Central Exon Fragment
    of Globins and to Corresponding Fragments in Cytochromes *by*
    S. Beychok 175
    Discussion 200

Heme Configuration in Cytochrome c *by Y.P. Myer* 203
    Discussion 225

Correlation between Near-Infrared, ESR Spectra and Oxidation-
    Reduction Potentials in Low-Spin Heme Proteins *by*
    A. Schejter, I. Vig and W.A. Eaton 227

Circular Dichroism Contributions of Heme-Heme and Chlorophyll-
    Chlorophyll Interactions *by R.W. Woody* 239
    Discussion 256

Optical Properties of Conjugated Proteins which Transport Haem,
    Porphyrins and Bilirubin *by B. Ketterer* 261
    Discussion 276

SECTION III. BILE PIGMENTS
            Chairmen: R.P.F. Gregory, H. Sund and R.W. Woody

Light Absorption of Bilatrienes-abc and 2,3-Dihydrobilatrienes-abc
    by H. Falk 281
    Discussion 293

Mechanism of Phototherapy of Neonatal Jaundice. Regiospecific
    Photoisomerization of Bilirubins *by A.F. McDonagh and*
    D.A. Lightner 297
    Discussion 308

Effects of Environment on Photophysical Processes of Bilirubin
    by A.A. Lamola 311
    Discussion 327

The Molecular Model of Phytochrome Deduced from Optical Probes
    by P.-S. Song 331
    Discussion 345

Phytochrome, the Visual Pigment of Plants: Chromophore Structure
    and Chemistry of Photoconversion *by W. Rüdiger, P. Eilfeld*
    and F. Thümmler 349
    Discussion 364

Recent Advances in the Photophysics and Photochemistry of Small,
    Large, and Native Oat Phytochromes *by K. Schaffner,*
    S.E. Braslavsky and A.R. Holzwarth 367
    Discussion 380

Contents

Time-Resolved Fluorescence Depolarisation of Phycocyanins in
Different States of Aggregation by *S. Schneider, P. Geiselhart,
T. Mindl, P. Hefferle, F. Dörr, W. John and H. Scheer* 383

The Influence of an External Electrical Charge on the Circular
Dichroism of Chromopeptides from Phycocyanin by *Ch. Scharnagl,
E. Köst-Reyes, S. Schneider and J. Otto* 397
Discussion 407

Bilins and Bilin-Protein Linkages in Phycobiliproteins: Structural
and Spectroscopic Studies by *H. Rapoport and A.N. Glazer* 411
Discussion 421

Structural Organization of Tetrapyrrole Pigments in Light-
Harvesting Pigment-Protein Complexes by *H. Zuber* 425
Discussion 442

The Crystal and Molecular Structure of C-Phycocyanin from *Mastigocladus laminosus* and its Implications for Function and Evolution
by *T. Schirmer, W. Bode, R. Huber, W. Sidler and H. Zuber* 445
Discussion 449

SECTION IV.   CHLOROPHYLLS
         Chairmen: M. Gouterman and A.A. Lamola

Optical Spectra of Photosynthetic Reaction Centers. Theoretical
and Experimental Aspects by *A.J. Hoff* 453
Discussion 472

Circular Dichroism of Chlorophylls in Protein Complexes from
Chloroplasts of Higher Plants: A Critical Assessment by
R.P. F. Gregory 475
Discussion 488

Optical Properties and Structure of Chlorophyll RC I by
D. Dörnemann and H. Senger 489
Discussion 505

Long-Wavelength Absorbing Forms of Bacteriochlorophylls. II. Structural Requirements for Formation in Triton X-100 Micelles and
in Aqueous Methanol and Acetone by *H. Scheer, B. Paulke and
J. Gottstein* 507
Discussion 521

Concluding Remarks by *M. Gouterman* 523

Index of Contributors 525

Subject Index 527

List of Contributors

A.D. Adler, Department of Chemistry, Western Connecticut State University, Danbury, Connecticut, USA

A. Andreoni, Istituto di Fisica del Politecnico, Centro di Elettronica Quantistica e Strumentazione Elettronica del C.N.R., Milano, Italy

S. Beychok, Departments of Biological Sciences and Chemistry, Columbia University, New York, New York, USA

G. Blauer, Department of Biological Chemistry, The Hebrew University, Jerusalem, Israel

W. Bode, Max-Planck-Institut für Biochemie, Martinsried, Germany

S.E. Braslavsky, Max-Planck-Institut für Strahlenchemie, Mülheim a.d. Ruhr, Germany

J.W. Buchler, Institut für Anorganische Chemie der Technischen Hochschule Darmstadt, Darmstadt, Germany

R. Cubeddu, Istituto di Fisica del Politecnico, Centro di Elettronica Quantistica e Strumentazione Elettronica del C.N.R., Milano, Italy

D. Dörnemann, Fachbereich Biologie/Botanik der Philipps-Universität Marburg, Marburg, Germany

F. Dörr, Institut für Physikalische und Theoretische Chemie, Technische Universität München, Garching, Germany

W.A. Eaton, Laboratory of Chemical Physics, National Institute of Arthritis, Diabetes, Digestive and Kidney Diseases, National Institute of Health, Bethesda, Maryland, USA

P. Eilfeld, Botanisches Institut der Universität München, München, Germany

H. Falk, Institut für Analytische, Organische und Physikalische Chemie der Johannes-Kepler-Universität, Linz, Austria

J.-H. Fuhrhop, Institut für Organische Chemie der Freien Universität Berlin, Berlin, Germany

P. Geiselhart, Institut für Physikalische und Theoretische Chemie, Technische Universität München, Garching, Germany

A.N. Glazer, Department of Microbiology and Immunology, University of California, Berkeley, California, USA

J. Gottstein, Botanisches Institut der Universität München, München, Germany

M. Gouterman, Department of Chemistry, University of Washington, Seattle, Washington, USA

R.P.F. Gregory, Department of Biochemistry, The Medical School, University of Manchester, Manchester, England

P. Hefferle, Institut für Physikalische und Theoretische Chemie, Technische Universität München, Garching, Germany

A.J. Hoff, Department of Biophysics, Huygens Laboratory of the State University, Leiden, The Netherlands

D. Holten, Department of Chemistry, Washington University, St. Louis, Missouri, USA

A.R. Holzwarth, Max-Planck-Institut für Strahlenchemie, Mülheim a.d. Ruhr, Germany

R. Huber, Max-Planck-Institut für Biochemie, Martinsried, Germany

W. John, Botanisches Institut der Universität München, München, Germany

B. Ketterer, Courtauld Institute of Biochemistry, The Middlesex Hospital Medical School, London, England

J. Kisslinger, Institut für Anorganische Chemie, Universität Regensburg, Regensburg, Germany

M. Knoff, Institut für Anorganische Chemie, Technische Hochschule Darmstadt, Darmstadt, Germany

H.-P. Köst, Botanisches Institut der Universität München, München, Germany

E. Köst-Reyes, Institut für Physikalische und Theoretische Chemie, Technische Universität München, Garching, Germany

A.A. Lamola, AT&T Bell Labs, Murray Hill, New Jersey, USA

T. Lehmann, Institut für Organische Chemie der Freien Universität Berlin, Berlin, Germany

D.A. Lightner, Department of Chemistry and The Cellular and Molecular Biology Program, University of Nevada, Reno, Nevada, USA

A.F. McDonagh, Department of Medicine and the Liver Center, University of California, San Francisco, California, USA

T. Mindl, Institut für Physikalische und Theoretische Chemie, Technische Universität München, Garching, Germany

Y.P. Myer, Department of Chemistry, State University of New York at Albany, Albany, New York, USA

K. Nagai, MRC Laboratory of Molecular Biology, Cambridge, England

J. Otto, Institut für Physiologische Chemie der Universität München, München, Germany

B. Paulke, Botanisches Institut der Universität München, München, Germany

H. Rapoport, Department of Chemistry, University of California, Berkeley, California, USA

W. Rüdiger, Botanisches Institut der Universität München, München, Germany

K. Schaffner, Max-Planck-Institut für Strahlenchemie, Mülheim a.d. Ruhr, Germany

Ch. Scharnagl, Institut für Physikalische und Theoretische Chemie, Technische Universität München, Garching, Germany

H. Scheer, Botanisches Institut der Universität München, München, Germany

A. Schejter, Sackler Institute of Molecular Medicine, Sackler Medical School, Tel-Aviv University, Tel-Aviv, Israel

T. Schirmer, Max-Planck-Institut für Biochemie, Martinsried, Germany

S. Schneider, Institut für Physikalische und Theoretische Chemie, Technische Universität München, Garching, Germany

H. Senger, Fachbereich Biologie/Botanik der Philipps-Universität Marburg, Marburg, Germany

W. Sidler, Institut für Molekularbiologie und Biophysik, ETH-Hönggerberg, Zürich, Switzerland

P.-S. Song, Department of Chemistry, Texas Tech University, Lubbock, Texas, USA

H. Sund, Fakultät für Biologie, Universität Konstanz, Konstanz, Germany

F. Thümmler, Botanisches Institut der Universität München, München, Germany

I. Vig, Sackler Institute of Molecular Medicine, Sackler Medical School, Tel-Aviv University, Tel-Aviv, Israel

A. Vogler, Institut für Anorganische Chemie der Universität Regensburg, Regensburg, Germany

R.W. Woody, Department of Biochemistry, Colorado State University, Fort Collins, Colorado, USA

H. Zuber, Institut für Molekularbiologie und Biophysik, ETH-Hönggerberg, Zürich, Switzerland

Section I
General

Chairman: S. Beychok

AN HISTORICAL SURVEY OF PORPHYRINS, THEIR RELATED SPECTRA, AND THEIR INTERPRETATION

Alan D. Adler

Department of Chemistry, Western Connecticut State University
Danbury, CT   06810   USA

Martin Gouterman

Department of Chemistry, University of Washington
Seattle, WA   98195   USA

Introduction

Man has long been interested in the colors (spectra) of
porphyrin related materials. The red color of blood, i.e.,
heme derivatives, and the green color of plants, i.e., chlorophyll derivatives, were integral components in the reli-
gion, art, and culture of prehistoric man and remain so,
even today, for primitive hunter-gatherer, pastoral, and
agricultural societies. Nevertheless, it is only in the
last and present century that such spectra have been delineated, associated with known chemical structures, and their
biological roles elucidated. In the past, the usual historical course of events was the discovery of a physiological
phenomenon, association of this physiology with some spectral characteristics, isolation of a chemical material dis-
playing that spectral character, and then determination of
that material's structure and chemical/physical properties.
However, today various types of spectra as a component in
modern analytical methods play a more key role. Spectral
methods often reveal new structural modifications, new physiological phenomena, and are employed in determining the
molecular structure and following its activity and function.
The spectra are also frequently of prime interest in the

growing technological importance of porphyrinic materials.

Historical surveys of these matters have been abetted not
only by the increasing number of pertinent monographs,
reviews, and biographies related to these materials that
have appeared in the past three decades or so, but by two
reviews devoted specifically to historical matters (1,2).
These reflect the concomitant growth of the porphyrin-
related literature in particular and, also, the increasing
interest in the history of biochemistry in general (3,4,5).

Survey of Structures, Spectra, and Spectral Interpretation.

Selected key events associated with physiological roles of
porphyrinic materials, those associated with the elucidation of structure, with spectra, and with spectral interpretation are given in Table 1. Sources are described as
to their utility below.

Porphyrins and hemeproteins. Drabkin (1) gives a comprehensive listing of the pertinent hemeprotein literature to
1978. Adler (2) provides a more varied, but less comprehensive source for the general field prior to 1950. Gamgee
(6) is strongly recommended as a source to the early hemoglobin literature, particularly for citations of the earliest
spectroscopic work. A concise history of the determination
of the structure of porphyrins is supplied by Corwin (7).
Gouterman has furnished a comprehensive review (8) of the
general porphyrin spectroscopic literature. Recent reviews
of the heme spectroscopic literature specifically are,
also, available (9,10). Chemical Abstracts Service now
maintains a CA Selects for Porphyrins as a source to the
most recent publications.

Many of the general monographs are also useful in tracing
historical matters (11-20). A biography of Hans Fischer is
available (21) and, also, an excellent semi-autobiographi-
cal account of the early history of cell respiration by
Keilin (22). A number of more specialized monographs are
also useful (23-32). The natural occurrence of porphyrins
has been treated in detail (33). A large literature also
exists for the medical and biosynthetic related aspects of
porphyrins (34-43). A monograph is now available for the
porphyrin photosensitization literature (44). Forensic
aspects of porphyrins, mainly blood detection methods, have
also been detailed (45,46).

Chlorins. The earlier work on chlorophyll is given in
Willstätter and Stoll (47). An autobiography by Willstatter
is, also, of interest (48). A monograph on chlorophyll (49)
is also available, but most information on chlorins is in
the porphyrin literature already cited or/ in the far more
extensive literature on photosynthesis. Several of these
monographs are useful for historical purposes (50-53).
Rabinowitch contains an excellent history of photosynthesis
(51). Connolly provides a source for the recent literature
on the employment of porphyrins/chlorins in the area of
solar energy (54). The spectroscopy of chlorophyll has
also been detailed (8,55,56).

Bile Pigments. Historical information on bile pigments can
be found in Lemberg and Legge (12), Gray (57), and Petryka
(58). More extensive information on bile pigments, including spectral data, is detailed in With (59) and in the
Dolphin series (18).

Corrins. These materials have recently been treated in an
extensive monograph (60). The first volume of this monograph contains an excellent history of vitamin B12 by

Folkers. Historical information may also be gleaned from
earlier sources (61,62). The Merck book (61) is of especial
interest.

Phthalocyanines. Moser and Thomas have provided two monographs on these materials and other porphyrin-like analog
structures (63,64). A good source for the spectroscopic
information is the review by Lever (65). See, also,
Gouterman (8). Discussion of other analog structures such
as porphyrazines can also be found in the Dolphin series (18).

Spectral Interpretations. The interpretation of the spectra
of porphyrinic materials are thoroughly reviewed in
Gurinovitch (55), Platt (56), and Gouterman (8). A number
of empirical correlations, also, exist (66). They may be
summarized by noting that addition of electrons to an
orbital tends to red shift and intensify the peaks associated with it.

Table 1
Selected Historical Events for Porphyrinic Materials

400	BC	Hippocrates – described symptoms of jaundice
1000	AD	Avicenna – distinguished obstructive jaundice
1628		Harvey – established circulation of the blood
1674		Leuwenhoek – observed and described erythrocytes
1684		Boyle – combusted blood to red powder, but did not identify as an iron compound
1747		Menghini – demonstrated presence of iron in blood
1772		Priestley – noted green plants revivified air
1774		Priestley, Hooke, Mayow – all noted color changes in blood
1779		Ingen-Housz – found light required for photosyn-thesis

1782   Senebier - observed carbon dioxide required for
               photosynthesis
1794   Marabelli - discovered "Gmelin" reaction for bili-
               verdin
1795   Fontana - noted effects of cyanides on organisms
1797   Wells - proposed that iron in blood was "organized"
               and not present in simple form
1801   Fourcroy - maintained iron in blood was a phosphate
1804   DeSaussure - established role of water in photo-
               synthesis
1807   Senebier, Spallanzini - studied tissue respiration
1810   Planche - discovered horse radish roots and milk
               produced guaiacum "blue" (peroxidase test)
1812   Brande - confirmed Wells, disproved Fourcroy (cf.
               above)
1817   Pelletier and Caventou - named chlorophyll
1818   Stokes - spectroscopic studies on chlorophylls
1818   Thenard - observed certain tissues decomposed
               peroxide
1824   Combe - tentatively described pernicious anemia
1826   Tiedemann, Gmelin - rediscovered "Gmelin" reaction
               (cf. above)
1838   Lecanu - extracted hemin from blood with acid
1838   Berzelius - unsuccessful attempt to isolate chloro-
               phyll
1845   Scherer - isolated biliverdin
1845   Schönbein - initiated studies on peroxidase
1849   Addison - gave detailed description of pernicious
               anemia
1853   Teichman - crystallized hemin (microscopic forensic
               test)
1864   Staedler - isolated bilirubin from gallstones
1864   Sachs - showed carbohydrates formed by photosyn-
               thesis
1864   Berthelet - proposed water splitting in photosynthesis

1864   Hoppe-Seyler - named hemoglobin
1864   Stokes - spectroscopic studies on gas interactions
                with hemoglobin
1866   Korber - distinguished fetal from adult hemoglobin
1867   Thudichum - discovered fluoresence of porphyrins
1867   Lankester - named chlorocruorin
1868   Jaffe - described urobilin
1868   Hoppe-Seyler - developed a preparation for hemin
1868   Gamgee - produced methemoglobin by nitrite reaction
1871   Hoppe-Seyler - prepared porphyrins from blood and
                demonstrated they were pyrrole
                derivatives
1872   Pfluger - described respiratory role of blood
1873   Sorby - extracted chlorophyll from other plant
                pigments
1876   Hoppe-Seyler - correct explanation of cyanide
                toxicity
1879   Hoppe-Seyler - prepared porphyrins from chlorophyll
                and demonstrated structural simi-
                larity to heme
1881   Borodin - crystallized chlorophylls in microscopic
                studies on leaves (described as mixtures)
1884   MacMunn - described myohematin (i.e., cytochromes)
1886   Knorr - developed a pyrrole synthesis
1890   Ranking, Pardington - described symptoms of
                porphyria
1895   Stokvis - induced an animal porphyria
1896   Gamgee - first observed "Soret" bands
1896   Haldane - demonstrated photodissociation of HbCO
1897   Bertrand - introduced the term oxidase
1898   Linossier - named peroxidase
1901   Nencki, Zaleski - degraded chlorophyll to pyrroles
1901   Gamgee - demonstrated magnetic properties of various
                hemoglobin derivatives
1901   Loew - named and specified catalase

1906	Willistätter - initiated chlorophyll characterization work
1907	Braun, Tchernias - accidentally synthesized phthalocyanine
1907	Sauer - patented a vinaceous beverage from blood
1908	Hausmann - observed porphyrin photosensitization with mice
1908	Tswett - chromatography of chlorophylls
1908	Warburg - initiated his cellular respiration work
1909	Barcroft - determined Hb oxygen dissociation curves
1912	Gunther - distinguished various clinical forms of porphyria
1912	Kuster - proposed a tetrapyrrolic macrocyclic structure for porphyrins (which was not accepted)
1913	Willstätter - proposed a tetrapyrrylethylene structure for porphyrins (which was accepted)
1915	Willstätter - awarded Nobel prize for chlorophyll work
1916	Van den Berge - discovered diazo reaction of bilirubin
1918	Willstätter - isolated horse radish peroxidase
1918	Milroy - described first general metalloporphyrin synthesis
1922	Hill, Meyerhof - awarded Nobel prizes for muscle studies
1923	Hartridge, Roughton - studied kinetics of Hb oxygen reaction
1923	DeDiesbach - synthesized phthalocyanines from dinitriles
1925	Keilin - rediscovered cytochromes (cf. MacMunn)
1925	Conant, Fieser - potentiometrically established the ferri- and ferro-states of Hb (confirming Kuster)
1926	Minot, Murphy - liver theraphy for pernicious anemia

1926   Hill, Holden - reconstituted hemoglobin
1926   Fischer - synthesized porphyrins from dipyrryl-
                methanes
1926   Fox - investigated properties of chlorocruorin
1926   Fischer - synthesized octamethylporphin
1928   Warburg, Negelein - spectra of respiratory ferment
                    (oxidase)
1929   Dandridge, Drescher, Thomas - patented phthalo-
                            cyanine synthesis
1929   Fischer, Zeile - synthesis of hemin (confirming
                Kuster)
1929   Fischer - demonstrated similarities of side chains
                of hemin and chlorophyll
1929   Conant - established phytyl ester group in chloro-
                phyll
1930   Fischer - awarded Nobel prize for porphyrin work
1931   Warburg - awarded Nobel prize for respiration work
1931   Fischer - determined structure of bilirubin
1931   Van Neel - generalized photosynthesis reaction
                (in microbes)
1933   Linstead - named phthalocyanines
1933   Watson - identified stercobilins
1934   Whipple, Minot, Murphy - awarded Nobel prize for
                    anemia work
1935   Klemm - investigated magnetic properties of metallo-
                phthalocyanines
1935   Fischer, Gleim - synthesized porphin
1936   Stern, Pruckner - initiated studies on porphyrin
                spectra
1936   Robertson - Xray structure of phthalocyanine
1936   Calvin, Cookbain, Polyani - studied catalytic prop-
                            erties of phthalo-
                            cyanines
1937   Turner - observed porphyria in fox squirrels

1938   Linstead - correlation of porphyrin and phthalo-
                  cyanine type spectra
1938   Helberger - chemiluminescence of phthalocyanines
1939   Waldenstrom - isolated porphobilinogen
1939   Rothemund - bomb synthesis of TPP
1939   Clark - reviewed extensive potentiometric work on
                  metalloporphyrins
1941   Kamen - isotopically demonstrated oxygen from water
                  in photosynthesis
1941   Chance - fast flow kinetic studies on peroxidase to
                  demonstrate intermediates
1941   Drabkin - precision iron methods for hemeproteins
1942   Fischer - synthesis of bilirubin
1944   Rabinowitch - extensive review of porphyrin spectra
1944   Goddard - demonstrated cytochromes in higher plants
1946   Shemin, Rittenberg - isotopic study of porphyrin
                  biosynthesis
1946   Granick, Gilder - vinyl side chain required for iron
                  insertion into heme in microbes
1946   Corwin, Erdman - IR investigation of NH tautomerism
                  in porphyrins
1948   Horlein, Weber - associated abnormal Hb spectra
                  with various diseases
1948   Folkers, Smith - isolation and characterization of
                  B12
1948   Eley, Vartanyan - semiconductivity of phthalo-
                  cyanines
1949   Putseiko - photoelectric effect in phthalocyanines
1949   Pauling, Itano - characterized sickle cell Hb
1949   Simpson - employed free electron model for spectra
                  of porphyrins
1950   Lounget-Higgins - LCAO MO model for porphyrin
                  spectra
1951   Dorough, Miller, Huennekens - metal TPP spectra
1952   Dorough, Huennekens - metal TPC spectra

1953   Shemin, Russell – established DALA in biosynthetic
                 path
1955   Theorell – awarded Nobel prize for oxidase work
1955   Hodgkin – Xray structure of vitamin B12
1956   Platt – extensive review of theoretical work on
                 porphyrin spectra
1958   Linshitz – studied triplet decay by flash photolysis
1959   Gouterman – included configuration interaction in
                 MO models
1961   Calvin – awarded Nobel prize for photosynthesis
                 work
1962   Perutz, Kendrew – awarded Nobel prizes for X-ray
                    structure of proteins (Hb and Mb)
1964   Hodgkin – awarded Nobel prize for B12 structure
1965   Hoard, Caughey – X-ray structure of a natural
                 porphyrin
1965   Fleischer – X-ray structure of porphin
1965   Mullins, Adler, Hochstrasser – gas spectrum of TPP
1967   Adler, Longo – improved synthesis of TPP's
1969   Dolphin, Fajer, Felton, Borg – spectra of π–cation
                    radicals
1970   Gouterman, et.al. – gas spectra of porphyrins and
                    phthalocyanines
1970   Longo, Adler – heat of combustion of porphin
1970   Hoffman – preparation of coboglobins
1973   Buchler – extensive review of improved methods for
                 metallation of porphyrins
1973   Siegal, Kamin – sirohemes as prosthetic groups for
                    sulfite and nitrite reductases
1974   Horrocks – synthesis and spectra of lanthanide
                 porphyrins
1975   Horrocks – synthesis and spectra of actinide
                 porphyrins
1975   Meot-Ner, Adler – Hammett correlation to ms-
                 porphyrin behavior

1977   Gouterman, Sayre - synthesis of phosphorus porphy-
                          rins and phthalocyanines

References

1. Drabkin, D.: "Selected Landmarks in the History of Por-
    phyrins and Their Biologically Functional Derivatives"
    in The Porphyrins (D. Dolphin, editor), Vol. I, Academic
    Press, New York 1978.

2. Adler, A.: "Some Landmarks in the History of Porphyrin
    Related Materials to 1950 A.D." in Porphyrin Chemistry
    Advances (F. Longo, editor), Ann Arbor Science, Ann
    Arbor 1979.

3. Fruton, J.: Molecules and Life, Wiley-Interscience, New
    York 1972.

4. Leicester, H.: Development of Biochemical Concepts from
    Ancient to Modern Times, Harvard University Press,
    Cambridge 1974.

5. Fruton, J.: A Bio-Bibliography for the History of the
    Biochemical Sciences since 1800, American Philosophical
    Society, Philadelphia 1982.

6. Gamgee, A.: "Haemoglobin: Its Compounds and the Principal
    Products of its Decomposition" in Text-Book of Physiology
    (E. Schafer, editor), Vol. 1, Macmillan, New York 1898.

7. Corwin, A.: "The Chemistry of the Porphyrins" in Organic
    Chemistry (H. Gilman, editor), Wiley, New York 1943.

8. Gouterman, M.: "Optical Spectra and Electronic Structure
    of Porphyrins and Related Rings" in The Porphyrins (D.
    Dolphin, editor), Vol. III, Academic Press, New York 1978.

9. Adar, F.: "Electronic Absorption Spectra of Hemes and
    Hemoproteins" in The Porphyrins (D. Dolphin, editor),
    Vol. III, Academic Press, New York 1978.

10. Lever, A. and Gray, H.: Iron Porphyrins, Part 1 and 2,
    Addison-Wesley, Reading 1983.

11. Fischer, H. and Orth, H.: Die Chemie des Pyrrols, Vol. I,
    II, and III, Akademische Verlagsgesellschaft, Leipzig
    1934 (also, available from Johnson Reprint Corporation,
    New York 1968).

12. Lemberg, R. and Legge, J.: Hematin Compounds and Bile
    Pigments, Interscience, New York 1949.

13. Falk, J.: "Porphyrins and Metalloporphyrins", Elsevier,
    Amsterdam 1964.

14. Goodwin, T.: Porphyrins and Related Compounds ,
    Academic Press, London 1968.

15. Adler, A.: The Chemical and Physical Behavior of Porphyrin
    Compounds, New York Academy of Sciences, New York 1973

16. Adler, A.: The Biological Role of Porphyrins and Related
    Structures, New York Academy of Sciences, New York 1975

17. Smith, K.: Porphyrins and Metalloporphyrins, Elsevier,
    Amsterdam 1975.

18. Dolphin, D.: The Porphyrins, Vol. I-VII, Academic Press,
    New York 1978.

19. Longo, F.: Porphyrin Chemistry Advances, Ann Arbor Press,
    Ann Arbor 1979

20. Berezin, B.: Coordination Compounds of Porphyrins and
    Phthalocyanines, Wiley, New York 1981.

21. Treibs, A.: Das Leben und Wirken von Hans Fischer,
    Technische Universität München, Munich 1971.

22. Keilin, D.: The History of Cell Respiration and Cyto-
    chrome, Cambridge University Press, Cambridge 1966.

23. Roughton, F. and Kendrew, J.: Haemoglobin, Butterworths,
    London 1949.

24. Falk, J., Lemberg, R., and Morton, R.: Hematin Enzymes,
    Pergamon Press, Oxford 1961.

25. Chance, B., Estabrook, R., and Yonetani, T.: Hemes and
    Hemoproteins, Academic Press, New York 1966.

26. Marks, G.: Heme and Chlorophyll, Van Nostrand, Princeton
    1969.

27. Hayaishi, O.: Oxygenases, Academic Press, New York 1962.

28. King, T., Mason, H., and Morrison, M.: Oxidases and
    Related Redox Systems, Part I and II, Wiley, New York 1965.

29. Wikstrom, M., Krab, K., and Saraste, M.: Cytochrome
    Oxidase, Academic Press 1981.

30. Okonuki, K., Kamen, M., and Sekuzu, I.: Structure and
    Function of Cytochromes, University of Tokyo Press,
    Tokyo 1968.

31. Lemberg, R. and Barrett, J.: Cytochromes, Academic Press,
    New York 1973.

32. Sato, R. and Omura, T.: Cytochrome P-450, Academic
    Press, New York 1978.

33. Rimington, C. and Kennedy, G.: "Porphyrins: Structure,
    Distribution, and Metabolism" in Comparative Biochem-
    istry (M. Florkin and H. Mason, editors), Vol. IV,
    Academic Press, New York 1962.

34. Heilmeyer, L.: Spectrophotometry in Medicine (translated by A. Jordan and T. Tippel), Hilger, London 1943.

35. Vanotti, A.: Porphyrins (translated by C. Rimington), Hilger and Watts, London 1954.

36. Jonxis, J. and Delafresnaye, J.: Abnormal Haemoglobins, Blackwell, Oxford 1959.

37. Goldberg, A. and Rimington, C.: Diseases of Porphyrin Metabolism, Thomas, Springfield 1962.

38. Lascelles, J.: Tetrapyrrole Biosynthesis and its Regulation, Benjamin, New York 1964.

39. Heilmeyer, L.: Disturbances in Heme Synthesis (translated by M. Steiner), Thomas, Springfield 1966.

40. Mare, M.: Porphyria, British Medical Association, London 1968.

41. Doss, M.: Regulation of Porphyrin and Heme Biosynthesis, Karger, Basel 1974.

42. Doss, M.: Porphyrins in Human Disease, Karger, Basel 1976.

43. Caughey, W.: Biochemical and Clinical Aspects of Hemoglobin Abnormalities, Academic Press, New York 1978.

44. Kessel, D. and Dougherty, T.: Porphyrin Photosensitization, Plenum Press, New York 1983.

45. Allen, A.: Commercial Organic Analysis, Vol. IV, Churchill, London 1898.

46. Lee, H.: "Identification and Grouping of Bloodstains" in Forensic Science Handbook (R. Saferstein, editor), Prentice-Hall, Englewood Cliffs, 1982.

47. Willstätter, R. and Stoll, A.: Untersuchungen uber Chlorophyll, Springer, Berlin 1913. (A translation is also available: Schertz, F. and Merz, A. Science Press Printing Company, Lancaster 1928).

48. Willstätter, R.: From My Life (translated by L. Hornig), Benjamin, New York 1965.

49. Vernon, L. and Seely, G.: The Chlorophylls, Academic Press, New York 1966.

50. Baly, E.: Photosynthesis, Methuen, London 1940.

51. Rabinowitch, E.: Photosynthesis, Vol. I and II, Interscience, New York 1951.

52. Clayton, R. and Sistrom, W.: The Photosynthetic Bacteria, Plenum Press, New York 1978.

53. Clayton, R.: Photosynthesis: Physical Mechanisms and Chemical Patterns, Cambridge University Press, Cambridge 1980.

54. Connolly, J.: Photochemical Conversion and Storage of
    Solar Energy, Academic Press, New York 1981.

55. Gurinovich, G., Sevchenko, A., and Solov'ev, K.: Spec-
    troscopy of Chlorophyll and Related Compounds, U.S.
    Atomic Energy Commission Translation Series, Oak Ridge
    1971.

56. Platt, J.: "Electronic Structure and Excitation of Poly-
    enes and Porphyrins" in Radiation Biology (A. Hollaender,
    editor), McGraw-Hill, New York 1956.

57. Gray, C.: The Bile Pigments, Methuen, London 1953.

58. Petryka, Z. and Howe, R.: "Historical and Clinical
    Aspects of Bile Pigments" in The Porphyrins (D. Dolphin,
    editor), Vol. VI, Academic Press, New York 1978.

59. With, T.: Bile Pigments, Academic Press, New York 1968.

60. Dolphin, D.: Vitamin B-12, Vol. 1 and 2, Wiley-Inter-
    science, New York 1982.

61. Vitamin B-12, Merck Service Bulletin, Merck and Company,
    Rahway 1958.

62. Smith, E.L.: Vitamin B-12, Methuen, London 1960.

63. Moser, F. and Thomas, A.: Phthalocyanine Compounds,
    Reinhold, New York 1963 (ACS Monograph No. 157).

64. Moser, F. and Thomas, A.: The Phthalocyanines, Vol. 1
    and 2, CRC Press, Boca Raton 1983.

65. Lever, A.: "The Phthalocyanines" in Advances in Inorganic
    Chemistry and Radiochemistry, Vol. 7, Academic Press,
    New York 1965.

66. Meot-Ner, M. and Adler, A.: J. Amer. Chem. Soc., 97,
    5107-5111 (1975).

Received July 16, 1984

Discussion

Fuhrhop: Would you please be so kind to comment on a strong 800 nm-band
in the heme-pyrazine-heme complex?

Would the charge transfer which takes place fit your picture on innocent
and non-innocent metals in porphyrins? As I understood intermolecular
charge transfer should only occur with non-innocent metals in the porphyrin centre.

Gouterman: It would be hard for me to answer without carefully considering the orbitals of the dimer system. As a guess I would suggest a transition from the metal d to the π^* orbital of the pyrazine.

Vogler: I would like to comment on Professor Fuhrhop's remark. The 800 nm
absorption band of a Fe^{II} porphyrin dimer bridged by pyrazine (porphyrin
Fe^{II} pyrazine Fe^{II} porphyrin) is most likely due to a CT ($Fe^{II} \rightarrow$ pyrazine)
transition (see: A. Vogler, H. Kunkely, Inorg. Chim. Acta 44, L 211 (1980).

Lamola: The presence of degeneracy or nondegeneracy of the 'x' and 'y'
transitions in porphyrins and chlorins should, of course, be reflected in
the absorption and emission band polarizations. Would you comment on the
contributions of polarization measurements to the understanding of electronic transition in cyclic tetrapyrroles.

Gouterman: One of the bases for establishing the four orbital model was
the study of fluorescence polarization. I neglected to mention this in my
talk. The bands predicted to be origin bands of x and y polarization were
found to be perpendicular to one another. The bands identified as vibronic
were of mixed polarization. The molecules predicted to have x,y degeneracy
also showed this in polarization studies.

Woody: The absolute transition moment direction of the long-wavelength
band in chlorins has been determined by Steven Boxer and coworkers. They
have measured energy transfer rates between chlorins in apohemoglobin substituted by a chlorin on all four subunits or by chlorin substituted in
either the α or β subunits with heme in the other type of subunit. Their
results agree with theory within a few degrees.

Song: Is the use of the porphyrin spectroscopic notations, $Q_{x,y}$ and $B_{x,y}$,
justified for open tetrapyrroles?

Gouterman: The use of this notation is probably not justified. But it may
be convenient and causes no harm.

Song: In one of the view graphs, you used a non-degenerate energy level
diagram (for $a_{1u,2u}$ and e_g orbitals) to explain the higher intensity of
Soret bands   relative to the red bands. Is the higher Soret
intensity not explainable regardless of the degeneracy of the porphyrin
orbitals?

Gouterman: The general form for the two states can be written as:

$$B_X = \cos\theta \; \bar{X} + \sin\theta \; \bar{X}'$$
$$Q_X = -\sin\theta \; \bar{X} + \cos\theta \; \bar{X}'$$
$$R^2(B_X) = (\cos\theta + \sin\theta)^2 R^2 = (1 + \sin2\theta)R^2$$
$$R^2(Q_X) = (-\sin\theta + \cos\theta)^2 R^2 = (1 - \sin2\theta)R^2$$

In this notation, $\bar{X} = (a_{2u}, e_{gx})$ and $\bar{X}' = (a_{1u}, e_{gy})$, which are supposed to
have equal transition   dipoles R.   When a_{1u} and a_{2u} are
degenerate, $\theta = 45°$ and θ deviates from 45° as they become less   degenerate. Since $0 \leq \theta \leq 90°$, the intensity of B_X is always larger than that of Q_X.

Hoff: I would like to come back to the question why Mg is the central metal in photosynthetic pigments. It cannot be the lifetime argument you used, as charge-separation and energy transfer is much faster than the fluorescence lifetime of H_2P, etc. There are certainly a number of other reasons, perhaps you can comment on this.

Gouterman: I think a metal such as Fe, which tends to lose energy from S_1 within a few picoseconds, would never be possible. Nor would Cu, which degrades from S_1 to T_1 within a few picoseconds. But other metals such as Zn or Cd, that have S_1 lifetimes above 100 ps, would also be possible. However, the free base is probably not possible, for the central protons might do unwanted acid-base chemistry when the initial cation radical is formed. Also, the Mg plays an important rôle with ligand bonding and dimer formation.

Sund: In addition to the events on the synthesis of porphyrins, hemin, bilirubin and phthalocyanines which are listed in your table, I would like to mention the synthetic work of Woodward and Inhoffen on the synthesis of chlorophylls a and b published in 1960 and 1971: R.B. Woodward et al., J.Am.Chem.Soc. 82, 3800 (1960); H.H. Inhoffen, P. Jäger and R. Mählhop, Liebigs Ann.Chem. 749, 109 (1971).

Gouterman: Thank you for pointing these out.

ROUTES TO FUNCTIONAL PORPHYRIN ASSEMBLIES
(MOLECULAR COMPLEXES, COVALENT REDOX PAIRS,AND REACTIVE
DOMAINS IN VESICLES)

Jürgen-Hinrich Fuhrhop and Thomas Lehmann
Institut für Organische Chemie der Freien Universität Berlin
Takustr. 3, D-1000 Berlin 33

Introduction

The motivation to synthesize organized assemblies which contain photochemically and/or redox-active porphyrins can be
exemplified with an elementary reaction in photosynthesis. In
photosystem II a chlorophyll molecule is localized close to
the inner surface of a thylakoid membrane and lies there parallel to the membrane plane. A pheophytin molecule is situated
in the environment of the chlorophyll. The π-systems of both
porphyrin-type chromophores do not interact directly, since
the magnesium-free chlorin is dissolved within the hydrophobic
membrane and lies orthogonal to the membrane plane (1,2). A
third component of a redox-chain,a plastoquinone (X 320) molecule,is bound to the outer surface of the thylakoid membrane,
approximately 30 Å apart from the in -side chlorophyll (3). In
the bacterial reaction center the bacterio pheophytin and ubi-
quinone components are separated by an estimated distance of
9-12 Å (4). The alignment of this chromophore in respect to
the membrane plane is unknown. Since the quinone chromophore
is quite small, one may expect that its exact position is somewhat flexible. Irradiation of the biomembrane-bound chloro-
phyll-pheophytin-plastoquinone system (Fig. 1) with visible
light leads to a charge separation of the type $Chl\ a\ ^{+}\cdot\ -\ PQ\ ^{-}\cdot$
within less than a nanosecond (5). The pheophytin anion acts
as a bridge to the more stable quinone anion and thus contri-

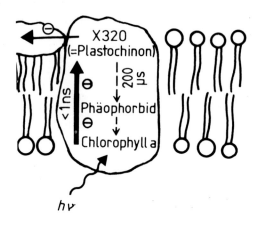

Fig. 1

Model of light induced chlorophyll⟶quinone electron transfer
in photosynthesis

butes a high-energy path through the membrane barrier.

The back reaction takes 200 µs, or 2×10^5 times longer (3). The
molecular causes of this vectorial proceeding of the reaction
are unknown. It is, however, of great interest to evaluate
these causes. The conversion of short-lived excited states into long-lived redox pairs is identical to the conversion of
light energy into storable chemical energy.

Several other applications of molecular aggregates containing
porphyrins are realized in nature: "Special pairs" of two partly overlapping chlorophyll chromophores are photoxidized in
photosystem I (6), porphyrin-steroid complexes are probably
involved in the stereoselective hydroxylation of steroids by
cytochrome P 450 (7), and heme-oxygen complexes play a prominent rôle in biological oxygen transport (8) and oxidation (9).
We shall, however, limit the discussion to systems relevant to
the problems delineated in the above paragraphs and Fig. 1.

Reported attempts to produce complex redox-systems of defined
stoichiometry and geometry fall into one of three categories:

(i) A porphyrin molecule, usually dissolved in an organic
    solvent, aggregates spontaneously with a second porphy-
    rin or a large-surface hydrocarbon, e.g. a steroid, or
    an electron-acceptor chromophore, e.g. a nitroaromat.

(ii) A porphyrin is covalently bound to a second porphyrin,
     or to a redox chromophore, e.g. a quinone, or to a redox-
     active metal complex.

(iii) A porphyrin is dissolved in micelles or bilayer lipid
      membranes, and the other redox components are dissolved
      in the aqueous phases or localized on the surface of the
      hydrophobic aggregate as head groups.

A fourth possibility, namely to arrange porphyrins and other
reactants, e.g. electron acceptors, in an artificial and well-
defined order in natural proteins, has to our knowledge not
been realized.

The given categories (i)-(iii) of organized redox-systems containing porphyrins also constitute the approximate chronologi-
cal order in the development of model systems. In the sixties
and seventies the molecular complexes and aggregates of porphyrins and chlorophylls were investigated in great detail,
mostly by spectroscopic means. The present times sees an ex-
plosion of the syntheses of covalent pairs and kinetic measurements of charge separation reactions. Experiments which try to
arrange porphyrins and other components regioselectively in
synthetic membranes are just beginning.

We shall shortly review and evaluate the results obtained with
molecular complexes and covalent pairs, and then concentrate
on new developments of synthetic membranes.

Dimers, Molecular Complexes and Aggregates

The knowledge on the geometrical and electronic structures of
these porphyrin assemblies has been summarized in the years
1976-1978 in three reviews (7, 10, 11). We point to a few general properties of these assemblies and to some more recent
results.

(i) Porphyrins of the etio-type (eight ß-pyrrolic substituents,
no methine bridge substituents) tend to form cofacial dimers
and molecular complexes with steroids. The binding force is
somewhat less than 40 kJ mol^{-1} and is mainly caused by van der
Waals type interactions (12). Steroids may block the catalytic
action of metalloporphyrins in autoxidations of olefins (13).
Other effects of complex formation on the chemical reactivity
of the components are not known to the authors.

More pronounced and therefore better known are physical changes
after the formation of molecular complexes of metalloporphyrins. NMR ring current effects and ESR-dectable metal interactions have often been applied in structural analyses. Long
wavelength-shifts in visible spectra are observed in dimers
of non-symmetric dipolar chromophores, such as chlorophylls
and protochlorophylls. An even stronger electronic interaction
is observed, if both porphyrin chromophores are oxidized to the
cation radical state. The dimer is then hold together by a

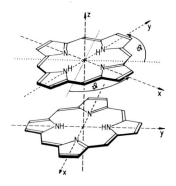

Fig. 2

Conformation of the
most stable porphin
dimer (calculated 12)

binding force of 70 kJ mol^{-1}, the unpaired electrons of the
radicals become paired and a strong, new absorption band close
to 900 nm is observed (7, 10, 11).

Light-induced charge-separation has been found, when the excited triplet and the ground states of zinc octaethylporphyrin
form an encounter complex. The initial ion-pair, ZnOEP $\cdot^+$ -
ZnOEP $\cdot^-$, recombines with rate constants in excess to 2 x 10^{11}
M^{-1} s^{-1}. It is therefore not possible to couple chemical redox
reactions to the photo-initiated electron transfer (14).

(ii) Small molecules are only tightly bound to porphyrin chromophores, if they act as electron acceptors in organic charge
transfer complexes. The organic charge transfer complexes have
again a face-to-face structure. Examples are complexes with
flavins (7), quinones (7), and nitroaromatics (15). The distance between the chromophore planes is again in the range
4-5 Å. There has, however, been one report in which a very
efficient charge transfer occurs over a distance of approximately 10 Å. In a pyrazine-bridged heme dimer light-induced
electron transfers were made responsible for a strong absorption band near 800 nm (16).

A dilute chlorophyll-quinone solution in ethanol showed the
photoinduced electron transfer from excited chlorophyll triplet states to quinone (17).

(iii) Electropositive metal ions, e.g. Mg^{2+}, bind to polarized
carbonyl substituents on the periphery of a second porphyrin
molecule. Dimers with an almost orthogonal orientation of the
porphyrin planes are thus formed (11).

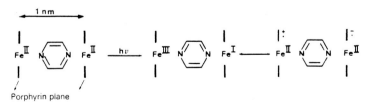

Fig. 3   Schematic structure of the pyrazine bridged heme
         dimer[16)].

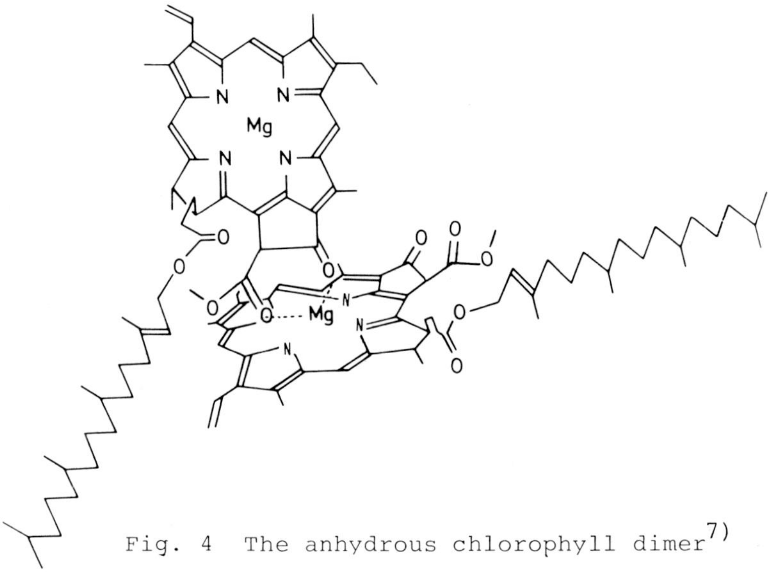

Fig. 4   The anhydrous chlorophyll dimer[7]

Such dimers can be detected and analyzed in organic solvents
by concentration dependent NMR-measurements. Addition of water
leads to extensive re-orientations of both chlorophyll chromophores in respect to each other.

Covalent Redox Pairs

In this section we shall discuss some recent experiments concerning light-induced charge transfers in bis-porphyrins and
porphyrin-quinone adducts. The interested reader will find
more detailed discussions and references to the photosynthesis
model work in the original papers.

Katz et al. have synthesized a photochlorophyllide dimer connected by an ethylene glycol unit (6). This dimer is in an
open configuration A in methylene chloride (preferably with
L = pyridine) and folds to conformation B if a hydrogen bonding
ligand such as ethanol is present.

Photoexcitation with laser pulses gave excited S_1 states in
high yields and significant triplet populations in the open
conformer A. In contrast the folded pair B exhibited an un-

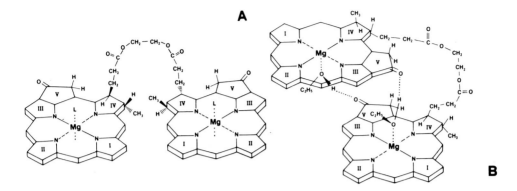

1

usually short fluorescence life time and a correspondingly low
quantum yield. One may therefore conclude that dimers with
strong electronic interactions and short porphyrin-porphyrin
distance are not useful in charge-separation experiments. The
same conclusion has obviously been reached by several other
workers, who have prepared different cofacial bis-porphyrins
with short porphyrin-porphyrin distances. No extraordinary
photochemical properties have been reported for the dimers 2
(18) and 3 (19).

anti-

2

3

It is, however, remarkable that dimer $\underline{2}$ exhibits a red-shift
of the Soret and α-bands (420 and 670 nm; explained as due to
weak incipient charge transfer interactions), whereas the
Soret band in dimer $\underline{3}$ is shifted to the blue (λ_{max} 376 nm
(M = 2H); probably due to excitonic interaction of both chromophores).

This shows that the differences in the electronic structures
of etio-type and meso-tetraphenyl porphyrins (20) also play a
role in the electronic interactions in dimers. Collman has also
used bis-metallic dimers for the 4-electron reduction of
oxygen. Accordingly dimers with short distances between the
porphyrins can be very useful in the promotion of redox-re-
actions (= electron transfer), but photo reactions are rather
quenched than promoted.

Porphyrin dimers with a long distance ($>$ 10 Å) between the
porphyrins or dimers with two perpendicular porphyrin planes
have, to our knowledge, not been described.

Experiments with covalently connected porphyrins and quinones,
e.g. $\underline{4}$, $\underline{5}$ and $\underline{6}$ have been more successful. In the cofacial
adduct $\underline{4}$ both components are held rigidly at a center-to-center
distance of 10 Å. Irradiation at 470 nm produced a porphyrin
triplet state which was quantitatively converted to a charge
separated ($P^{+}\cdot - Q^{-}\cdot$) -state ($\tau \sim$ 30 ns). The decay was about
a hundred fold slower ($\tau \sim$ 1.4 µs), (21).

The quinone adducts $\underline{5}$ and $\underline{6}$ did not show long-lived charge
separation in irradiation at room temperature. If the diamide
$\underline{5}$ and diester $\underline{6}$ however, are irradiated in frozen solvents an
ESR-spectrum arises, which corresponds to a 1:1 mixture of
porphyrin and quinone radicals. This signal is formed essentially irreversible in the diamide linked compound $\underline{5}$ whereas in
the diester linked analogon $\underline{6}$ rapid decay of the ESR signal

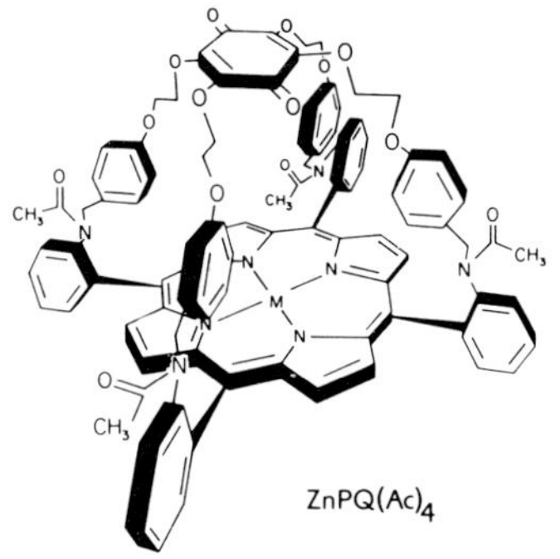

ZnPQ(Ac)$_4$

4

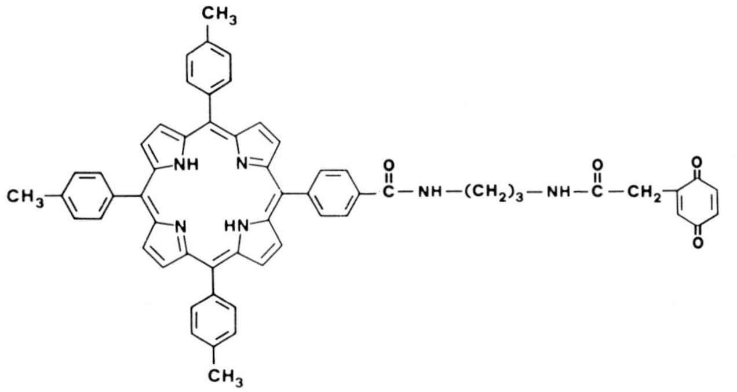

5

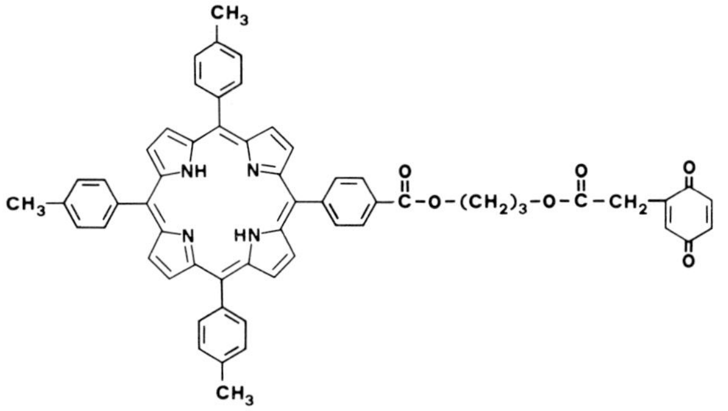

6

after cessation of the irradiation is observed. From the dipolar splitting of the signal ($\sim$ 2.0 mT) an average distance of
10 - 12 Å between the two unpaired electrons was estimated
(22).

7

The carotene-porphyrin-quinone (C-P-Q) adduct 7 is a concise structural sign for biological photosynthesis: electron donor (carotene), photo catalyst (porphyrin) and electron acceptor (quinone) are closely attached to each other. Irradiation with visible light leads to a very fast charge separation (30 ps), whereas recombination of charges is 2×10^5 times slower (60 μs) (23). This compares well with the natural system (see Fig. 1). The key to obtaining long lifetimes of the charge-separated state appears to be the interposition of a neutral porphyrin and perhaps amide bonds between widely separated ions.

Porphyrins and Reactive Domains in Vesicles

Long-lived vesicle membranes constitute a very useful frame-work for the construction of complex molecular assemblies. Nine different regions can be separately occupied by reactive molecules (24).

The water phases 1 and 9 are not organized and can only be used as reservoirs for water-soluble reagents, such as redox-active metal ions. The head group regions 3 and 7 have been

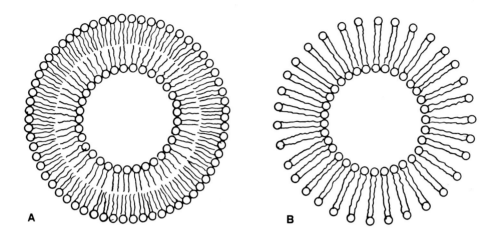

Fig. 5 Cuts through bilayer (A) and monolayer (B) membranes.
Amphiphiles with two head groups are called bolaamphiphiles.

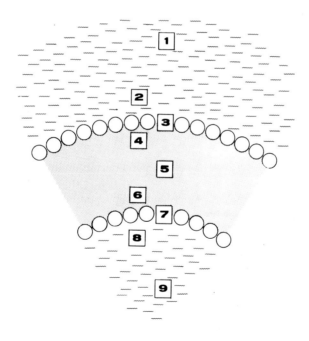

Fig. 6 The nine regions in vesicular solutions (1,9 aqueous
phases, 2,8 Gouy-Chapman layers, 3,7 head groups, 4,6 polar
part of the hydrophobic membrane, 5 hydrophobic membrane).

functionalized with several redox active chromophores, e.g.
phenylene diamine, viologen, diazonium benzene and quinone
groupings. Since the outside region 3 can be regioselectively
oxidized or reduced with water-soluble, membrane-impermeable
reagents, one can easily produce unsymmetrical membranes, e.g.
with an electron acceptor on the outside surface and a donator
on the inside.

The distance of both head group regions can be varied from
18 Å for the thinnest known monolayer membrane, to about 70 Å
for the thickest monolayer membrane. Bilayer lipid membranes
have been made up to 100 Å thick (25).

The charge of the head groups may be positive, negative or
neutral. Charged solutes in the aqueous phase may therefore be
adsorbed and localized in regions 2 and 8. Such adsorbed molecules are characterized by a tight adherence to the vesicle
membrane, e.g. in Sephadex chromatography, and by the accessibility to water-soluble reagents. An example is the adsorption
of meso-tetra (methylpyridinium)porphyrin $\underline{8}$ to dihexadecyl-
phosphate vesicles (26). The particle load on the vesicle sur-

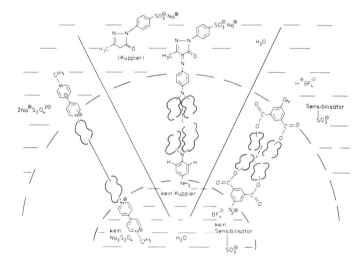

Fig. 7   Three examples of reactive head groups in vesicle
         membranes. The outside has reacted with water-soluble
         reagents, the inside has not.

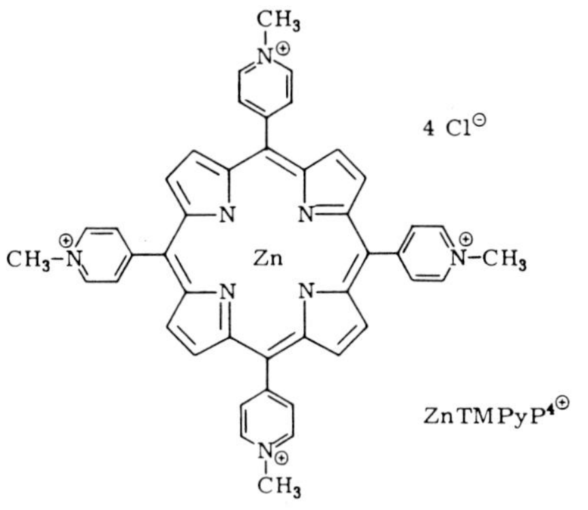

8

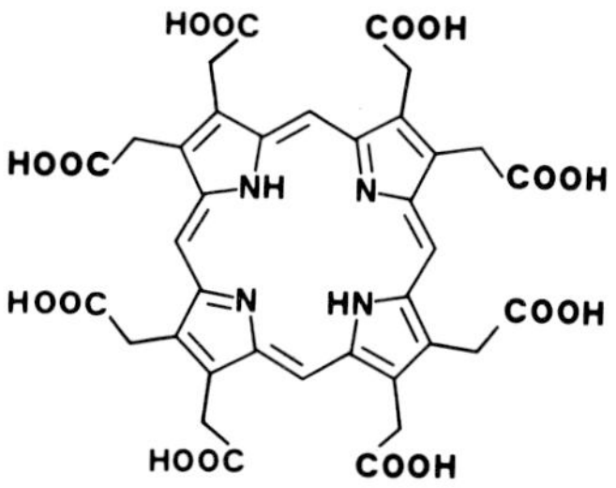

9

face should, however, not exceed a few percent, since vesicle
precipitation is observed at high porphyrin concentrations.

Regions 4 and 6 have also been be selectively occupied. This
has been demonstrated with octaacetic acid porphyrin 9 , an
analogon of uro- and coproporphyrins (24,27). If this porphyrin is dissolved in vesicular solutions at pH 7 it is located in the bulk water phase as an oligoacetate. Acidification
to pH = 5 leads to a half-neutralization of the acetate side-
chains and monoprotonation of the porphyrin centre. Further
acidification, which would lead to precipitation in pure water,
drives the neutral porphyrin base into the vesicle membrane.
From a bathochromic shift of the α-band (616$\rightarrow$ 625 nm) and
from the fact that the porphyrin base is not protonated even
at pH 1, it is concluded that porphyrin 9 has migrated into
the hydrophobic membrane. Rising the pH in the bulk aqueous
phase, leads to a quantitative migration of the porphyrin back
into the aqueous phase. The side-chains are therefore accessible to hydroxyl ions, the porphyrin's center is not accessible to protons. It should be noted, that acidification of a
porphyrin in the aqueous phase leads to deprotonation of the
porphyrin cation, which then migrates into the membrane. The
described phenomena taken together clearly indicate, that the

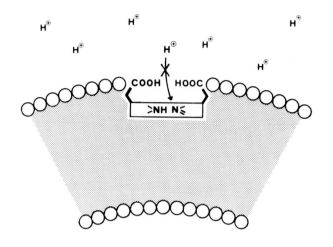

Fig. 8

The porphyrin
plane of por-
phyrin 9 is lo-
cated in the
hydrophobic
region 4 of
vesicles and
its nitrogen
atoms cannot
be protonated.

porphyrin is localized in region 4.

Region 6 can be occupied, if one sonicates porphyrin 9 and
vesicles together at pH = 7, separates the outside porphyrin
by gel chromatography and acidifies the solution.

The selective occupation of the hydrophobic region with porphyrins can be achieved in a trivial, non-specific way and in
a more sophisticated manner. The trivial kind has often be realized with water-insoluble porphyrins, such as magnesium octaethylporphyrin 10 (MgOEP). This porphyrin can be co-sonicated
with any vesicle forming amphiphile and is then dissolved in
any part of the hydrophobic membrane (region 4-6). Since MgOEP
is well soluble in petrol ether, one may assume that it is
evenly distributed over the whole width of the membrane. Experiments on black lipid membranes have indeed shown, that
MgOEP and its cation radical freely diffuse through the membrane (28-30). If such a porphyrin is dissolved in a vesicle
membrane with reactive head groups, one can assume that (i)
the average porphyrin-head group distance is about half of the
thickness of the membrane (ii) the porphyrin and the head
groups are mostly aligned perpendicular to the membrane plane.
Both generalizations will only describe an average situation.

10   11   12

Large fluctuations and overall deviations may occur.

The locality of a hydrophobic porphyrin chromophore gets more
restricted, if it bears one or two charged substituents in one
hemisphere, e.g. positively charged viologen (11) or cholyl
hydrazone (e.g. 12) derivatives (32,33).If the porphyrin 11 is
dissolved in positively charged or neutral vesicles, sodium
dithionite reduces the substituent to the radical. In negatively charged vesicles this becomes impossible. The porphyrin
ligand of 11 is totally inaccessible to protons or other water-
soluble reagents. One may conclude that the porphyrin is buried
within the hydrophobic membrane (regions 4 and 5), whereas the
bipyridinium substituent is in the head group region 3 (31).
The non-flexible bond between porphyrin and substituent should
lead to a defined orientation of the porphyrin perpendicular
to the membrane plane and a distance of about 4 Å between the
porphyrin centre and vesicle surface. Porphyrin 1 can also be
dissolved in vesicles which contain reactive head groups (see
Fig. 6). If all the outer head groups of the vesicle are active, as electron acceptors, e.g. viologens or quinones, then
the distance to the porphyrin donor would be as short as
2 Å. If the porphyrin is located at the" outside region 5"

and the reactive head groups are on the inside surface (region 7), then the porphyrin acceptor distance has a length equal to the membrane thickness minus 7 Å. It has been shown, that totally unsymmetric monolayer vesicle membranes can be prepared either from bolaamphiphiles with one large and one small head group or by regioselective precipitation. The small head groups are all at the vesicle at the small, inner surface, the large head groups are all outside (34). A water-soluble viologen bolaamphiphile vesiculates spontaneously, if titrated with perchlorate (35).

A porphyrin-loaded vesicle such as given in Fig. 9 can therefore be realistically visualized and will be a starting point for complex systems modelled after the photosystems of nature.

If one wants to work with a vesicle, in which only a few percent of the head groups are reactive, then the problem of domain formation becomes important. This problem is solved in nature by the application of protein complexes, in which all redox- and photo-reactive components are concentrated. In artificial membranes without proteins one may enforce domain formation by coulombic or charge transfer interactions at the surface. An example is the polymeric, blue charge transfer complex between viologen head groups of bolaamphiphiles and

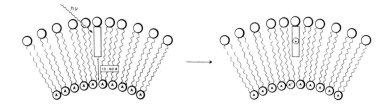

Fig. 9 A porphyrin with a hydrophilic substituent is localized
       at the outer surface of vesicle membranes (regions 3-5).
       The inner head groups (region 7) are electron acceptors
       A. Irradiation with visible light leads to charge sepa-
       ration. The electron distributes over the inner head
       group region (hypothetical).

benzidine. The formation and dissociation of this "Scheibe-type complex is highly dependent of temperature, counter-ions and pH and can thus be regulated (36). The formation of por-phyrin domains, e.g. with negatively charged porphyrin 8 parallel and positively charged porphyrins 11 or 12 perpen-dicular to the membrane should also be possible. If the lat-ter porphyrins were connected via a long chain to a polar quinone molecule on the other side of the membrane, the orga-nization of a synthetic analogue to Fig. 1 would be perfect. Another possibility is the introduction of bolaamphiphiles with one polar and one apolar chain. The apolar chain will dis-solve within the membrane, the polar chain will aggregate and form pores (37).

All presuppositions to synthesize complex assemblies on the basis of well organized vesicle membranes are nowadays ful-filled. It is only a question of time, until all the ready-made parts can be assembled in rational (and reproducible) ways. Is there any chance to make these reactive vesicles work as light-driven vectorial redox-chains and to run interesting irreversible chemical reactions with solar energy?

Multicomponent systems of a relatively low degree of organi-zation have been obtained and tested. Mauzerall has reviewed in 1975 the work with black lipid membranes, where dissolved chlorophyll produces photopotentials of a few millivolts. More recent work shows that chlorophyll derivatives, with posi-tively charged substituents give higher photovoltages (18 compared to 11 mV), because the charged chlorophyll molecules are more concentrated in the surface area than the neutral chlorophyll (30).

Calvin has extensively discussed vesicles which have EDTA as reducing agent entrapped in the vesicle aqueous volume (region 9), hydrophobic ruthenium complexes as electron carriers in regions 4-6 and methylviologen as electron acceptors together with water-soluble zinc porphyrins as sensitizers in the bulk aqueous phase. Although this arrangement with the final accep-

tor in the bulk phase should not be very efficient one obtained
some photoreduction of the viologen, indicating light-induced
electron transfer through the membrane (38). Fendler has recently given more direct evidence for this phenomenon (39).

Tollin et al. have studied the photoreaction of chlorophyll
and quinone in vesicle membranes. They found that the lifetime of the charge separated state Chl^{+} - Q^{-} can be extended
if negatively charged amphiphiles are mixed into electroneutral
vesicles. The negative surface charge presumably leads to an
expulsion of the negatively charged quinone radicals into the
water-phase. The back-reaction is significantly retarded. In
another system the chlorophyll sensitized vectorial electron
transport across a bilayer membrane was investigated. A water-
soluble quinone was entrapped in region 9, chlorophyll was
dissolved in the vesicle membrane and glutathione was added as
reducing agent to the bulk water phase (region 1). It was
stated that 20 % of all the photons absorbed by the vesicle
system resulted in electron transfer across the membrane from
glutathione to the entrapped quinone (40, 41).

Hurst et al. have adsorbed the zinc complex of the tetra-
cationic porphyrin 7 on negatively charged vesicles and
claimed the formation of a ZnP^{+} - ZnP^{-} pair with a decay
rate of $\tau = 20$ µs (26). This compares very favorably to the
20 ps observed in homogeneous solution and is even ten times
longer then with the carotene porphyrin-quinone triad.

Fendler has arranged all parts of the Shilov system for the
photoreduction of water (reducing agent, sensitizer, redox
catalyst and colloidal platinum in different compartments of
vesicles. He found again that separated units are cooperating
well through the membrane, if membrane soluble electron transporting molecules are present (42).

None of the examined systems has a high quantum yield or is
very long-lived. Better organized systems should improve the
efficiency. More stable dyes in connection with inorganic

systems preferably bound to regions 2 and 8, should provide
longevity.

Because of space and time - limitations we have only discussed
organized porphyrin systems in respect to photoinduced electron
transfer reactions. Since vesicles have also extensively been
applied in work on oxygen reactions with metalloporphyrins, we
shall give two hints on very recent work on this closely related topic. Tsuchida has synthesized vesicles with artificial
hemes which reversibly bind molecular oxygen in aqueous media
(8).   Whitten has shown that photooxidations of porphyrins in
vesicles are predominantly operating with the intermediacy
of superoxide, whereas in organic solvents most of the products arise from singlet oxygen attack (43). These examples
indicate that slow, secondary reactions of porphyrins can be
controlled as well by vesicles as the fast primary electron
transfer reactions.

Acknowledgement

The authors are grateful to the Deutsche Forschungsgemeinschaft,
the Fonds der Chemischen Industrie and the Förderungskommission
der Freien Universität für Forschung und wissenschaftlichen
Nachwuchs for current support of our research. Thanks are due
to the many graduate students who worked on our own projects,
most of whom may be identified as co-authors of papers from
this laboratory.

References

1. Mathis, P., Breton, J., Vermeglio, A., and Yates, M.:
   FEBS Lett. 63, 171-173 (1976)
2. Ganago, I.B., Klimov, V.V., Ganago, A.O., Shuvalov, V.A.,
   and Erokhin, Y.E.: FEBS Lett. 140, 127-130 (1982)
3. Witt, H.T.: Biochim. Biophys. Acta 505, 355-427 (1979)

4. Clayton, R.K., in "The Photosynthetic Bacteria";
   Clayton, R.K., Sistrom, W.R., Eds.; Plenum Press, New York,
   387-396 (1978)

5. Trissl, H.W., Gräber, P.: Biochim. Biophys. Acta 595,
   96-108 (1980)

6. Hunt, J.E., Katz, J.J., Svirmickas, A., Hindman, J.C.:
   J.Am.Chem.Soc. 106, 2242-2250 (1984)

7. Fuhrhop, J.-H.: Angew. Chem. Intern.Ed.Engl. 15, 648-659
   (1976)

8. Tsuchida, E., Nishide, H., Juasa, M.: J. Chem. Soc.,
   Chem. Commun., 96-98 (1984)

9. White, R.E., Coon, M.J.: Annu.Rev.Biochem. 49, 315-356
   (1980)

10. White, W.I.: D. Dolphin (Ed), The Porphyrins, Vol. V,
    Academic Press, New York, 303-341 (1978)

11. Katz, J.J., Shipman, L.L., Cotton, T.M., Janson, T.R.:
    J.Am.Chem.Soc., 106, 402-458 (1984)

12. Sudhindra, B.S., Fuhrhop, J.-H.: Internat. J. Quantum Chem.
    20, 747-753 (1981)

13. Fuhrhop, J.-H., Baccouche, M., Grabow, H.: J.Mol. Catal.
    7, 245-256 (1980)

14. Ballard, S.G., Mauzerall, D.C.: J.Chem.Phys. 72, 933-947
    (1980)

15. Chandrashekar, T.K., Krishnan, V.:Inorg. Chim.Acta 62, 259-
    264 (1982)

16. Fuhrhop, J.-H., Baccouche, M., Bünzel, M.: Angew. Chem.
    Internat. Ed. 19, 322-323 (1980)

17. Tollin, G.F., Castelli, G., Cheddar, G., Rizzuto, F.:
    Photochem.Photobiol. 29, 147-152 (1979)

18. Kagan, N.E., Mauzerall, D., Merrifield, R.B.: J.Am.Chem.
    Soc. 99, 5484-5486 (1977)

19. Collman, J.P., Bencosme, C.S., Branes, C.E., Miller, B.D.:
    J. Am. Chem. Soc. 105, 2704 - 2710 (1983)

20. Felton, R.H., Dolphin, D., Borg, D.C., Fajer, J.: J. Am.
    Chem. Soc. 91, 196 (1969)

21. Lindsey, J.S., Mauzerall, D., Linschitz, H.: J. Am. Chem.
    Soc. 105, 6528 - 6529 (1983)

22. Mc Intosh, A.R., Siemiarczuk, A., Bolton, J.R., Stillman,
    M.J., Ho, T.-F., Weedon, A.C.: J. Am. Chem. Soc. 105,
    7215-7223 (1983)

23. Moore, T.A., Gust, D., Mathis, P., Mialocq, J.-C.,
    Chachaty, C., Bensasson, R.V., Land, E.J., Doizi, D.,
    Liddell, P.A., Lehman, W.R., Nemeth, G.A., Moore, A.L.:
    Nature 307, 630-632 (1984)

24. Fuhrhop, J.-H., Mathieu, J.: Angew. Chem. Internat. Ed.
    Engl. 23, 100-113 (1984)

25. Fuhrhop, J.-H.: unpublished results

26. Hurst, J.K., Lee, L.Y.C., Grätzel, M.: J. Am. Chem. Soc.
    105, 7048-7056 (1983)

27. Fuhrhop, J.-H., Lehmann, T., Meding-Angrick, M.: sub-
    mitted for publication

28. Mauzerall, D., Hong, F.T.: in Smith, K. (ed), Porphyrins
    and Metalloporphyrins, Elsevier, Amsterdam 701-728 (1975)

29. Ilani, A., Mauzerall, D.: Biophys. J. 35, 79-92 (1981)

30. Krakover, T., Ilani, A., Mauzerall, D.: Biophys. J. 35,
    93-98 (1981)

31. Fuhrhop, J.-H., Wanja, U., Bünzel, M.: Liebigs Ann.Chem.
    426-432 (1984)

32. Loser, A., Mauzerall, D.: Photochem. Photobiol. 38, 355-
    361 (1983)

33. Fuhrhop, J.-H., Lehmann, T.: Liebigs Ann. Chem.
    in press

34. Fuhrhop, J.-H., Mathieu, J.: J. Chem. Soc.
    Commun. 144-145, (1983)

35. Fuhrhop, J.-H., Fritsch, D., Schmiady, H., Tesche, B.:
    J. Am. Chem. Soc. 166, 1998-2001 (1984)

36. Fuhrhop, J.-H., Fritsch, D.: J.Am.Chem.Soc., in press

37. Fuhrhop, J.-H., Liman, U.: J.Am.Chem.Soc., in press

38. Calvin, M.: Photochem. Photobiol. 37, 349-360 (1983)

39. Tricot, Y.M., Fendler, J.H.: J.Am.Chem.Soc. 106, 2477
    (1984)

40. Hurley, J.K., Castelli, F., Tollin, G.: Photochem. Photo-
    biol. 32, 79-86 (1980)

41. Ford, W.E., Tollin, G.: Photochem. Photobiol. 38, 441-449
    (1983)

42. Tunuli, M.S., Fendler, J.H.: J.Am.Chem.Soc. 103, 2507
    (1981)

43. Krieg, M., Whitten, D.G.: J.Am.Chem.Soc. 106, 2477-2479,
    (1984)

Received July 5, 1984

Discussion

Scheer: What is your evidence that the porphyrin plane in the octaacetic acid porphyrin at low pH is parallel to the plane of the membrane?

Fuhrhop: So far we have only chemical evidence. We have a C_4-symmetric porphyrin with 8 polar groups at the perimeter, and the most likely arrangement then seems to be the one shown. Anisotropic fluorescence measurements also prove that the porphyrin binds to the vesicle below pH 5 and is in solution at pH 7. Such reversibility is, in our experience, not observed, if the porphyrin ligand is buried in the hydrophobic membrane. Cooperation with physico-chemical groups on this point is planned.

Myer: I have three questions with respect to the orientation of porphyrin in the membrane:

a) How is the orientation related to structural asymmetry of the porphyrin moiety?

Fuhrhop: Asymmetry of the head group distribution does not affect the porphyrin orientation, if the porphyrin cannot enter the hydrophobic membrane region. This remark relates to porphyrins with charged side-chains and to experiments with water-soluble reagents, which are only tests to prove whether or not a porphyrin appears in the aqueous phase or is tightly bound to the hydrophobic membrane. In cases where the porphyrin can move freely in the membrane, e.g. magnesium octaethylporphyrin, we do not know anything about distribution or orientation with respect to the head groups. Experiments on light induced charge separation with different acceptor head groups may tell.

Myer: All COOH groups do not have heme pK_a. pK_a-values in membranes are quite different from those in solution.

Fuhrhop: The octaacetic acid porphyrin separates from electroneutral or negatively charged membranes at pH 6.5. It is rebound at pH 5. From this one may conclude that the average pK_a of the acetic acid side-chains is around 5 in both cases. Since we do not know how many charges are needed to make the porphyrin water-soluble, we assume that at pH 5 each pyrrole ring bears a negative charge. This assumption then implies that the pK_a is not changed dramatically on the membrane surface.

Myer: What effect, if any, has an additional substituent?

Fuhrhop: We did not synthesize derivatives of octaacetic acid porphyrin.

Falk: What is the evidence that in case of a "half polar" porphyrin embedded inside the membrane, oriented in a parallel fashion, this porphyrin comes out of the apolar layer to catch cations?

Fuhrhop: A porphyrin which is located in the inner aqueous phase of the vesicle cannot be metalated with zinc ions in the bulk aqueous phase. This indicates that zinc ions do not penetrate the membrane. Porphyrins with

two flexible polar side-chains, e.g. the Girard porphyrin, can move to the
vesicle surface and are then metalated by zinc ions in the aqueous phase.

Woody: Have you explored the limits to which the octaacetic acid porphyrin derivative can be loaded in your vesicles?

Fuhrhop: In a 10^{-3} molar dipalmitoyl phosphatidylcholate vesicular solution, about 1.5×10^{-4} moles of octaacetic acid porphyrin can be added to
the vesicle's surface at pH 4.5. This corresponds to a quantitative cover-
age of the outer surface. This is based on rough calculations comparing the
vesicle surface with porphyrin surface areas.

Woody: Do you observe any spectral changes when the system is fully
loaded?

Fuhrhop: No. Only when you go beyond the indicated limit, line broadening
occurs. It must, however, be noted that porphyrin-covered vesicles have a
shorter life time than vesicles without such cover. After a few hours one
usually observes the slow formation of precipitates.

Song: How flexible are your monomolecular layer vesicles as compared to
bilayer type vesicles? In this regard, does a porphyrin in the MLM show
a sharp NMR spectrum?

Fuhrhop: ^{1}H-NMR experiments show that the oligomethylene chains are quite
flexible at room temperature in symmetric monolayer vesicle membranes (both
end groups identical). In non-symmetric monolayer membranes (large and
small end groups) the mobility is reduced, but it is still higher than in
comparable bilayer membranes. It should be noted that monolayer membranes
did not melt below 70°C.

Porphyrins up to 10 mole per cent have only very small effects on the
^{1}H-NMR spectrum of the vesicle membrane, although about half of the oligomethylene chains should be in contact with the porphyrin. The mobility of
the porphyrin itself has not been investigated so far, because the concentration ($\leq 10^{-4}$ molar) is relatively low in our preparations. Preliminary
measurements indicated sharp signals for the methylene protons of magnesium octaethylporphyrin in vesicular solutions. The methine bridge protons
have not been detected so far.

ORGANISM DEPENDENT PATHWAYS OF TETRAPYRROLE BIOSYNTHESIS

Dieter Dörnemann and Horst Senger
Fachbereich Biologie/Botanik der Philipps-Universität
Marburg, Lahnberge, D-3550 Marburg, F.R.G.

1. The two pathways to tetrapyrroles

Tetrapyrrole biosynthesis comprises as well the formation
of non-cyclic, open chain tetrapyrroles like bile pigments
in animal tissue and phycobilins and phytochrome in the
plant kingdom, as the synthesis of cyclic tetrapyrroles
like heme in all kind of living cells, chlorophylls in
plants and corrins in certain microorganisms.

From the early studies on heme biosynthesis the intermediates of the heme pathway and the corresponding enzymes
are well known. The pathway starts with the condensation
of glycin and succinyl-CoA by ALA-synthase (Neuberger,
1961; Jordan and Shemin, 1972) to form ALA, the first
specific intermediate of all tetrapyrroles. This pathway
is called the Shemin-pathway. Subsequent condensation of
two molecules of ALA leads to porphobilinogene, a step
catalyzed by ALA-dehydratase (Shemin, 1972).

The formation of the first cyclic tetrapyrrole structure
is then performed by two enzymes, porphobilinogen deaminase
and uroporphyrinogen III cosynthase (Bogorad, 1958 a,b)
leading to uroporphyrinogen III, one of the most common
intermediates in the field of tetrapyrrole synthesis.

The next step is then the decarboxylation of uroporphyrinogen III to coproporphyrinogen III by uroporphyrinogen
decarboxylase. Oxidation of coproporphyrinogen III by
coproporphyrinogen oxidase leads to protoporphyrinogen IX,
which is subsequently oxidized by protoporphyrinogen

oxidase, under certain circumstances also nonenzymatically
by oxygen. The last step before the branching into different heme-chromophores is the incorporation of Fe^{2+} into
the molecule by heme synthase, also called ferrochelatase.

All intermediates of heme biosynthesis and the enzymes
catalyzing the interconversions are shown in Fig. 1. The
branching of this pathway, leading to the formation of
corrins and phycobilins will be discussed in detail in
one of the following chapters.

As chlorophyll, the most spread biomolecule on earth, is
also a cyclic tetrapyrrole and resembles hemes very much it
was assumed for a long time that the pathway to chlorophyll
was identical with the Shemin-pathway leading to hemes.
But when attempts were made to demonstrate the presence
of ALA-synthase in plants, the key-enzyme of the Shemin-
pathway, could not be detected there.

This lead to the conclusion that another pathway for ALA-
and thus chlorophyll-formation must be working in plants.
The existence of this new pathway to ALA and chlorophyll
was first proven by Beale and Castelfranco (1973, 1974a,b).
In these papers they clearly demonstrated by ^{14}C-labelling
experiments that the intact C-5 skeleton of glutamate or
2-oxoglutarate is incorporated into ALA without the loss
or exchange of one of the C-atoms.

From this it was generally concluded that the C-5-pathway
is restricted to plants whereas the classical Shemin-
pathway only existed in animals and bacteria. This would
mean that on one hand in plants the total amount of tetrapyrroles derives from the ALA formed via the C-5-pathway
and that on the other hand animals and bacteria cover
their tetrapyrrole requirement via the Shemin-pathway.

This assumption was first ruled out by Klein and Senger
(1978) who could demonstrate that in the pigment mutant

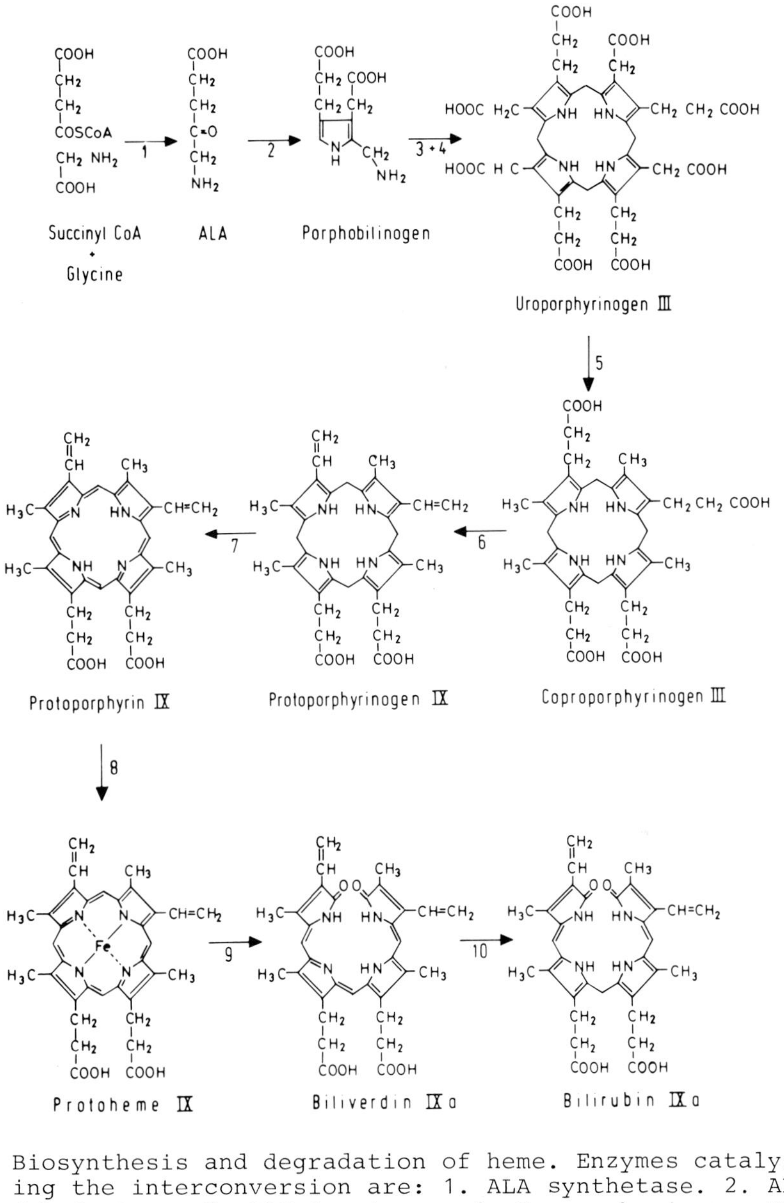

Fig.1 Biosynthesis and degradation of heme. Enzymes catalyzing the interconversion are: 1. ALA synthetase. 2. ALA
dehydratase. 3. PBG deaminase. 4. Uroporphyrinogen III
cosynthetase. 5. Uroporphyrinogen decarboxylase. 6. Coproporphyrinogen oxidase. 7. Protoporphyrinogen oxidase. 8. Heme synthetase. 9. Heme oxygenase 10. Biliverdin reductase.
(From: Heme and Hemoproteins, G.H. Tait)

C-2A' of the green alga <u>Scenedesmus obliquus</u> both pathways are working. Under the influence of levulinic acid,
a competitive inhibitor of ALA-dehydratase, great amounts
of ALA are accumulated. The accumulated intermediates are
both labelled by glutamate and 2-oxoglutarate via the
C-5 pathway and by glycin and succinate via the Shemin
pathway.

Both pathways are shown in Fig. 2 including the compartimentation of the reactions, but this will be discussed in
one of the following chapters.

By specific labelling experiments with 1- and $5-^{14}C$-gluta-
mate and identically labelled 2-oxoglutarate and subsequent cleavage of the accumulated ALA it could be excluded
that the label was incorporated via the transformation of
the C-5 compounds to succinate. In the case of $1-^{14}C$-
glutamate and $1-^{14}C$ 2-oxoglutarate the specific radioactivity in ALA compared to labelling by $5-^{14}C$-compounds
should be very low and occur by random label only, when
ALA derives from succinate. When ALA formed from $1-^{14}C$-
precursors was cleaved by periodate the label was exclusively found in the C-5 of ALA which was the C-1 of the
precursors. The presence of the Shemin-pathway in the same
organism could also be reconfirmed by labelling experiments and <u>in vitro</u> enzyme assays.

The coexistence of the two pathways in one organism lead
to reconsideration of the assumption that the pathways
are restricted to only one kind of organism. By now it is
generally accepted that both pathways possibly are present
in animals, plants and facultative photosynthetic organisms (Klein and Porra, 1982). When chlorophyll formation
was not inhibited by levulinic acid labelling of the chlorophylls via both pathways was found (Klein and Senger,
1978). Thus it was concluded that both pathways contribute
to chlorophyll formation, the intensity of contribution

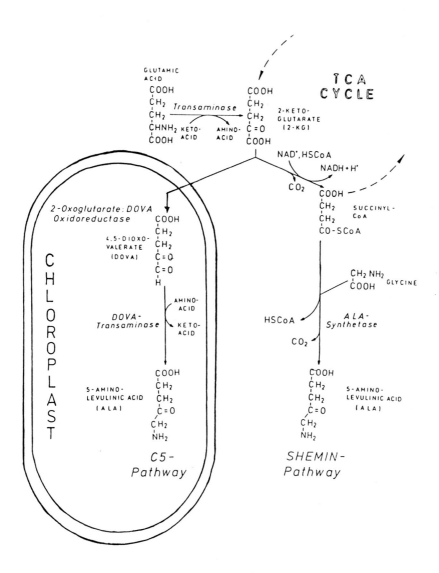

Fig.2 Compartimentation of the two pathways to ALA in plants. The C-5 pathway leads to chlorophyll formation and is locallized in the chloroplast.

being regulated by internal factors. When Oh-Hama et al.
(1982) however cultured green algae with ^{13}C-labelled precursors of both pathways, they could clearly demonstrate
by ^{13}C-NMR spectroscopy that only the C-5 pathway labelled
at the expected positions in the porphyrin skeleton of
chlorophyll, whereas glycin labelled chlorophyll only at
the methyl group of the methylester adjacent to the isocyclic pentanone ring via the normal methyl metabolism.
This result obtained with green algae could be reconfirmed
by Porra et al. (1982) for higher plants. As well as in
algae in higher plants glycin labels only via 5-adenosyl-
methionine at the methylester group attached to C-13^2.

Further evidence for the presence of both pathways was
given by the observed labelling pattern (Rühl, 1984). By
radioactive labelling with different precursors of both
pathways the results of Oh-Hama et al. (1982) and Porra
et al. (1983) could be reconfirmed and quantitatively
reinsured by this different method. The effectiveness of
labelling via the C-5 pathways could as well be demonstrated as the fact that there is no real label via the
Shemin pathway into chlorophyll, but that this incorporation occurs via the methylmetabolism. In this case the
presence of label in only the methylester position at
C-13^2 was shown by reesterification with unlabelled methanol, resulting in a strong loss of radioactive label in
the formed methylpheophorbide molecule.

A summary of these results is given in Tab. 1

Further steps of chlorophyll biosynthesis from the first
unequivocally identified precursor of chlorophyll formation, ALA, seem to be much more clear. Fig. 3 shows the
pathway from ALA to chlorophyll with the principal intermediates of this pathway. It is taken from Castelfranco
and Beale (1983).

Table I

radioactive precursors	biosynthetic pathway	Organism	ALA-accumulation (+LA) in light	Phaeophytin a (-LA)	Methyl-Phaeophorbide a (-LA)
$(1-{}^{14}C)$-Glycin	unspecific	Maize	2,36	0,105	0,101
$(2-{}^{14}C)$-Glycin	Shemin	Maize	12,23	0,346	0,135
$(1-{}^{14}C)$-Oxoglutarate	C-5	Maize	0,669	0,620	0,744
$(1-{}^{14}C)$-Glycin	unspecific	C-2A'	$0,48 \cdot 10^{-3}$	$1,95 \cdot 10^{-3}$	
$(2-{}^{14}C)$-Glycin	Shemin	C-2A'	$14,4 \cdot 10^{-3}$	$4,61 \cdot 10^{-3}$	
$(1-{}^{14}C)$-Oxoglutarate	C-5	C-2A'	$22,18 \cdot 10^{-3}$	$8,27 \cdot 10^{-3}$	

values represent µ Ci/µ mole

Tab.I Labelling of ALA, phaeophytin a and methylphaeophor-
       bide a by different ^{14}C-labelled precursors of both
       C-5- and Shemin-pathway.

After stepwise condensation of two molecules ALA to porphobilinogen (Jordan and Seehra, 1980) an instable hy-
droxymethylbilane (Battersby et al., 1979; Scott et al.,
1980) is formed by the enzyme porphobilinogen deaminase.
Inversion of ring D of the linear bilane and following
closure of the tetrapyrrole cycle to yield uroporphyrinogen III is performed by uroporphyrinogen III cosynthase
(Battersby et al., 1981). Decarboxylation of uroporphyrinogen III to coproporphyrinogen III by uroporphyrinogen
decarboxylase (Jackson et al., 1976) is followed by oxidative decarboxylation of two propionic acid groups to
vinyl groups by coproporphyrinogen III oxidase to form
protoporphyrinogen IX (Games et al., 1976). Finally protoporphyrin IX is formed by protoporphyrinogen oxidase by
the removal of six electrons (Jacobs et al., 1982). The
next steps lead to magnesium protoporphyrin IX monomethyl-
ester and/or magnesium protoporphyrin IX, catalyzed by
magnesium chelatase (Pardo et al., 1980; Richter and
Rienits, 1980, 1982,and Fuesler et al., 1981).

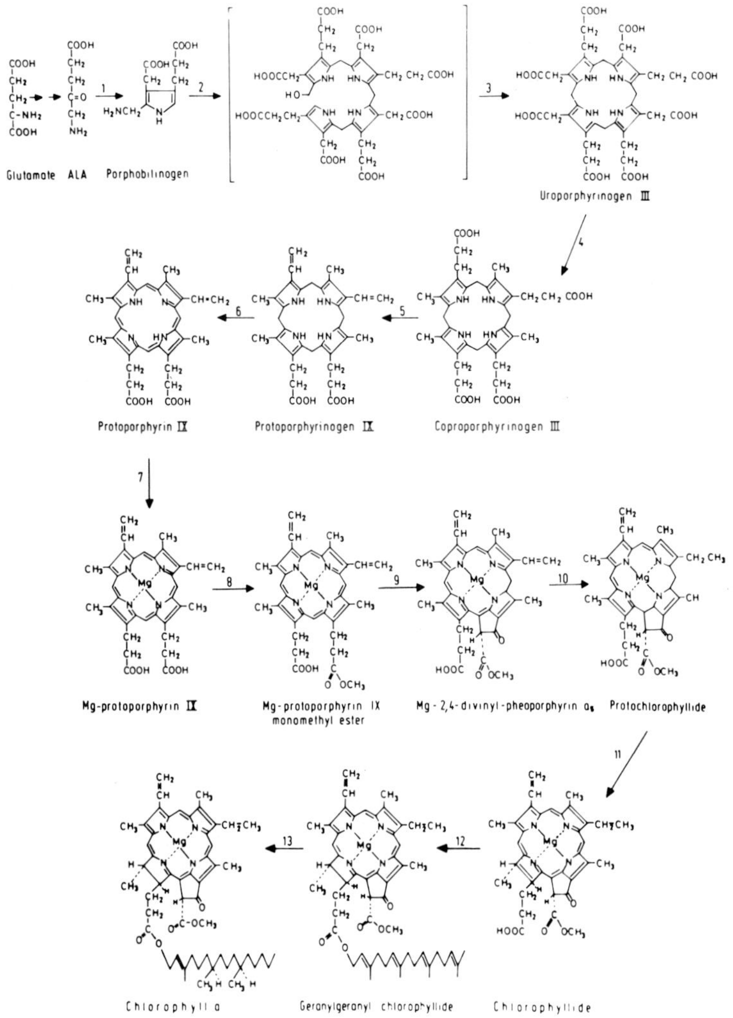

Fig.3 The intermediates of chlorophyll biosynthesis. Enzymes
      catalyzing the interconversion are: 1. ALA-dehydratase,
      2. PBG-deaminase, 3. Uroporphyrinogen cosynthase,
      4. Uroporphyrinogen decarboxylase, 5. Coproporphyrino-
      gen oxidase, 6. Protoporphyrinogen oxidase, 7./8. Mg-
      chelatase.
      The last steps are not discussed in detail.

      (From: Castelfranco and Beale in Ann.Rev. Plant
      Physiol., 1983)

The next steps, isocyclic ring formation, protochlorophyllide reduction and phytylation will not be discussed here
in detail. For review see Castelfranco and Beale (1983).

2. Intermediates of the C-5-pathway

As the formation of ALA via the Shemin-pathway is already
elucidated and there is no doubt about further steps of
tetrapyrrole biosynthesis via this pathway focus now lies
on the early intermediates of the C-5 pathway which are
still in discussion. In Fig. 4 (Dörnemann and Senger,
1980) all possible intermediates of the C-5-pathway are
presented as earlier proposed by Beale et al. (1975) but
today discussion focusses on only two of the possibilities:
the route via glutamate-1-semialdehyde as proposed by
Kannangara and Gough (1978) and the route via 4,5-dioxo-
valerate (DOVA) as demonstrated by Dörnemann and Senger
(1980).

Kannangara and Gough (1978) could show that plastids isolated from barley can transfer glutamate to ALA. 2-Oxo-
glutarate was converted at lower rate. From the ATP dependence of this conversion the authors concluded that via
glutamate- 1- phosphate glutamate-1-semialdehyde should be
formed and thus be the missing intermediate in the conversion of glutamate to ALA, although it must be stated
that glutamate-1-semialdehyde could never be isolated
from any organism.

The chemical synthesis of this compound, as described
by Kannangara and Gough (1978) was found to be impossible
by this and many other methods (Meisch et al., 1983; Kah,
1984).

From our experiments it became clear that glutamate-
semialdehyde consisted mainly of glutamate, thus only
demonstrating the conversion of glutamate to ALA by

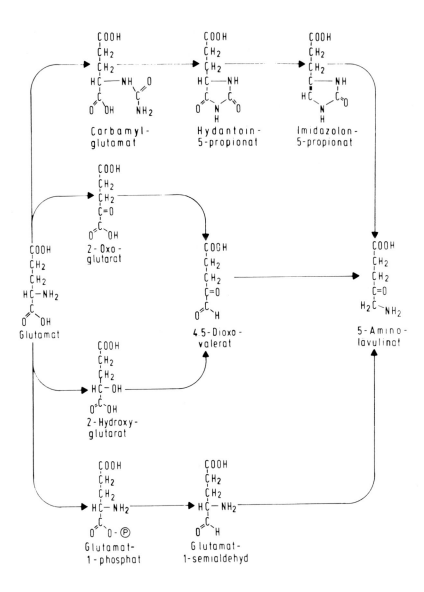

Fig.4 Various reaction sequences for the biosynthesis of 5-
aminolevulinic acid from glutamate via the C-5 pathway
as proposed by Beale et al.
(From: Dörnemann and Senger, 1980)

plastids, but not the intermediate role of glutamate-1-semialdehyde in the C-5-pathway. Harel and co-workers (1983a,b) could also demonstrate the formation of ALA in isolated plastids from etiolated maize which was before isolation illuminated for 90 min, from glutamate, 2-oxo-glutarate and 4,5-dioxovalerate. From their experiments they conclude that DOVA is formed and may be an inter-mediate with the diversion of ALA to respiratory meta-bolism.

Evidence for the intermediate role of DOVA in the C-5-path-way was given by Dörnemann and Senger (1978, 1980), who could demonstrate that under the competitive inhibition of ALA the accumulation of DOVA was paralleling the accumulation of ALA by a factor of 1/5 over a period of 24 hours in green algae. Radioactive label in DOVA was found to be 18 times higher via the C-5-pathway than by a reversible transamination of ALA formed from 2-^{14}C-la-belled glycin via the Shemin-pathway (Klein et al., 1978). Formerly DOVA-transaminase was discussed to be more or less reversible (Klein, 1978), but these experiments were per-formed with crude cell homogenates and might have included an unspecific transaminase reaction. Recent results how-ever (Kah, 1984) showed that DOVA-transaminase, purified to homogeneity, was only working in the direction of ALA, reconfirming the former results of DOVA labelling.

In these experiments it could also be demonstrated by in vitro labelling with 1-^{14}C-glutamate and 1-^{14}C-2-Oxo-glutarate that 5-^{14}C-labelled DOVA and ALA were formed. The ATP- and NADPH-dependence of the reactions was also demonstrated, NADH giving only 40% of NADPH response. Thus it has to be concluded that DOVA is the intermediate in the C-5-pathway in algae. Corresponding experiments in maize seem to point out in the same direction and will be reported soon as well.

3. Branching of tetrapyrrole biosynthesis

Formerly branching of tetrapyrrole biosynthesis was understood without any problem because only one pool of ALA
formed via the Shemin-pathway had to be reconsidered,
followed by a chain of reactions to yield all types of
tetrapyrroles. Under the viewpoint of two pathways to ALA
and the assumption of their compartimentation problems of
branching became more differentiated and questions of
possible pool interchanges had to be discussed.

A general scheme of branching of tetrapyrrole biosynthesis
is given in Fig. 5.

A first intercrossing of both pathways seems to be possible
on the level of ALA. Under certain conditions which will
be discussed in chapter 4 the ALA-pools are able to interchange, but these conditions are more or less unphysio-
logical or stress situations (Rühl, 1984). Thus the first
branching point in the prolonged Shemin-pathway is uroporphyrinogen III, the compound from which corrin biosynthesis starts (for review see Battersby and McDonald,
1982).

This very important group of tetrapyrroles includes a
great variety of very important compounds in regulation
of cell metabolism like vitamin B_{12}, also known as cyano-
cobalamine, methylcobalamin or adenosyl cobalamin, all
these compounds being derivatives of cobyrinic acid, the
key substance in cobalamin biosynthesis.

The next point of branching is reached at the level of
protoporphyrin IX. From this compound in tetrapyrrole
biosynthesis again a great variety of biologically important chromophores is derived: the group of phycobilins
(for review see McDonagh, 1979 and Bennet and Siegelman,
1979) and phytochrome (for review see Kasemir, H., 1983).

Phycobilins, non cyclic tetrapyrroles, are accessory pig-

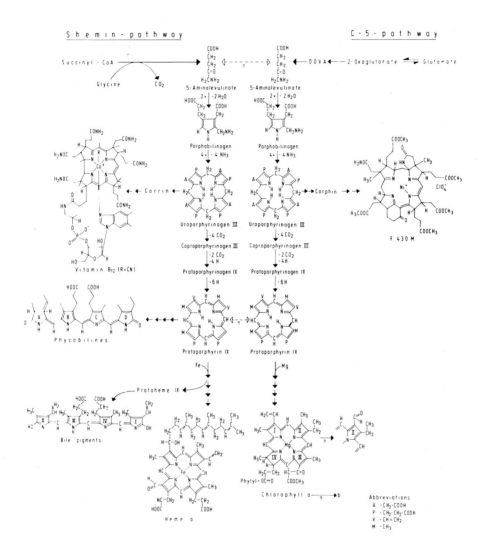

<u>Fig. 5</u> Branching of tetrapyrrole biosynthesis. Deriving from
the classical Shemin-pathway are corrins like vitamin
B$_{12}$ as well as phycobilins, bile pigments and hemes.
From the C-5 pathway F 430 in methanogenic bacteria
and chlorophylls derive.

ments in chlorophyll-b-less blue green- and red algae and
form the "light-harvesting-complex" of the algae, the phycobilisomes. Phytochrome plays a great role in photomorphogenesis of higher plants, whereas in algae blue light
effects seem to control this phenomenon. Some other lectures in this symposium will deal with these chromophores.

The C-5 pathway, in plants mainly leading to chlorophylls
shows in methanogenic bacteria a very interesting branching
to a nickel-containing tetrapyrrole. This chromophore was
called F_{430} and is directly involved in the formation of
methane in these organisms. The structure of the chromophore was clarified by the groups of Thauer and Eschenmoser (1983).

That uroporphyrinogen is the point of branching could be
shown by Gilles and Thauer (1983) by enzyme studies. It
became also clear that the carrinoid pathway up to Siro-
hydrochlorin is followed and that afterwards the corphi-
noid structure of this nickel porphyrin is build up.
Which early precursor to ALA is involved in this pathway
was also clarified by Gilles et al. (1983) who could
demonstrate that labelled succinyl-CoA via reductive
carboxylation to 2-oxoglutarate is incorporated into ALA
in this archae bacterium and is then further metabolized to
uroporphyrinogen III. As glycine did not yield significant
label in ALA and uroporphyrinogen III it has to be concluded that this new tetrapyrrole is formed via the C-5
pathway, demonstrating, that also in non photosynthetic
bacteria the C-5 pathway is working.

4. Compartimentation

As already mentioned in the previous chapters the existence
of two pathways and their branching to yield the great
variety of tetrapyrroles in all kind of living cells

suggest a compartimentation of at least the two pathways.
For the first time Kannangara and Gough (1977)could show
that isolated chloroplasts were able to form ALA and
chlorophyll from glutamate, indicating that the C-5 pathway in higher plants is located in the chloroplast. These
results from barley chloroplasts could be reconfirmed by
Harel and Ne'eman (1983) with maize plastids and with green
algae and maize chloroplasts by Rühl (1984).

As in cell-free extracts from both maize and Scenedesmus
ALA-synthase activity could be measured (Rühl, 1984) the
question of compartimentation of this ALA providing system
arose. It could be shown that in isolated maize plastids
no ALA-synthase activity was present. The assumption that
this enzyme could in plants be located in the mitochondria
could not be verified (Rühl, 1984). When mitochondria were
isolated from both organisms even by radioactive assay no
ALA-synthase activity could be detected in the mitochondria
so that it seems to be reasonable that the Shemin-pathway
is located in the cytoplasm of plant cells (Fig. 2). This
is in contradiction to animal cells where ALA-synthase is
described to be a mitochondrial enzyme (Granick and Sassa,
1971). The further enzymes up to Coproporphyrinogen III
oxidase are located in the cytoplasm. The final steps up
to heme are located again in the mitochondria.

Control and compartimentation are shown in Fig. 6, which
is taken from Tait (1979).

In plants the transport metabolite into the organelles
seems to be ALA itself, but it cannot be excluded that
also protoporphyrin IX can cross organelle membranes.

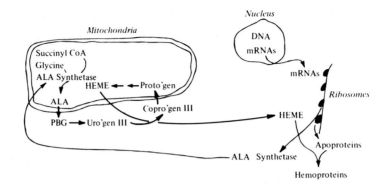

<u>Fig.6</u> Schematic representation of heme and hemoprotein bio-
synthesis showing how heme may regulate activity of
ALA synthase, formation of ALA synthase and apoprotein
moities of hemoproteins.
(From: Heme and Hemoproteins, G.H. Tait)

References

 1. Battersby, A.R., Fookes, C.J.R., McDonald, E., Matcham,
      G.W.J. (1979): Bioorg.Chem. 8,451-463.
 2. Battersby, A.R., Fookes, C.J.R., Matcham, G.W.J., Pandey,
      P.S. (1981): Angew.Chem.Int.Ed.Engl. 20,293-295.
 3. Battersby, A.R., McDonald, E. (1982): In: B$_{12}$ Vol 1,
      D. Dolphin, ed., Wiley Interscience Publication, New
      York, pp. 107-145.
 4. Beale, S.I., Castelfranco, P.A. (1973): Biochem.Biophys.
      Res.Commun. 52,143-149. Corrected in 53.
 5. Beale, S.I., Castelfranco, P.A. (1974a): Plant Physiol.
      53,291-296.
 6. Beale, S.I., Castelfranco, P.A. (1974b): Plant Physiol.
      53,297-303.
 7. Beale, S.I., Gough, S.P., Granick, S. (1975): Proc.Nat.
      Acad.Sci. USA 72,2719-2723.
 8. Bennett, A., Siegelman, H.W. (1979): In: The Porphyrins,
      D. Dolphin ed., Vol. VI, pp. 493-520. Wiley-Interscience
      Publ.
 9. Bogorad, L. (1958a): J. Biol.Chem. 233,501-509.
10. Bogorad, L. (1958b): J. Biol.Chem. 233,510-515.
11. Castelfranco, P.A., Beale, S.I. (1983): Ann.Rev.Plant
      Physiol. 34,241-278.
12. Dörnemann, D., Senger, H. (1980): Biochim.Biophys.Acta
      628,35-45.
13. Fuesler, T.P., Wright, L.A.Jr., Castelfranco, P.A. (1981):
      Plant Physiol. 67,246-249.

14. Games, D.E., Jackson, A.H., Jackson, J.R., Belcher, R.V.
    Smith, S.G. (1976): J.Chem.Soc.Chem.Commun. pp. 187-189.
15. Gilles, H., Thauer, R.K. (1983): Eur.J.Biochem. 135,109-
    112.
16. Gilles, H., Jaenchen, R., Thauer, R.K. (1983): Arch.Micro-
    biol. 135,237-240.
17. Granick, S., Sassa, S. (1971): In: Metabolic Regulation
    (Vol. V of Metabolic Pathways, 3rd ed., Vogel, H.J.,
    ed), pp. 77-141, Academic Press, New York.
18. Harel, E., Ne'eman, E., Meller, E. (1983a): Plant Physiol.
    72,1056-1061.
19. Harel, E., Ne'eman, E. (1983b): Plant Physiol. 72,1062-
    1067.
20. Jackson, A.H., Sancovich, H.A., Ferramola, A.M., Evans,
    N., Games, D.E., et al. (1976): Philos.Trans.R.Soc.
    London Ser. B 273, 191-206.
21. Jacobs, J.M., Jacoby, N.J., De Maggio, A.E. (1982): Arch.
    Biochem.Biophys. 218,233-239.
22. Jordan, P.M., Shemin, D. (1972): In: The Enzymes, Boger,
    P.D., ed., 3rd. ed., Vol. VII, pp. 339-356, Academic
    Press, New York-London.
23. Jordan, P.M., Seehra, J.S. (1980): J.Chem.Soc.Chem.Commun.
    pp. 240-242.
24. Kah, A. (1984): PhD Thesis: Philipps-Universität Marburg,
    FB Biologie/Botanik.
25. Kannangara, C.G., Gough, S.P. (1977): Carlsberg Res.
    Commun. 42,441-457.
26. Kannangara, C.G., Gough, S.P. (1978): Carlsberg Res.
    Commun. 43,185-194.
27. Kasemir, H. (1983): In Encyclopedia of Plant Physiol.,
    Shropshire, W. and Mohr, H., ed., Vol 15, pp
28. Klein, O., Senger, H. (1978): Plant Physiol. 62,10-13.
29. Klein, O., Dörnemann, D., Senger, H. (1978): In: Chloro-
    plast Development, Vol. 2, Akoyunoglou, G. and Argy-
    roudi-Akoyunoglou, J.H., eds., Elsevier/North-Holland
    Biomedical Press, Amsterdam, pp. 45-50.
30. Klein, O., Porra, R.J. (1982): Hoppe-Seyler's Z. Physiol.
    Chem. 363,551-562.
31. McDonagh, A.F. (1979): In the Porphyrins, D. Dolphin,
    ed., Vol. VI, pp. 293-491, Wiley-Interscience Publ.
32. Meisch, H.U., Maus, R. (1983): Z. Naturforsch. 38c, 563-
    570.
33. Neuberger, A. (1961): Biochem.J. 78,1-10.
34. Oh-Hama, T., Seto, H., Otake, N., Miyachi, S. (1982):
    Biochem.Biophys.Res.Comm. 105,647-652.
35. Pardo, A.D., Chereskin, B.M., Castelfranco, P.A.,
    Francheschi, V.R., Wezelman, B.E, (1980): Plant Physiol.
    56,956-960.
36. Pfalz, A., Jaun, B., Fässler, A., Eschenmoser, A. and
    Jaenchen, R., Gilles, H., Dickert, G., Thauer, R.K.
    (1983): Helv.Chim.Acta 65,828-865.
37. Porra, R.J., Klein, O., Wright, P.E. (1983): Eur.J.
    Biochem. 130,509-516.

38. Richter, M.L., Rienits, K.G. (1980): FEBS Lett. 116,211-
    216.
39. Richter, M.L., Rienits, K.G. (1982): Biochim.Biophys.Acta
    717,255-264.
40. Rühl, D. (1984): PhD Thesis: Philipps Universität Marburg,
    FB Biologie/Botanik.
41. Scott, A.I., Burton, G., Jordon, P.M., Matsumoto, H.,
    Fagerness, P.E., Pryde, L.M. (1980): J.Chem.Soc.Chem.
    Commun. pp. 384-387.
42. Shemin, A. (1972): In: The Enzymes, Boger, P.D., ed.,
    3rd ed., Vol. VII, pp. 323-337. Academic Press, New
    York-London.

Received August 12, 1984

Discussion

Scheer: There is now very good evidence that chlorophylls in higher
plants and green algae are biosynthesized by the C-5 pathway. Is there an
equally good, or any, evidence that the hemes in plants are synthesized
by the Shemin paythway?

Dörnemann: As far as I know, no. We have shown that the Shemin pathway
operates in higher plants, too, by isolation of the properly labeled Ala,
but we do not yet know the further fate of this intermediate with respect
to the formation of hemes.

Rapoport: You showed several arrows indicating phycocyanobilin biosynthesis. Is this conjecture or fact?

Dörnemann: More or less a conjecture. There is a report by McColl and
Burns which claims the proposed branching from the normal heme pathway.

Beychok: Do you know whether there is a free heme-pool and what is the
state of iron in unbound heme of the reticulocyte? Is the iron ferrous
or ferric?

Dörnemann: In a review by Tait (see Fig. 1 of the lecture) the regulation of heme biosynthesis in reticulocytes is discussed. It is described
that ALA-synthase activity is feedback-regulated by hemin which also controls the uptake of iron and thus both mechanisms possibly work together
in regulating the pool sizes of intermediates and final product. From this
it can be concluded that there should be no or only very small pools of
free heme. This is in accordance with plant regulation where also intermediates have an inhibitory effect on ALA-synthesis. When not enough
chromophore apoprotein is synthesized for some reason, chlorophyll and
heme biosynthesis are stopped by feedback inhibition, thus avoiding pools
of free pigment. In the same review iron is reported to be in the Fe^{2+} form.

Section II
Porphyrins and Metalloporphyrins

*Chairmen: H. Rapoport, P.-S. Song
and H. Zuber*

TRANSIENT ABSORPTION SPECTRA AND EXCITED STATE KINETICS OF TRANSITION METAL PORPHYRINS

Dewey Holten
Department of Chemistry, Washington University, St. Louis, MO 63130, USA

Martin Gouterman
Department of Chemistry, University of Washington, Seattle, WA 98195 USA

Introduction

Extensive experimental and theoretical investigations of the electronic
structure of metalloporphyrins, their excited states and interconversion
processes have been carried out over the past three decades (1,2). Much of
the experimental work has focused on the ground state absorption,
fluorescence emission from the lowest excited singlet, $^1Q(\pi,\pi^*)$, and
phosphorescence emission from the lowest excited triplet, $^3T(\pi,\pi^*)$.
However, there are a rather large number of transition-metal porphyrins
that can be characterized as "nonemitters". It is thought that in these
compounds the normally-emissive ring (π,π^*) states decay rapidly via low-
lying ligand-field (d,d) or ring $\longleftrightarrow$ metal charge transfer (CT) excited
states, (π,d) or (d,π^*), predicted by iterative extended Hückel (IEH)
theory (1-6). Recently an electronic taxonomy for metalloporphyrin absorption and emission behavior has been developed, that relates the optical
properties to electronic structure and excited state manifolds (1).

Evidence for ligand field and CT states of transition-metal porphyrins has
been found previously in many complexes in the ground state absorption
spectra. However, transitions to such states are usually weak or forbidden, particularly when they are in the near-infrared. Picosecond transient
absorption measurements on a number of transition-metal porphyrins have
been undertaken to help characterize these predicted low-lying ligand field

and CT excited states, both spectrally and kinetically (7-32). Some of the
most interesting findings have included (1) the implication of these states
in the photodissociation and photoassociation of basic axial ligands in
Ni^{II} (15,17,18), Co^{II} (19,26), Co^{III} (19,26), Cu^{II} (16,24) and Fe^{II} (24,30)
complexes and (2) the observation that the excited state dynamics of Os^{II}
(14,22) and Ru^{II} (28) porphyrins are strongly dependent on the σ-donating
and π-backbonding characteristics of the two axially-coordinated ligands.

In this article we give an overview of this field. Spectral and kinetic
results from our work are presented as examples of data to be found in the
literature. We point out some expectations and trends observed in the
transient absorption spectra for the (π,π) (d,d), (d,π*) and (π,d) excited
states. Some working hypotheses on the binding and release of axial
ligands are outlined. Correlations between the picosecond results and
earlier experimental and theoretical measurements are pointed out. We hope
that this discussion will be of assistance in the identification of
metalloporphyrin transient states and in the planning of future experiments.

Table 1 summarizes the metalloporphyrins that have been investigated with
picosecond transient absorption spectroscopy. Details about the porphyrin
macrocycle, axial ligands, and solvent will be mentioned where appropriate
in the discussion that follows, or in the literature cited. Picosecond
studies on the photosynthetic pigments _in vivo_ and _in vitro_ and on hemoglobin and myoglobin have been reviewed elsewhere (33).

Instrumentation

The picosecond transient absorption spectrometers used in the various
laboratories cited in Table I are conceptually similar, but differ in
detail. The apparatus used in our studies has been described elsewhere
(17). Basically, a pump pulse of ∼ 35 ps duration at 532 nm or 355 nm
excites a small region in a 2 mm pathlength optical cell through which the
sample is flowed. New wavelengths can be generated by stimulated Raman

Table I: Summary of Transition Metal Porphyrins Investigated by Picosecond
Spectroscopy

Metal ion	d^n	$[L]^*$	Reference
Mo(V)	1		31
Cr(III)	3		30
Mn(III)	4		30,32
Fe(III)	5		8,13,20,27,29,30
Ru(III)	5		28
Fe(II)	6	*	8,20,27
Ru(II)	6	*	23,28
Os(II)	6	*	14,22
Co(III)	6		19,26
Pt(IV)	6		25
Co(II)	7		11,19,26
Ni(II)	8	*	9,12,15,17,18
Pd(II)	8		7,9,25
Pt(II)	8		9,22,25,32
Cu(II)	9	*	7,10,12,16,21,24
Ag(II)	9		10,12
Zn(II)	10		9
Sn(IV)	10		7

*Evidence for either photodissociation/photoassociation of axial ligands
or of dependence of the photophysics on axial ligands.

scattering. At a time "delay" adjustable from about -500 ps to about 10 ns
a weaker ~ 35 ps "white-light" (450-900 nm) probe pulse passes through
excited and unexcited regions of the sample and is dispersed by a poly-
chromator onto two 500-channel tracks of a vidicon detector coupled to a
computer-controlled optical multichannel analyzer. The digitized information is used to construct a "transient absorption difference" spectrum or
"change in absorbance" spectrum. The spectrum represents the difference in
absorption (ΔA) between the excited or transient state and the ground
state, at the preset pump-probe time delay. Flashes of 5-10 ps duration
have been employed in many of the studies from other laboratories listed in
Table I.

General Background on Metalloporphyrin Excited States

Figure 1 presents the ground-state near-UV/visible absorption spectrum of
Ni^{II} protoporphyrin IX dimethylester in toluene. The inset and legend to
Fig. 1 give substituents and abbreviations for some common porphyrin macro-
cycles.

Porphyrin electronic states and spectra generally can be discussed within
the framework of the four-orbital model developed by Gouterman and
coworkers (1,34,35). Transitions occur from the two ring highest-filled
molecular orbitals $a_{1u}(\pi)$ and $a_{2u}(\pi)$ to the lowest-empty orbitals $e_g(\pi*)$.
Configuration interaction between the excited singlets $^1(a_{1u}(\pi), e_g(\pi*))$
and $^1(a_{2u}(\pi), e_g(\pi*))$ gives rise to the intense, near-UV $^1B(\pi,\pi*)$ band
(380-420 nm), also called the Soret band, and to the weaker, visible
$^1Q(\pi,\pi*)$ bands (500-600 nm). The two triplets don't interact and remain
relatively pure states. The lowest of them will be referred to as
$^3T(\pi,\pi*)$.

It has been observed that the ground-state absorption spectra and emission
spectra, lifetimes, and yields in metalloporphyrins are strongly influenced
by the nature of the central metal ion, the macrocycle and substituents,
and axial ligands coordinated to the metal (1-6). These factors and others

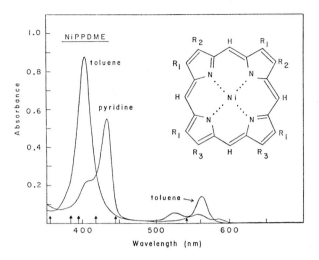

Figure 1. Ground-state absorption spectra for Ni[II] protoporphyrin IX dimethyl ester (NiPPDME) in tolune (four-coordinate Ni) and in pyridine (six-coordinate Ni). The inset shows structures for several "octaalkyl" substituted porphyrins: octaethylporphyrin (OEP), $R_1 = R_2 = R_3$ = ethyl; PPDME, R_1 = methyl, R_2 = vinyl, $R_3 = CH_2CH_2COCH_3$; mesoporphyrin IX dimethyl ester, same as PPDME, but with R_2 = ethyl; deuteroporphyrin IX dimethyl ester, as as PPDME, but with R_2 = H; etioporphyrin (Etio) R's alternating methyl and ethyl. Tetraphenylporphyrin (TPP) has $R_1 = R_2 = R_3$ = H and phenyl groups replacing H at the four meso positions. Ref. (17).

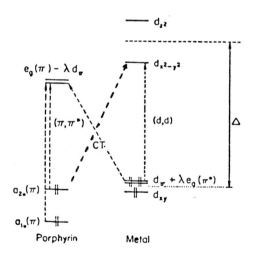

Figure 2. Schematic orbital diagram for a d^6 metalloporphyrin, Ru[II] or Os[II], showing (π,π^*) transitions on the ring, (d,d) transitions on the metal, and ring $\longleftrightarrow$ metal charge-transfer (CT) transitions. Ref. (46).

affect the relative energies of the filled and empty metal d orbitals with
respect to each other and to the highest-filled π and lowest-empty $\pi*$ orbitals on the ring. Figure 2 shows a molecular orbital diagram for a d^6
metalloporphyrin, Ru^{II} or Os^{II} (3). Gouterman and coworkers have used such
diagrams obtained from iterative extended Hückel (IEH) calculations to
rationalize variation in observed absorption and emission behavior with
metal ion, axial ligands, etc. (1,3-6). The calculations have been used in
this semiempirical spirit because they generally do not directly give an
accurate measure of excited-state energies, particularly of CT states (1).
Figure 2 illustrates $(\pi,\pi*)$ transitions on the porphyrin ring, a (d,d)
transition on the metal, and ring $\longleftrightarrow$ metal (π,d) and $(d,\pi*)$ CT transitions
(3). Knowledge of such states is crucial for understanding the picosecond
excited state dynamics of transition-metal porphyrins.

Absorption Difference Spectra of the Excited States

One of the most important ways by which metalloporphyrin excited states can
be characterized is through their optical absorption difference spectra
(the difference in absorption between the excited or transient state and
the ground state). Absorption difference spectra for each type of metalloporphyrin excited state - $(\pi,\pi*)$, (d,d), (π,d) CT, and $(d,\pi*)$ CT - appear
to exhibit, or is expected to exhibit, features that hopefully can be used
to identify them and, most importantly, to help elucidate their photophysical and photochemical behavior.

<u>$(\pi,\pi*)$ States of the Ring</u>: Metalloporphyrin $(\pi,\pi*)$ excited state spectra
appear to exhibit the following common features on the basis of picosecond
(7-32) and slower time scale [see 36-38 for examples] studies on complexes
in which the $(\pi,\pi*)$ excited states are lowest: (1) net bleaching in the
Soret or $^1B(\pi,\pi*)$ band; (2) in the 430-650 nm region, OEP complexes usually
show broad featureless absorption, broken by bleaching of the ground state
Q-bands, while TPP complexes should show a stronger new absorption with a
peak between 450 and 500 nm, tailing through the Q-band bleachings to near
650 nm; (3) one or two weak new absorption peaks between 700 and 900 nm

that may differ for singlet and triplet in a given complex.

Evidence for the second point can be seen in many of the figures shown
below. Distinction between $^1Q(\pi,\pi*)$ and $^3T(\pi,\pi*)$ in the near-infrared was
pointed out by Ponterini et al. (22) for $Pt^{II}(OEP)$, and in our recent
measurements on $Pt^{II}(TPP)$ (32) as discussed below.

(d,d) Ligand Field States: The absorption spectra of low-lying (d,d)
excited states of metalloporphyrins, formed by deactivation of higher-
energy states, appear to be similar to, but red-shifted from, the ground
state spectra. This can be rationalized because normally the porphyrin
ring has returned to its ground-state $a_{2u}(\pi)^2 a_{1u}(\pi)^2$ configuration.
Therefore, the strongest absorption by these (d,d) excited states will be
the normal four-orbital $\pi \rightarrow \pi*$ transitions. Picosecond studies on Ni^{II}
(9,12,15,17,18) and Co^{III} (19,26) porphyrins have shown absorption difference spectra in the visible, and in a few studies in the Soret region
(12,15), that have bleachings in the ground state bands with new bands of
similar shape 15-20 nm to the red. These new bands are assigned to the
corresponding transitions in the (d,d) excited state. Such low-lying (d,d)
transient states in these complexes were predicted earlier from theoretical
studies and the lack of emission from these complexes (1,6,39). They also
may be important in the picosecond photochemistry of six-coordinate Fe^{II}
porphyrins (8,27).

(π,d) and $(d,\pi*)$ Metal <-> Ring CT States: Low-lying (π,d) and $(d,\pi*)$ CT
states also are normally formed by radiationless deactivation of higher-
energy excited states; the net effect is an "internal" electron transfer
between ring and metal. Thus, the energies of these CT states and their
transient behavior are expected to be dependent on the electronic properties of the ring, central metal ion, and axial ligands. Again, most of the
oscillator strength for absorption by these excited states should come from
the porphyrin ring. Therefore, the absorption changes due to formation of
low-lying metal <-> ring (π,d) and $(d,\pi*)$ CT excited states should be similar to those observed upon production of metalloporphyrin ring $\underline{\pi\text{-cation}}$
radicals and $\underline{\pi\text{-anion}}$ radicals, respectively. Dolphin, Fajer, Felton, and

coworkers (see (40-42) and refs. therein) have studied extensively by optical and ESR techniques metalloprophyrin π-cation and π-anion radicals produced chemically or electrochemically. Important spectral features for
these states include possible strong absorption between bleachings in the
Soret and visible bands, the observation of broad absorption to the red of
the Q-band bleachings, and the distinct peaks between 650 and 950 nm found
in the π-radical spectra (40-42). Some variations occur in the π-radical
spectra depending on macrocycle, metal ion, and ground state ($^2A_{1u}$ vs. $^2A_{2u}$
for the cation radicals) (40-42). Spectral evidence from the picosecond
studies for predicted (1-6) metal ⟷ ring CT states has been reported for
Co^{II} (11,19,26), Co^{III}(19,26), Os^{II} (14,22), Ru^{II} (23,28), Pt^{IV} (25),
Ru^{III} (28), and Mo^V (31) porphryrins. Kinetic support has been given in
these cases and for the Fe^{II} and Fe^{III} (8,13,20,27,29,30), Ag^{II} (10,12) and
Cu^{II} (16,21,24) porphyrins.

Ligated and Deligated Transients: The spectra of transients resulting from
the binding or release of axial ligands upon excitation should resemble
those calculated from the ground state spectra of authentic four-, five-,
or six-coordinate species. Such results have been found for Ni^{II}
(15,17,18), Co^{II} and Co^{III} (19,26), and Fe^{II} (27) porphyrins.

Hypotheses on the Binding and Release of Axial Ligands

We wish to summarize some current views on the binding and release of axial
ligands by excited states of metalloporphyrins. These ideas help draw
together important aspects of the research and also provides a framework
for devising new experiments.
1) Photodissociation and photoassociation should occur predominantly from
the lowest excited states, because radiationless decay to them generally
appears to be extremely fast ($\leq$ 35 ps).
2) The (π,π*) and (d,π*) CT states should not be photoactive with respect
to release of σ-bonded axial ligands, but perhaps may participate in the
release of π-bonded species or those in which π-backbonding is substantial
(52). Most of the ligands we shall discuss are primarily σ-bonded.

3) Low-lying (π, d_{z^2}) CT and (d, d_{z^2}) excited states should be dissociative with respect to σ-bonded axial ligands because increasing electron
density in the d_{z^2} orbital will weaken the metal-axial-ligand bond.

4) The yields for ligand release from low-lying (d, d_{z^2}) excited states
may be larger than from (π, d_{z^2}) CT states. The latter appears to decay
more rapidly to the ground state than (π,π^*) or (d,d) states at comparable
energies, possibly because the shift in electron density from ring to metal
in the CT state may result in larger geometry changes (nuclear distortions)
and better Franck-Condon factors for radiationless deactivation.

5) Ligand binding may be enhanced following formation of (d_{z^2}, π^*) CT and
(d_{z^2}, d) states because of the removal of electron density from the d_{z^2}
orbital.

6) Wavelength-dependent photochemistry should be observed only when photoactive (d,d) or CT states lie between the ring (π,π^*) visible (Q) and
near-UV Soret (B) states.

These ideas are consistent with arguments based on calculated potential
curves (43,44) and earlier suggestions that the low (~ 5%) yield of O_2
release from hemoglobin and myoglobin as compared to the higher yield
(50-100%) of CO release observed in picosecond and slower time-scale studies may be due in part to the lowest states being CT and (d, d_{z^2}) in the
two cases, respectively (44-48,65). (Of course, in these systems geminate
recombination also appears to be important (45,46 and references therein).)
The ideas are consistent with reports of low dissociation yields of nitro-
geneous bases from simple Fe^{II} and Co^{II} porphyrins (49-52). We have
discussed some of these ideas in recent picosecond studies on the binding
and release of axially-coordinated nitrogenous bases from Ni^{II} (17,18),
Co^{II} and Co^{III} (19,26), and Fe^{II} (27) porphyrins and of CO from Ru^{II}
porphyrin π-cation radicals (23). Numerous other authors have rasied similar points.

Particular Systems

Ni^{II} complexes: Ni^{II} porphyrins are very interesting nonluminescent d^8
complexes that exhibit a variety of excited state behavior. These
complexes have been the most extensively studied by picosecond transient
absorption spectroscopy of any metalloporphyrin (9,12,15,17,18). Figure 3
shows a series of absorption difference spectra for $Ni^{II}TPP$ in three
solvents that are representative of the observations on these complexes.
We now briefly summarize these results.

In noncoordinating solvents, such as toluene, the ground state electronic
configuration for the diamagnetic four-coordinate Ni^{II} is $^1A_{1g}(d_{z^2})^2$. It
is believed that the normally-emissive lowest ring excited states, $^1Q(\pi,\pi^*)$
and $^3T(\pi,\pi^*)$, are quenched by rapid radiationless decay to low-lying
$B_{1g}(d_{z^2}, d_{x^2-y^2})$ excited states predicted by theory (6,39). Picosecond
studies on a number of Ni^{II} porphyrins in noncoordinating solvents have all
revealed absorption changes that decay with a 250-300 ps time constant
(Fig. 4A). Kobayashi et al. (9) assigned this transient as the $^1B_{1g}$, while
Chirvonyi et al. (12,15) and Kim et al. (17,18) assigned it as the $^3B_{1g}$.
These studies also have revealed an initial 10-20 ps step, the nature of
which is not fully clear.

In basic solvents such as pyridine and piperidine, Ni^{II} porphyrins can bind
two axial ligands to make a six-coordinate complex with paramagnetic $^3(d_{z^2},$
$d_{x^2-y^2})$ ground state (53-56). Evidence for a five-coordinate complex is
rarely found (53-56). The ratio of four-coordinate to six-coordinate
ground-state species depends on the ligand and its concentration, and on
the peripheral groups on the porphyrin macrocycle (17,18,53,55). For
example, $Ni^{II}TPP$ in pyridine is mainly four-coordinate, while in piperidine
it is predominantly six-coordinate, as evidenced by the ground state
absorption spectra in the visible region (Fig. 3D). Therefore, it is
essential to compare results with a variety of excitation wavelengths
where, to the extent possible, a single type of species can be pumped.
Room temperature picosecond measurements in basic solvents, such as pyridine or piperidine, have shown the following (15,17,18):

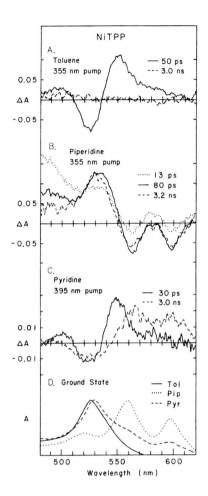

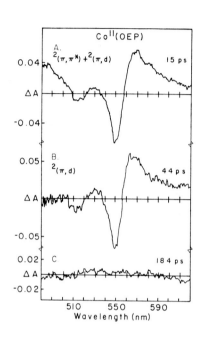

Figure 3 (left). Absorption difference spectra at various delay times following pulse excitation of NiTPP: (A) 355 nm pump in toluene, (B) 355 nm pump in piperidine, (C) 395 nm pump in pyridine. Ground-state spectra in the Q-band region are shown in panel (D). Soret-band maxima are at 416 and 433 nm for uncomplexed and complexed species, respectively. Each transient spectrum shown here represents the average of ≈ 300 spectra, with standard deviations in ΔA of < 0.01 over the wavelength regions shown. Ref. (18).

Figure 4 (right). Absorption difference spectra at three delay times following excitation of Co[II]OEP in toluene with 35 ps flashes at 355 nm. Ref. (19).

(1) A transient is observed at early delays following excitation of mainly four-coordinate Ni^{II} porphyrins (solid spectrum in Fig. 3C) that appears to be (d_{z^2}, $d_{x^2-y^2}$) on the basis of comparison with the spectrum so assigned in toluene (Fig. 3A). At longer delay times a second transient is observed (dashed spectrum in Fig. 3C) that is assigned to the six-coordinate complex on the basis of comparison with the ground state spectrum of this species in piperidine (Fig. 3D). This longer-lived transient persists for > 20 ns. Kim et al. (17,18) have interpreted these observations to mean that two ligands are bound in the $^3(d_{z^2}, d_{x^2-y^2})$ excited state of the four-coordinate species. Thus, on a 10 ns time scale the net conversion $^1A_{1g} \rightarrow {}^3B_{1g}(L)_2$ has occurred.

(2) Excitation of six-coordinate Ni^{II} porphyrins results in the rapid appearance ($\leq$ 35 ps) of a transient assigned to the four-coordinate species (Fig. 3B). However, an earlier transient can be resolved during the excitation flash that exhibits strong absorption increasing in strength below 500 nm expected for a (π,π^*) excited state (dotted spectrum in Fig. 3B). The spectrum due to the four-coordinate complex in neat piperidine persists for > 20 ns (dashed spectrum in Fig. 3B). Kim et al. (17,18) proposed that dissociation occurs mainly from the low-lying $^1(d_{z^2})^2$ excited state of the six-coordinate complex that is too short-lived to be observed. Perhaps this transient will be found at low temperature. Thus, the effect of pumping the six-coordinate species is the net conversion on the 10 ns time scale $^3B_{1g}(L)_2 \rightarrow {}^1A_{1g}$. On these interpretations, the absorption difference spectra at long delays in pyridine and piperidine should be mirror images (about $\Delta A = 0$) of each other, as is observed (dashed spectra in Figs. 3B and 3C).

Co^{II} complexes: The lowest (π,π^*) excited states of these d^7 complexes are $^2T(\pi,\pi^*)$ and $^4T(\pi,\pi^*)$. They do not emit (1,6). We have investigated the excited state relaxations of four-coordinate Co^{II}(OEP) and Co^{II}(TPP) in toluene (19,26). Absorption changes in the 480-630 nm region were assigned to a (π,π^*) feeding a $^2(a_{2u}(\pi), d_{z^2})$ CT (Fig. 4A), in agreement with recent IEH calculations (6). The decays are complete by 50 ps. The short lifetimes are in agreement with another picosecond study on Co^{II}(TPP) using 7 ps flashes (11). The lack of emission from these complexes, therefore, is

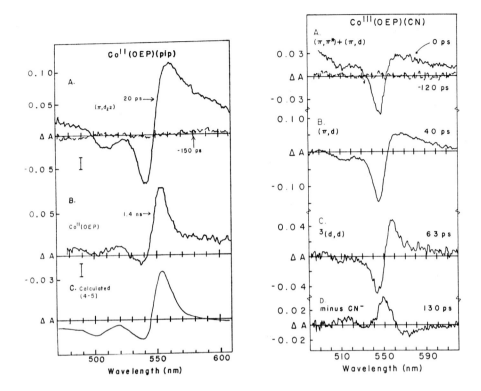

Figure 5 (left). Absorption difference spectra (A and B) at two delay times following excitation of Co[II](OEP)(pip) in 0.1 M piperidine in toluene with 35-ps flashes at 355 nm. The difference spectrum shown in (C) was constructed by substracting the ground state spectrum of the five-coordinate Co[II](OEP)(pip), obtained by dissolving Co[II](OEP) in 0.1 M piperidine in toluene, from the ground state spectrum of the four-coordinate Co[II](OEP) in toluene; both solutions had the same porphyrin con-centration. Ref. (26). The maximum error across the wavelength region shown is given by error bars at the left.

Figure 6 (right). Absorption difference spectra for Co[III](OEP)CN in CH_2Cl_2 produced by excitation with 35 ps flashes at 355 nm. Ref. (19).

attributed to rapid deactivation of the $^2T(\pi,\pi*)$ to the $^2(\pi,d)$ CT, which relaxes to the ground state in $\leq$ 35 ps.

We also have investigated $Co^{II}(OEP)$ and $Co^{II}(TPP)$ in the presence of piperidine (26). Predominantly five- or six-coordinate species can be obtained depending on the porphyrin and piperidine concentration. Figure 5 shows representative results obtained upon excitation of the pentacoordinate $Co^{II}(OEP)(pip)$ in 0.1 M piperidine. The absorption difference spectrum over the 480-630 nm region during the 35 ps, 355 nm excitation flash (Fig. 5A) was ascribed mainly to the $^2(\pi, d_{z^2})$ CT, possibly with some $(\pi,\pi*)$ contribution (compare with Fig. 4A). The spectrum after the flash (Fig. 5B) was assigned to the deligated four-coordinate complex, upon comparison with the difference spectrum calculated for this process from ground state absorptions of the authentic species (Fig. 5C). Photolysis of the hexacoordinate $Co^{II}(TPP)(pip)_2$ led to the release of one ligand to form the five-coordinate complex (26). In all cases the time constant for rebinding of the ligand to the dissociated species depended on the ligand concentration. These values gave bimolecular rate constants of $\sim 10^9$ M^{-1} s^{-1}, similar to those obtained in a previous slower time scale study of $Co^{II}(MPDME)$ in very dilute basic (ligand) solutions, in which photodissociation yields of $\sim$ 5% were reported (51). We attributed the low yields of ligand release from the Co^{II} complexes as being due to rapid deactivation competing with dissociation from the (π, d_{z^2}) CT. Rapid geminate recombination appeared less likely from the data, although some contribution from this process could not be ruled out.

Co^{III} Complexes: We have reported on the transient behavior of $Co^{III}(OEP)(CN)$ in CH_2Cl_2 excited with 35 ps, 355 nm flashes (19,26). Absorption changes shown in Fig. 6A-6C have been assigned as indicated to the chain of transient states $(\pi,\pi*) \rightarrow (a_{2u}(\pi), d_{z^2}) \rightarrow (d_\pi, d_{z^2})$ on the basis of resemblances with the absorption changes expected for these states, as discussed above, and predictions of the low-lying states from IEH calculations by Antipas and Gouterman (6). The (d_π, d_{z^2}) spectrum was found to give way with a time-constant of 20-40 ps to a long-lived (> 5 ns relaxation) transient species (Fig. 6D) that was ascribed to formation of the CN^--free $(Co^{III}(OEP))^+$. A quantum yield of > 0.2 has been reported previously for release of CN^- from Co^{III} horseradish peroxidase (52).

We have investigated the transient behavior of some six-coordinate Co^{III} complexes (19,26). We have studied mainly $Co^{III}(OEP)(L)_2$ with L = piperidine or DMSO, prepared by dissolving $Co^{III}(OEP)(CN)$ in concentrated solutions of the ligand (L). The authentic $Co^{III}(OEP)(CN)(\gamma\text{-picoline})$ complex in CH_2Cl_2 has been studied also. Excitation of a predominantly six-coordinate complex with either 355 or 532 nm flashes gave absorption difference spectra in the 480-630 nm region that were assigned to $(\pi,\pi^*) + (\pi, d_{z^2})$ excited states during the flash, and at successively longer delay times to five-coordinate and then four-coordinate species. Again the spectrum for the pentacoordinate transient agrees very well with that calculated from spectra of the authentic compounds. Rebinding of a ligand (DMSO or piperidine) to the five-coordinate complex occurs at $\sim 10^9$ M^{-1} s^{-1}, while rebinding of two ligands to the four-coordinate transient is much slower, possibly because the ground state of $(Co^{III}(OEP))^+$ may be a high-spin (d_π, d_{z^2}) triplet (26,43b). There was no clear spectral evidence for a (d,d) transient for any of these six-coordinate Co^{III} complexes as observed for $Co^{III}(OEP)(CN)$ in Fig. 6C. It was suggested that the dissociative state probably remained (d_π, d_{z^2}), but that release of the nonionic ligands may be too fast to prevent its observation under the conditions employed. Possibly the (d,d) will be resolved at low temperature.

Cu^{II} complexes: Cu^{II} has a d^9 electronic configuration and a half-filled $d_{x^2-y^2}$ orbital. In noncoordinating solvents these Cu^{II} porphyrins exhibit a temperature-dependent (spectra, lifetimes, and yields) luminescence from a "tripmultiplet" manifold of states that arises from coupling of the $d_{x^2-y^2}$ odd electron with the lowest ring excited $^3(\pi,\pi^*)$ state. The normal (π,π^*) ring singlets become doublets (1,4,57-59).

Picosecond studies on several Cu^{II} porphyrins in noncoordinating solvents indicate that deactivation of the singdoublet $^2Q(\pi,\pi^*)$ to the tripdoublet $^2T(\pi,\pi^*)$ occurs in < 10 ps followed by establishment of the $^2T(\pi,\pi^*) \longleftrightarrow$ $^4T(\pi,\pi^*)$ equilibrium in ≤ 500 ps (10,12,24). The latter levels are split

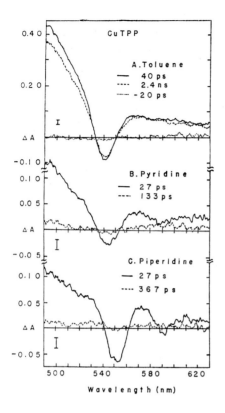

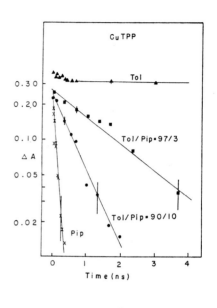

Figure 7 (left). Absorption difference spectra in the Q-band region for Cu(TPP) in toluene (A), pyridine (B), and piperidine (C) at short and long time delays following excitation with 355-nm flashes. Each spectrum is the average of approximately 500 spectra. Ref. (24).

Figure 8 (right). Kinetics for decay of transient absorption at 495 nm for Cu(TPP) in toluene (Δ), piperidine (X), toluene/piperidine = 90/10 (●) and toluene/piperidine = 97/3 (■). Note the ordinate is a log scale. Ref. (24).

by 200-700 cm^{-1}, depending on the macrocycle and lie ~ 14,400 cm^{-1} above
the ground doublet state (1,57-59). Spectra for the tripdoublet and quar-
tet appear to be similar, possibly because they differ only in how the
unpaired metal electron couples with the porphyrin π-system, as suggested
in early picosecond studies (7). The recent picosecond studies found that
deactivation of the $^2T(\pi,\pi*)$ <—> $^4T(\pi,\pi*)$ manifold takes > 10 ns
(10,12,24). Typical absorption difference spectra in the visible for
Cu^{II}(TPP) are shown in Fig. 7A, and the kinetics in Fig. 8 (triangles)
(24). Recent nanosecond transient absorption and emission studies in
toluene at room temperature give lifetimes for Cu^{II}(OEP), Cu^{II}(Etio), and
Cu^{II}(TPP) of 140 ns, 130 ns, and 40 ns, respectively (60).

A number of earlier observations have suggested that a CT state may lie
close in energy above the ^2T and ^4T (24). These include the finding that
the Cu^{II}(meso) emission is quenched above 77K and that the luminescence of
Cu^{II}(TPP) is quenched in pyridine (4,57,58).

Recently, it has been reported by Rentzepis' group that the transient lifetime of Cu^{II}(PPDME) in pyridine is reduced to 45 ps and that of Cu^{II}
cytochrome-c is pH dependent (16,21). We also found rapid (40-150 ps)
quenching following excitation of several five-coordinate Cu^{II} porphyrins
in neat piperidine (24). Spectra for Cu^{II}TPP in this solvent are shown in
Fig. 7C and the kinetics in Fig. 8 (X). That a five-coordiante ground
state complex is formed in strong base was shown previously (61). In pyridine and dilute piperidine, excitation of the four-coordinate complex gives
rise to similar spectra (Fig. 7B), but kinetics that depend on the ligand
concentration, reflecting diffusion and binding of a ligand (Fig. 8 circles
and squares). The spectra are similar in all solvents, except for shifts
in the bleaching wavelengths, and resembled those expected for a (π,π*)
excited state. On the basis of the observed behavior, the earlier work
described above and recent calculations (24,62), we ascribed this behavior
to decay of the $^2T(\pi,\pi*)$ <—> $^4T(\pi,\pi*)$ via a thermally acessible (a$_{2u}(\pi)$,
d$_{x^2-y^2}$) CT state that drops closer in energy to the $^2T(\pi,\pi*)$ in the five-
coordinate complex (24).

Os^{II} and Ru^{II} Complexes: In the d^6 Os^{II} and Ru^{II} porphyrins, the filled
metal d_{xy} and d_π orbitals have comparable energy to the highest filled
porphyrin ring orbitals $a_{1u}(\pi)$ and $a_{2u}(\pi)$ (3,5). Thus, these complexes are
expected to have $(d_\pi, e_g(\pi^*))$ metal $\rightarrow$ ring CT states in the same energy
range as the lowest ring (π,π^*) states. It has been argued on the basis of
calculations and absorption and emission studies that the relative energies
of these states should be influenced strongly by the σ-donating and π-
backbonding characteristics of the two axial ligands. These properties
affect the relative importance of equatorial π-backbonding between metal d_π
and empty $e_g(\pi^*)$ ring orbitals, and the axial π-backbonding between the
metal d_π and π^* orbitals on the axial ligand(s) (3,5).

Extensive picosecond studies have explored the effects of π-backbonding on
the photophysical properties of a series of Os^{II} porphyrins in which the
two axial ligands are varied (14,22). All of the complexes studied show
singlet lifetimes in the range $\leq$ 9 ps to 50 ps and triplet lifetimes in the
range 1 to 9 ns at room temperature. $Os^{II}(OEP)(L_1)(L_2)$ complexes in which
$^3(d,\pi^*)$ are thought to be lowest have the following axial ligands: $L_1 =$
$L_2 = $ py, $L_1 = L_2 = P(OMe)_3$ and $L_1 = $ py, $L_2 = $ CO (3,5,14,22). The first two
compounds do not emit and the third shows a peculiar luminescence spectrum.
Complexes having $^3T(\pi,\pi^*)$ lowest have $L_1 = L_2 = $ NO, $L_1 = L_2 = $ O, and $L_1 =$
NO, $L_2 = $ OMe. The first compound does not emit. (The metal in the second
to last compound formally has charge + 6.) It was noted that the excited
state absorptions in the 550-770 nm region in the compounds with (d,π^*)
lowest are shifted from those having (π,π^*) lowest (14,22). The bispyri-
dine complex gave evidence for a photochemical product, possibly ligand
loss (22).

We have recently examined a number of Ru^{II} porphyrins having different
axial ligands (28). The first series of compounds are carbonyl complexes
of the form $Ru^{II}(OEP)(CO)(L)$ with L = EtOH, pyridine or imidazole. These
complexes have $^3T(\pi,\pi^*)$ lifetimes of $\sim$ 75 µs as determined by slower time
scale transient absorption and emission studies (28). These lifetimes are
in the same range as previous studies on analogous TPP compounds (38). The
solid spectrum in Fig. 9 shows a difference spectrum at a 215 ps delay for

$Ru^{II}(OEP)(CO)(EtOH)$ in ethanol (28). (Ethanol is maintained as the second
axial ligand in the neat solvent (63).) The inset shows the absorption
changes in the Soret region 1 μs after 10 ns flash at 532 nm. This transient and that of the other carbonyl complexes are assigned as the
$^3T(\pi,\pi*)$, the lowest excited state assigned from earlier phosphorescence
measurements on $Ru^{II}(OEP)(CO)(py)$ (3). It should be noted that the weak
transient absorption peaks near 720 and 810 nm (Fig. 9 solid spectrum) are
consistent with the assignment as a ring $(\pi,\pi*)$ excited, as discussed
earlier in the text. The $^1Q(\pi,\pi*)$ lifetimes in these complexes appear to
be $\leq$ 35 ps. Some of these carbonyl complexes were found to be weakly pho-
todissociative, as noted previously by numerous groups.

Figure 9 also shows the transient spectrum for $Ru^{II}(OEP)[P(nBu)_3]_2$ in
toluene at 215 ps after excitation with a 35 ps 355 nm flash (dashed). The
absorption changes are clearly different from those just assigned to the
$^3T(\pi,\pi*)$ in the carbonyl-ethanol complex (solid spectrum). The lifetime is
12 ± 3 ns. This short relaxation time is suggestive of the transient being
a $(d,\pi*)$ metal → ring CT, as assigned for certain Os^{II} complexes (14,22).
The lack of a π-accepting ligand and $P(^nBu)_3$ being a reasonably strong σ-
doner would tend to push the $(d,\pi*)$ down in energy relative to the $(\pi,\pi*)$
state (3,28). The prominant peak near 700 nm in the absorption difference
spectrum (Fig. 11) also is expected for a $(d,\pi*)$ CT, on the basis of
spectra for π-anion radicals of other transition-metal complexes, such as
Zn^{II} (40). (The π-anion radical of the corresponding $Ru^{II}(TPP)$ complex
should show an additional peak near 900 nm not expected for the OEP macrocycle, and not observed in the present case.)

The $Ru^{II}(NO)(OMe)$ complex exhibits a very fast decay ($\tau \leq$ 50 ps) and a
longer-lived (> 5 ns) component. The two transients are assigned to
$^1Q(\pi,\pi*) \rightarrow {}^3T(\pi,\pi*)$ and decay of the $^3T(\pi,\pi*)$ respectively. Decay of the
triplet may proceed via $(\pi, d_\pi + NO(\pi*))$ CT states predicted by the calculations and lack of emission (3).

Picosecond flash photolysis studies also have been carried out on the π-
cation radicals of several $Ru^{II}(OEP)(CO)(L)$ complexes radicals in which the

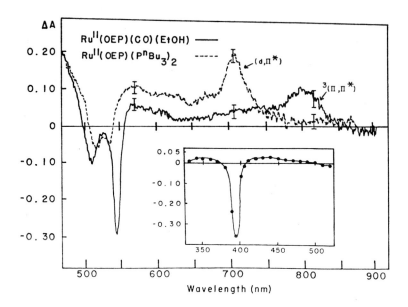

Figure 9. Absorption difference spectra observed at a 215 ps time delay following excitation with 35 ps, 355 nm flashes: $Ru^{II}(OEP)(CO)(EtOH)$ in ethanol (——); $Ru^{II}(OEP)[P(nBu)_3]_2$ in toluene (---). The <u>inset</u> shows the absorption changes in the Soret region for a more dilute solution of $Ru^{II}(OEP)(CO)(EtOH)$ in ethanol observed 1 μs after excitation with a 10-ns, 532-nm flash. Ref. (23).

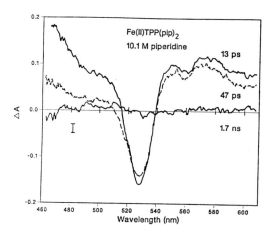

Figure 10. Absorption difference spectra at three time delays with respect to excitation of $Fe^{II}TPP(pip)_2$ in neat (10.1 M) piperidine with 35-ps flashes at 355 nm. Ref. (27).

ground state is either $^2A_{2u}$ (L = EtOH or py) or $^2A_{1u}$ (L = Br$^-$) or a mixture
(L = Im), depending on the axial ligand (23). In each case the decay was $\leq$
35 ps, but appeared to be faster for the $^2A_{1u}$ ground state π-cation radicals. The short lifetimes were attributed to rapid relaxation of the six-
coordinate complex via low-lying metal → ring (d,π) CT states, not present
in the neutral parent compounds. Excitation of the $^2A_{2u}$ ground state π-
cation radicals with 355 nm pulses gave a $\leq$ 35 ps decay, but also a long-
lived component that was attributed to release of the CO followed by
internal electron transfer from metal to ring. It was suggested that a
dissociative state such as (d$_\pi$, d$_{z^2}$) may be accessible with 355 nm pulses
but not with 532 nm flashes (23). Calculations on the neutral parent compounds indicate that this may be a reasonable possibility (3).

Fe^{II} and Fe^{III} Complexes: Numerous experimental and theoretical investigations, including picosecond transient absorption studies, have attempted to
understand the mechanism of photorelease of CO, O_2, and NO from the Fe^{II}
porphyrin active sites of hemoglobin and myoglobin, (see 45, 46 and refs.
therein). Here, we discuss investigations on a few simple ferrous and
ferric porphyrins.

Iron porphyrins are radiationless (1), exhibiting fluorescence quantum
yields < 10^{-6} (1,64,65), implying $^1Q(\pi,\pi^*)$ lifetimes < 1 ps. It is thought
that rapid deactivation of the (π,π*) occur via lower energy ring ⟷ metal
(d,π*) CT or ligand field (d,d) states predicted from calculations and
assigned in near-infrared ground state absorption spectra (1,3,47,65-68).
Rapid decays have been reported from picosecond transient absorption studies on iron porphyrins: < 2 ps for ferrous and ferric cytochrome-c (8), $\leq$
13 ps for Fe^{III}(OEP)(Cl) (20), 25-50 ps for Fe^{III}(TPP)(Cl) in benzene or
toluene (13,27,29) and $\leq$ 35 ps for Fe^{II}(TPP)(2-MeIm) (27). The last three
compounds have S = 5/2, S = 5/2, and 2, respectively. The half-filled d$_\pi$
and d$_{z^2}$ orbitals between the ring filled-π and empty-π* orbitals make a
number of low energy quenching states possible.

Photodissociation of hexacoordinate Fe^{II}(P)L$_2$ complexes also have been
investigated (27,30), with P = TPP, PPDME or DPDME and L = pip or 1-MeIm

(27). Figure 10 shows a spectrum for $Fe^{II}(TPP)(pip)_2$ excited with a 35 ps
355 nm flash in neat piperidine. A fast ($\leq$ 35 ps) component is observed in
all cases .that shows strong absorption in the blue expected for a $(\pi,\pi*)$
excited state. It may be due to a short-lived $^3T(\pi,\pi*)$ formed by recycling
through shorter lived excited states during the excitation flash (13,27).
The transient absorption to the red of the Q-band bleaching also forms
early in the excitation flash and is similar to that expected for the loss
of one ligand to make a five-coordinate complex. The time-constant for
recovery of this second component was found to depend on the axial ligand,
its concentration, and the porphyrin macrocycle (27). The bimolecular rate
constant for rebinding of the ligand was 0.6-3.5 x 10^8 M^{-1} s^{-1} in all
cases. These values agree well with those from slower time scale measurements on similar systems which showed also that quantum yields for photorelease of the ligand were low, ~ 5% (49,50). A four-coordinate transient
was not observed, nor was there any evidence for a high initial ligand
release yield followed by geminate (cage) recombination. Dissociation was
attributed (27) to a (d_π, d_{z^2}) excited state below the lowest $(\pi,\pi*)$, as
assigned by previous calculations and near infrared spectra (1,47,48,65-
68). Rapid radiationless decay evidently competes with ligand release,
accounting for the low photodissociation yield.

Pd^{II}, Pt^{II}, Pt^{IV}: Fluorescence studies on Pd^{II} and Pt^{II} porphyrins some
years ago found a weak prompt fluorescence from the former species,
implying a lifetime for $^1Q(\pi,\pi*)$ of about 20 ps, but only delayed
fluorescence from the latter, implying a shorter lifetime (69). Although
picosecond studies on several Pd^{II} and Pt^{II} porphyrin have been done
(7,9,22,25,32), only upper bounds of $\leq$ 13 ps have been reported for the
$^1Q(\pi,\pi*)$ of both species. Both metal complexes have $^3T(\pi,\pi*)$ decays of $>$
10 ns (7,9,22,25,32). Calculations indicate that there are no low energy
CT or (d,d) excited states (6). In the case of $Pt^{II}(TPP)$, transient absorbance peaks are observed at 740 nm and 840 nm (32). The latter becomes
more intense when the initial $^1Q(\pi,\pi*)$ gives way to the $^3T(\pi,\pi*)$. A similar phenomenon is observed for $Pt^{II}(OEP)$, with the two peaks at 680 and 750
nm (22). Such investigations suggest that the absorbance of the $^1Q(\pi,\pi*)$
may have consistent differences from that of $^3T(\pi,\pi*)$ that could prove use-

ful for determining singlet versus triplet decay routes.

The decay kinetics of $Pt^{IV}(TTP)Cl_2$ (TTP is tetratolyporphyrin) are entirely different from that of the Pt^{II} complexes (25). The excited state is entirely quenched in $\leq$ 45 ps. This rapid relaxation was predicted by the calculation of a low energy $(a_{2u}(\pi), d_{z^2})$ CT state (6) that would act to quench the $^3T(\pi,\pi^*)$ state.

$\underline{Ag^{II}, Mo^V, Mn^{II}}$: Picosecond studies on $Ag^{II}(PPDME)$ and $Ag^{II}(DPDME)$ have shown absorption changes that decay in $\leq$ 10 ps, taken to reflect mainly equilibration of the $^2T(\pi,\pi^*) \longleftrightarrow ^4T(\pi,\pi^*)$ states (10,12). A > 1 ns transient was reported for the first complex, but not for the second. These compounds are nonluminescent and the short lifetime probably reflect rapid relaxation via a low-lying metal $\rightarrow$ ring CT $(d_{x^2-y^2}, e_g(\pi^*))$ predicted by theory (4,10,12).

Picosecond studies on $Mo^V(TPP)(OCH_3)$ have revealed a $\leq$ 10 ps component attributed to establishment of the $^2T(\pi,\pi^*) \cdot ^4T(\pi,\pi^*)$ equilibrium, followed by a 4 ± 1 ns decay ascribed to relaxation of this equilibrium mixture via a lower energy (π, d_{xy}) ring $\rightarrow$ metal CT state (31). This compound also does not emit (1). Early picosecond results on the S = 1/2 metalloporphyrins have been tabulated by Serpone et al. (31).

Several Mn^{III} porphyrins have been exmained by picosecond transient absorption spectroscopy. These include the five-coordinate $Mn^{III}(OEP)(Cl)$, $Mn^{III}(TPP)(Cl)$ and $Mn^{III}(MPDME)(Cl)$ in toluene (30,32) and the six-coordinate $Mn^{III}(OEP)(py)_2$ and $Mn^{III}(TPP)(py)_2$ in pyridine (32). All complexes show broad transient absorption in the visible region and relaxation <35 ps. Calculations, assignments of near-infrared ground state absorption bands in these complexes, and low luminescence yields (1) indicate that the short lifetimes are probably due to relaxations through a manifold of low-lying metal $\rightarrow$ ring CT and ligand field (d,d) states.

Conclusions

In this paper we have attempted to summarize the current knowledge of the picosecond transient absorption and kinetic spectroscopy of transition-metal porphyrins. As shown above there are distinct spectral and kinetic behaviors. Quite generally the $(\pi,\pi*)$ states have absorption spectra like that of $^3T(\pi,\pi*)$ states measured some years ago (36). The ring $\longleftrightarrow$ metal CT states have excited state spectra like that of ring anion or cation radicals, depending on the direction of the charge transfer. These states tend to be short lived. The absorption spectra of (d,d) excited states tend to resemble the absorption of ground state porphyrin except shifted. Most of the ligation-deligation phenomena observed so far observed goes through these (d,d) states. This rich excited state chemistry provides new insights into how the electrons of transition metals interact with those of the porphyrin ring and how this determines the photokinetics.

Acknowledgements: DH wishes to thank the Camille and Henry Dreyfus Foundation for a grant for Newly Appointed Young Faculty in Chemistry and to the National Institutes of Health for support through the Biomedical Research Grant Program (BRSG SO7RR7054-17). We also wish to acknowledge students and colleagues who have collaborated in this research.

References

1. Gouterman, M.: The Porphyrins, Dolphin, E. Ed.; Vol. III, Academic Press: New York 1978.
2. Buchler, J.W.: "Porphyrins and Metalloporphyrins" edited by K.E. Smith, Elsevier, Amsterdam.
3. Antipas, A., Buchler, J.W., Gouterman, M., Smith, P.D.: J. Am. Chem. Soc. 100, 3015-3024 (1978).
4. Antipas, A., Dolphin, D. Gouterman, M., Johnson, E.C.: J. Am. Chem. Soc. 100, 7705-7709 (1978).
5. Antipas, A., Buchler, J.W., Gouterman, M. Smith, P.D.: J. Am. Chem. Soc. 102, 198-207 (1980).
6. Antipas, A., Gouterman, M.: J. Am. Chem. Soc. 105, 4896-4901 (1983).
7. Magde, D., Windsor, M.W., Holten, D., Gouterman, M.: Chem. Phys. Lett. 29, 183-188 (1974).
8. Huppert, D., Straub, K.D., Rentzepis, P.M.: Proc. Natl. Acad. Sci. USA 74, 4139-4143 (1977).
9. Kobayashi, T., Straub, K.D., Rentzepis, P.M.: Photochem. Photobiol. 29, 925-931 (1979).

10. Kobayashi, T., Huppert, D., Straub, K.D., Rentzepis, P.M.: J. Chem. Phys. _70_, 1720-1726 (1979).
11. Dzhagarov, B.M., Timinskii, Y.V., Chirvonyi, V.S., Gurinovich, G.P. Dokl. Biophys. _247_, 138-140 (1979).
12. Chirvonyi, V.W., Dzhagarov, B.M., Timinskii, Y.V., Gurinovich, G.P. Chem. Phys. Lett. _70_, 79-83 (1980).
13. Cornelius, P.A., Steele, A.W., Chernoff, D.A., Hochstrasser, R.M.: Chem. Phys. Lett. _82_, 9-14 (1981).
14. Serpone, M., Netzel, T.L., Gouterman, M.: J. Am. Chem. Soc. _104_, 246-252 (1982).
15. Chirvonyi, V.S., Dzhagarov, B.M., Shul'ga, A.M., Gurinovich, G.P.: Dokl. Biophys. _259_, 144-148 (1982).
16. Reynolds, A.H., Straub, K.D., Rentzepis, P.M.: Biophys. J. _40_, 27-31 (1982).
17. Kim, D-H., Kirmaier, C., Holten, D.: Chem. Phys. _75_, 305-322 (1983).
18. Kim, D-H., Holten, D.: Chem. Phys. Lett. _98_, 584-589 (1983).
19. Tait, C.D., Holten, D., Gouterman, M.: Chem. Phys. Lett. _100_, 268-272 (1983).
20. Fujita, I., Netzel, T.L., Chang, C.K., Wang, C.B.: Proc. Natl. Acad. Sci. USA _79_, 413-417 (1982).
21. Straub, K.D., Rentzepis, P.M.: Biophys. J. _41_, 411a (1983).
22. Ponterini, C., Serpone, N., Bergkamp, M.A., Netzel, T.L.: J. Am. Chem. Soc. _105_, 4639-4645 (1983).
23. Barley, M., Dolphin, D.H., James, B.R., Kirmaier, C., Holten, D.: J. Am. Chem. Soc., _106_, 3937-3943 (1984).
24. Kim, D-H., Holten, D., Gouterman, M.: J. Am. Chem. Soc., _106_, 2793-2798 (1984).
25. Kim, D-H., Holten, D., Gouterman, M., Buchler, J.: J. Am. Chem. Soc., _106_, 4015-4017 (1984).
26. Tait, C.D., Holten, D., Gouterman, M.: J. Am. Chem. Soc., in press, (1984).
27. Dixon, D.W., Kirmaier, C., Holten, D.: J. Am. Chem. Soc., submitted.
28. Tait, C.D., Holten, D., Barley, M., Dolphin, D., James, B.R. (manuscript in preparation).
29. Liang, Y., Negus, D.K., Hochstrasser, R.M., Grunner, M., Dutton, P.L.: Chem. Phys. Lett. _84_, 236-240 (1981).
30. Chirvonyi, V.S., Gurinovich, G.P.: Eest. NSV Teaduste Akadeemia Toimetised Koide Fuusika Matamaatika _31_, 129-132 (1982).
31. Serpone, N., Ledon, H., Netzel, T.: Inorg. Chem., _23_, 454-457 (1984).
32. Yan, X-W., Kirmaier, C., Holten, D., (unpublished results).
33. "Ultrafast Laser Spectroscopy" edited by R.R. Alfano (Academic Press, New York).
34. Gouterman, M.: J. Chem. Phys. _30_, 1139-1161 (1959).
35. a) Gouterman, M.: J. Mol. Spectrosc. _6_, 138-163 (1961). b) Gouterman, M., Wagniere, G.H.: J. Mol. Spectrosc. _11_, 108-127 (1963).
36. Pekkarinen, L., Linschitz, H.: J. Am. Chem. Soc. _82_, 2407-2411 (1960).
37. Kalyanasundaram, K.: Chem. Phys. Lett. _104_, 357-362 (1984).
38. Rillema, D.P., Nagle, J.K., Barringer, L.F., Meyer, T.J. J. Am. Chem. Soc. _103_, 56-62 (1981).
39. Ake, R.L., Gouterman, M.: Theor. Chim. Acta 15, 20-42 (1969).
40. Felton, R.H., "The Porphyrins"; Dolphin, D., Ed.; Academic Press: New York, 1978; Vol. V, pp. 53-125, and references therein.
41. Dolphin, D., Felton, R.H.: Accts. Chem. Res. _7_, 26-32 (1974).

42. Dolphin, D., Muljiani, Z., Rousseau, K., Borg, D.C., Fajer, J., Felton, R.H.: Annals. N.Y. Acad. Sci. 206, 177-200 (1973).
43. "Iron Porphyrins" edited by A.P.B. Lever and H.B. Gray, Addison-Wesley, Reading, MA, 1983. See (a) Chapter 1 by Loew, G.H. and (b) Chapter 2 by Scheidt, W., Gouterman, M.
44. Waleh, A., Loew, G.H.: J. Am. Chem. Soc. 104, 2346-2351, 2352-2356 (1982).
45. See Ref. 33 for reviews, Chapter 14 by Eisenstein, L., Frauenfelder, H., and Chapter 15 by Noe, L.J.
46. Chernoff, D.A., Hochstrasser, R.M., Steele, A.W.: Proc. Natl. Acad Sci. USA 77, 5606-5610 (1980).
47. Eaton, W.A., Hanson, L.K., Stephens, P.J., Sutherland, J.C., Dunn, J.B.R.: J. Am. Chem. Soc. 100, 4491-5003 (1978).
48. Zerner, M., Gouterman, M.: Theor. Chem. Acta 6, 363-400 (1966).
49. Momenteau, M., Lavelette, D.: J. Am. Chem. Soc. 100, 4322-4324 (1978).
50. Lavalette, D., Tetreau, C., Momenteau, M.: J. Am. Chem. Soc. 101, 5395-5401 (1979).
51. Tetreau, C., Lavelette, D., Momenteau, M.: J. Am. Chem. Soc. 104, 1506-1509 (1983).
52. Hoffman, B.M., Gibson, Q.H.: Proc. Natl. Acad Sci. USA 75, 21-25 (1978).
53. McLess, B.D., Caughey, W.S.: Biochemistry 7, 642-652 (1968).
54. Caughey, W.S., Fujimoto, W.Y., Johnson, B.P.: Biochemistry 5, 3830-3843 (1966).
55. Walker, F.A., Hui, E., Walker, M.: J. Am. Chem. Soc. 97, 2390-2397 (1975).
56. Baker, E.W., Brookhart, M.S., Corwin, A.H.: J. Am. Chem. Soc. 86, 4587-4590 (1964).
57. Gouterman, M., Mathies, R.A., Smith, B.E., Caughey, W.S.: J. Chem. Phys. 52, 3795-3802 (1970).
58. Eastwood, D., Gouterman, M.: J. Mol. Spectrosc. 30, 437-458 (1969).
59. Ake, R.L., Gouterman, M.: Theor. Chim. Acta 17, 408-416 (1970).
60. Russell, M.D., Gouterman, M., Kim, D-H., Holten, D.: (to be published).
61. Baker, M.S., Brookhart, M.S., Corwin, A.H.: J. Am. Chem. Soc. 86, 4587-4590 (1964).
62. Shelnutt, J.A., Straub, K.D., Rentzipis, P.M., Gouterman, M., Davidson, E.R.: Biochemistry (submitted).
63. Barley, M.H., Becker, J.Y., Domazetis, G., Dolphin, D., James, B.R.: Can. J. Chem. 61, 2389-2396 (1983).
64. Harriman, A.: J. C. S. Faraday I 76, 1978-1985 (1980).
65. Adar, F., Gouterman, M., Aronowitz, S.: J. Phys. Chem. 80, 2184-2190 (1976).
66. Eaton, W.A., Charney, E.: J. Chem. Phys. 51, 4502-4505 (1969).
67. Kobayashi, H.: Adv. Biophys. 8, 191, (1978) and references therein.
68. Rawlings, D., Davidson, E.R., Gouterman, M. (work in progress).
69. Callis, J.B., Gouterman, M., Jones, Y.M., Henderson, B.H.: J. Mol. Spectrosc. 39, 410-420 (1971).

Received August 9, 1984

Discussion

Buchler: The excitation of the copper porphyrin leads to an elevation of the reduced Cu^I ion from the porphyrin plane in the presence of axial donors. It should be easier to remove the bigger Cu^I ion from the porphyrin. Are there any experiments showing a photodemetallation of metalloporphyrins?

Gouterman: This is a good suggestion that we never thought of. Under the right conditions it should be possible to demetallate Cu porphyrin in the light when it is not demetalled in the dark.

Vogler: I would like to comment on the remark of Prof. Buchler concerning the possibility of photodemetallation of Cu^{II} porphyrins. Even in the released CT excited state, containing Cu^I and a porphyrin radical cation, back electron transfer (which regenerates the starting Cu^{II} porphyrin) may be too fast to compete with "photodemetallation".

Gouterman: In a weak acid, such as acetic acid which normally does not demetallate Cu, we might get a fast reaction:

$$Cu^I(P\cdot) + HAc \longrightarrow Cu^I Ac + (HP\cdot)$$

Here $(HP\cdot)$ is a monoprotonated free base π"cation" radical. Subsequent dark reactions may lead to an overall reaction:

$$CuP + h\nu + 2HAc \longrightarrow H_2P + Cu(Ac)_2$$

Song: Is it possible to observe an enhanced luminescence due to the recombination of a photodissociated pair of metalloporphyrin and ligand, analogous to the enhanced phosphorescence of organic molecules via recombination of the photoionized pairs?

Gouterman: I think that in general the answer is no. The reason is that we believe that the dissociation occurs from states such as (d_π, d_{z^2}) or (π, d_{z^2}), which are at lower energy than (π, π^*). So that even if these states reformed by reversing the dissociation, they would generally deactivate rather than radiate.

There may be cases of a dissociation due to such states at higher energy than the lowest energy (π, π^*), so that in these cases radiation after recombination may be possible.

Lamola: I would assign the $\sim$10 ps transient you observed for iron porphyrin as a triplet π, π^* state since if it were the singlet π, π^* state, one would predict an observable fluorescence $(\phi_f \approx 10^{-4})$ and it is well known that iron porphyrins have fluorescence yields smaller than 10^{-6}. Do you agree?

Gouterman: Yes.

Scheer: You mentioned the formation of Pd^{II}-porphyrins chlorinated at the macrocycle, when you start with $Pd^{IV}-Cl_4$. Is this chlorination a photochemical process?

Gouterman: No, it is formed during the attempted synthesis of Pd^{IV} porphyrin dichloride from Pd^{II} porphyrin.

METAL COMPLEXES OF TETRAPYRROLE LIGANDS

XXXVII. Structure and Optical Spectra of Sandwich-Like Lanthanoid Porphyrins (1)

Johann W. Buchler, Martina Knoff
Institut für Anorganische Chemie, Technische Hochschule Darmstadt, D-6100 Darmstadt, Germany (F.R.)

1. Introduction

Sandwich-type lanthanoid bisphthalocyaninates, $Ln(Pc)_2$
$[(Pc)^{2-}$: phthalocyaninate dianion], have been extensively studied (for review articles see 2)   because of the intense colour changes accompanied with redox or protonation-deproto-
nation processes (2, 3) which may be associated with the elec-
trochromic properties of these substances (4). Not only the
interest in this kind of new materials, but also a biophysical
aspect renders a study of the corresponding porphyrins worthwhile: compounds in which two porphyrin disks held together
at a fixed distance by a large rare earth cation could serve
as models for energy transfer in biological systems; the lanthanoid ion is not expected to have much electronic interaction with the porphyrin systems. For this reason, we had prepared the lanthanoid bisporphyrinates, $Ce(TTP)_2$ and
$PrH(TTP)_2$ $[(TTP)^{2-}$: dianion of 5,10,15,20-tetra(p-tolyl)por-
phyrin] (5). This study describes the structural and spectral
properties of similar compounds obtained from 2,3,7,8,12,13,
17,18-octaethylporphyrin $[H_2(OEP)]$; the pattern of the peripheral substituents in this ligand is more closely related to
hemes or chlorophylls.

2. Experimental

2.1 Materials and methods

The instruments, materials and methods used were essentially
the same as indicated in our previous paper (5). $H_2(OEP)$ was
purchased from Strem Chemicals/M. Braunagel, Karlsruhe. The
UV/Vis spectra were taken in diethyleneglycoldiethylether
(DEDE, series A̱) or in cyclohexane (series Ḇ). The near infrared spectra were measured with a Zeiss spectrophotometer
DMR 21. Correct elemental analyses were obtained for all compounds. Magnetic measurements (3 - 300 K) were performed with
an automatic Faraday balance in the Institut für Festkörper-
physik, TH Darmstadt.

2.2 Synthesis of the lanthanoid porphyrins A̱ [$Ln_2(OEP)_3$] and Ḇ [$Ln(OEP)_2$] (Ln = Ce, Pr, Eu)

300 mg (0.56 mmol) $H_2(OEP)$ and 1 g of the trisacetylacetonate
of the required lanthanoid [$Ln(acac)_3 \cdot H_2O$; Ln = Ce,Pr,Eu] are
heated under reflux in 1,2,4-trichlorobenzene (TCB) and an
inert gas for 20 h. A colour change from red to brown occurs
in the course of the reaction, and according the UV/Vis-spec-
trum, all $H_2(OEP)$ is consumed. After removing the solvent by
distillation in vacuo the remaining oil is chromatographed at
alumina (basic, grade I). Toluene elutes the brown compounds
A̱ [$Ln_2(OEP)_3$], toluene/methanol (100:1) the brown compounds Ḇ
[$Ln(OEP)_2$]. After removal of the solvents in vacuo, the remaining oils containing A̱ or Ḇ besides TCB and some unidentified
acetylacetone derivatives are dissolved in benzene, concentrated and set aside for crystallization. The resulting dark blue
crystals are dried in vacuo at 50°C. Yields of crystalline
compounds vary between 10 and 34 % for A̱ or 65 and 83 % for Ḇ.

2.3 Properties of individual compounds

<u>A1</u>: $Ce_2(OEP)_3 \cdot 2TCB$, $C_{120}H_{138}N_{12}Ce_2Cl_6$.

$\lambda_{max}(\log \varepsilon)$ = 755 (2.86), 639 (3.40), 571 (3.70), 531 (3.76),
387 (5.27) nm. - IR: 832, 841 cm^{-1}. - Lattice parameters:
P 2_1/c, monoclinic, a = 13.94, b = 22.76, c = 18.51 Å, β =
110.25°, V = 5507 $Å^3$ (9).

<u>A2</u>: $Pr_2(OEP)_3 \cdot 2TCB$, $C_{120}H_{138}N_{12}Pr_2Cl_6$.

$\lambda_{max}(\log \varepsilon)$ = 775 (3.12), 645 (3.42), 572 (3.72), 529 (3.81),
383 (5.28) nm. - IR: 832, 839 cm^{-1}. - Lattice parameters:
tetragonal body centered, a = 14.80, c = 28.81 Å, V = 6307 $Å^3$
(9). - $\mu_{obs} \sim$ 4.8 μ_B (300 K).

<u>A3</u>: $Eu_2(OEP)_3 \cdot 2TCB$, $C_{120}H_{138}N_{12}Eu_2Cl_6$.

$\lambda_{max}(\log \varepsilon)$ = 800 (3.03), 664 (3.32), 574 (3.74), 532 (3.75),
381 (5.27) nm. - IR: 831, 842 cm^{-1}. - Lattice parameters:
tetragonal body centered, a = 14.77, c = 28.82 Å, V = 6290 $Å^3$
(9). $\mu_{obs} \sim$ 4.8 μ_B (300 K).

<u>B1</u>: $Ce(OEP)_2$, $C_{72}H_{88}N_8Ce$.

$\lambda_{max}(\log \varepsilon)$ = 661 (3.16), 573 (4.24), 530 (3.77), 467 (3.92),
378 (5.03) nm. - Diamagnetic; molecular ion at A = 1204; IR:
841 cm^{-1}.

<u>B2</u>: $Pr(OEP)_2$, $C_{72}H_{88}N_8Pr$.

$\lambda_{max}(\log \varepsilon)$ = 1400 (3.50), 670 (3.21), 540 (3.80), 490 (3.72),
391 (5.18) nm. - $\mu_{obs} \sim$ 3.3 μ_B (300 K); molecular ion at A =
1206; IR: 835 cm^{-1}.

<u>B3</u>: $Eu(OEP)_2$, $C_{72}H_{88}N_8Eu$.

$\lambda_{max}(\log \varepsilon)$ = 1280 (3.89), 676 (3.36), 543 (3.87), 376 (5.04). -
μ_{obs}: 1.28 - 3.54 μ_B (3 - 300 K); molecular ion at A = 1215;
IR: 836 cm^{-1}. - Lattice parameters: tetragonal, P $4_1 2_1 2$ or
P $4_3 2_1 2$, a = 15.27, c = 40.98 A, V = 9555 A^3 (9).

Table 1: ^{1}H-NMR data of the bisporphyrinates B (δ [ppm] vs.
int. TMS, 300 MHz; half widths in brackets)

Complex	HC⟨	H_2C⟨(exo)	H_2C⟨(endo)	H_3C-
Ce(OEP)$_2$	9.11	4.18 a)	3.84 a)	1.68 b)
		4.23 a)	3.89 a)	
Pr(OEP)$_2$	-14.5 (17.8)	13.25 (28)	9.55 (27.4)	1.26 (15)
Eu(OEP)$_2$	22.6 c)	22.62	14.6	3.41

a) Centres of quadruplets (see Fig. 3a); J_{gem} = 14.2 Hz. –
b) Centre of triplet. – c) According integral (see Fig. 3b),
methine signal might be buried by H_2C⟨(exo) signal. –

3.   Discussion

3.1 Formation of the compounds

As described in the experimental section, two sets of sandwich-type porphyrins, A and B, are obtained in the octaethylporphyrin series. Compounds A are practically insoluble after
crystallization and probably all have a tripledecker structure; compounds B are soluble in organic solvents and have a
doubledecker structure. The formation of these compounds is
tentatively explained in scheme 1 (overleaf).

The primary product, a monoporphyrin (C), can be isolated when
Ln = Lu (6) and is obtained for any Ln when imidazole serves
as a solvent in the metal insertion (7). While the common precursor of A and B, the "acid" bisporphyrinate LnH(OEP)$_2$, is
not observed for Ln = Ce, Pr, and Eu, it is the main product
in the tetra(p-tolyl)porphyrin series for Ln = Pr (5).

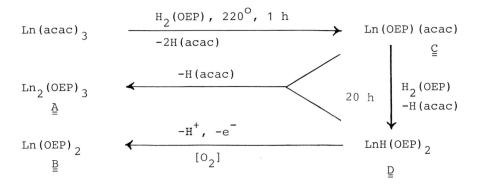

Scheme 1: Formation of Compounds A and B

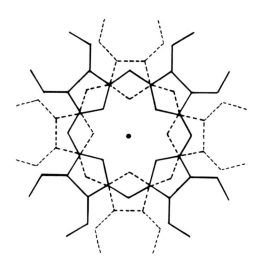

Figure 1: Schematic representation of the doubledecker (bi-plane) structure of $Ln(OEP)_2$ (Ln = Ce, Pr, Eu). The dot in the centre indicates the metal.

3.2 Structure of the compounds

The absence of $\underline{C}$ and $\underline{D}$ in the sets of compounds $\underline{A}$ and $\underline{B}$ can be seen from the infrared spectra which do not show any bands of acetylacetonate or NH groups. Neither $\underline{A}$ nor $\underline{B}$ migrate when a thin layer electrophoresis is attempted [$PrH(TTP)_2$ is mobile in an electrical field (5)].

It is suggested that all compounds $\underline{A}$ have a tripledecker structure, i.e., two Ln^{III} ions are held together by a common, internal, porphyrinate dianion between them, and the molecule is completed by two external porphyrinate dianions at each Ln^{III} ion opposite to the internal porphyrinate; each Ln^{III} ion is octacoordinate. Such a structure may be assumed for a compound, $Nd_2(Pc)_3$, known in the phthalocyanine series (8) and has been proved for $Ce_2(OEP)_3 \cdot 2TCB$ by x-ray diffraction (9). As the elementary cells of all three compounds of series $\underline{A}$ have a rather large volume of about 6000 $Å^3$, and $Pr_2(OEP)_3 \cdot 2TCB$ shows a doubly charged molecular ion at A = 940 in the mass spectrum, they very probably have all the tripledecker structure. All compounds $\underline{A}$ show two prominent bands at 830 and 840 cm^{-1} in their IR spectra, whereas all compounds $\underline{B}$ have only a single band at 836 - 841 cm^{-1}. Thus, the series $\underline{A}$ and $\underline{B}$ are easily distinguished.

The composition $Ln(OEP)_2$ of the compounds $\underline{B}$ (Ln = Ce, Pr, Eu) is proved by their mass spectra (see 2.3). The doubledecker structure of the diamagnetic $Ce(OEP)_2$ is unambiguously derived from its ^1H-NMR-spectrum as in the case of $Ce(TTP)_2$ (5) and shown in Fig. 1. The NMR spectrum (see Fig. 2a and Table 1) shows the expected pattern for equivalent methine and methyl protons and diastereotopic methylene protons. The latter appear as four quadruplets because the protons at the endo position have a distinctly different chemical shift as compared with those at the exo positions and a coupling arises between

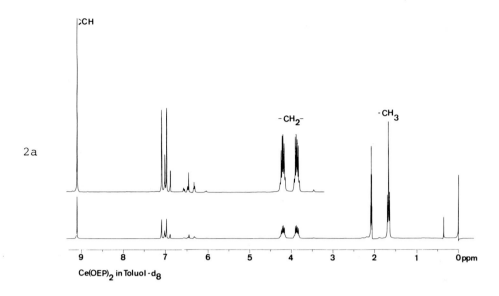

Ce(OEP)$_2$ in Toluol-d$_8$

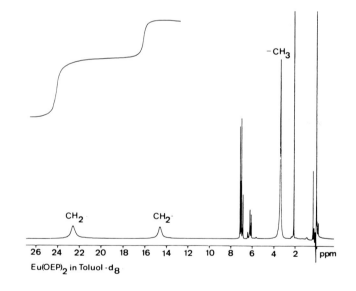

Eu(OEP)$_2$ in Toluol-d$_8$

Figure 2: NMR spectra of lanthanoid octaethylporphyrinates Ln(OEP)$_2$. a): Ln = Ce, b): Ln = Eu. The additional peaks are due to TMS, TCB, incompletely deuterated toluene, and minor impurities.

them (see Table 1). The methine protons appear at an unusual high field (9.12 ppm) as compared with other octaethylporphyrin complexes of octacoordinate tetravalent metals, e.g. Zr^{IV} or Hf^{IV} (10.4 ppm) (10). This is typical for two porphyrin ligands linked together, e.g. by a μ-oxo bridge in $OSc_2(OEP)_2$ (11).

The ^{1}H-NMR spectra of the compounds $Pr(OEP)_2$ (Fig. 2b) and $Eu(OEP)_2$ show broad peaks with large shifts as compared with $Ce(OEP)_2$. This indicates paramagnetism of the former compounds. The appearence of four groups of peaks and the integration accords with the sandwich structure of Fig. 1 (assignments see Table 1) if one assumes that the four partially superimposed quartets of Fig. 2a collapse to two broad signals (Fig. 2b). These NMR spectra mean that both porphyrin systems are identical (at least on the NMR time scale and under the assumption that the spectra of both rings are observable, see 3.3).

3.3 Electronic configuration of $Pr(OEP)_2$ and $Eu(OEP)_2$

While the composition $Ce(OEP)_2 = [Ce^{4+}/2(OEP)^{2-}]$ poses no question for the inorganic chemist, the other compositions, $Pr(OEP)_2 = [Pr^{4+}/2(OEP)^{2-}]$ or $Eu(OEP)_2 = [Eu^{4+}/2(OEP)^{2-}]$ seem very improbable, Pr^{4+} being a very strong oxidant (12) and Eu^{4+} being unknown in coordination chemistry. A hint to an understanding of these compounds is the observation of Marchon et al. (3) that the green lutetium bis(phthalocyaninate), $Lu(Pc)_2$, does not contain acidic hydrogen and shows the paramagnetism of an organic radical (1.7 B.M.; ESR: g = 2). This was accomodated with a composition $[Lu^{3+}/(Pc)^{2-}/(Pc)^{\overline{\cdot}}]$, i.e., the charge balance is achieved by abstracting one electron from the delocalized π-orbitals of the phthalocyanine ring. Such a situation is suggested to exist as well in the complexes $Pr(OEP)_2 = [Pr^{3+}/(OEP)^{2-}/(OEP)^{\overline{\cdot}}]$ and $Eu(OEP)_2 = [Eu^{3+}/(OEP)^{2-}/(OEP)^{\overline{\cdot}}]$.

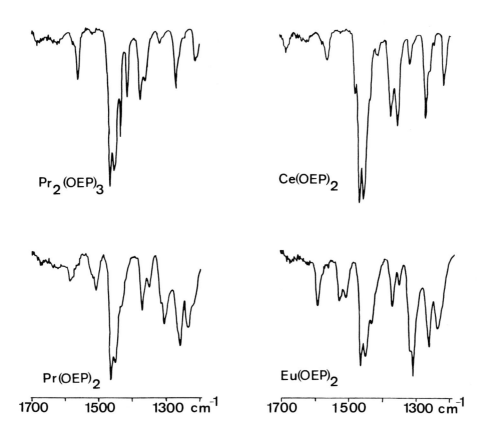

Figure 3: Infrared spectra (1200 - 1700 cm^{-1}) of the lanthanoid octaethylporphyrinates $Ce_2(OEP)_3$ and $Ln(OEP)_2$ (Ln = Ce, Pr, Eu; see text).

Evidence for such compositions comes from the observation that both the Pr and Eu sandwiches show so-called "oxidation state marker bands" in their IR spectra (see Fig. 3), i.e. extra bands in the regions 1200 - 1350 and 1500 - 1550 cm^{-1}, respectively, as compared with $Ce_2(OEP)_3$ and the other compounds of series A and $Ce(OEP)_2$ in which all bonding molecular orbitals of the porphyrin π-systems are filled. Such additional

bands have been previously found in zinc-, iron-, and cobalt
porphyrins that contain uninegative porphyrin π-radicals as
equatorial ligands (13).

Detailed analyses of the magnetic data and electron spin resonance spectra of these complexes are necessary to prove
these ideas and are under way. Eu(OEP)$_2$ shows a temperature
dependent ESR spectrum with several resonances at g ~ 2 that
cannot simply be attributed to the Eu^{III} ion. At 300 K this
complex has μ_{obs} ~ 3.5 μ_B which decreases in a manner typical
for Eu^{III} to about 1.3 μ_B at 3 K; the comparatively high moment at 3 K seems to indicate that a paramagnetic contribution
of the porphyrin π-radical exists besides the temperature-in-
dependent paramagnetism of the Eu^{III} ion (14).

3.4 Electronic absorption spectra

While the monoporphyrinates of all the Ln^{III} ions show normal
electronic spectra with only negligible effects of the metal
ion (15), the sandwich-type porphyrinates of series A and B
show extra bands besides the usual B (Soret) and Q bands at
~ 400, 530 and 570 nm. All compounds A show a similar pattern
(see 2.3) in the visible region with two weak extra bands at
639 - 664 nm and 755 - 800 nm and B bands that are hypsochro-
mically shifted as compared with octacoordinate monoporphyrinates, e.g. Hf(OEP)(acac)$_2$ where the B band appears at 400 nm
(10). Such a shift has been observed in a less pronounced manner in the condensed scandium porphyrin, $OSc_2(OEP)_2$, and hence
is typical for two porphyrin disks in close vicinity (11).
Compounds A do not show near infrared absorption between 800
and 2000 nm.

Three special features can be seen in the visible region of
the compounds B (Fig. 4a): 1. As expected from the situation
found with $OSc_2(OEP)_2$ (11) or $Ln_2(OEP)_3$ (see 2.3), the B band

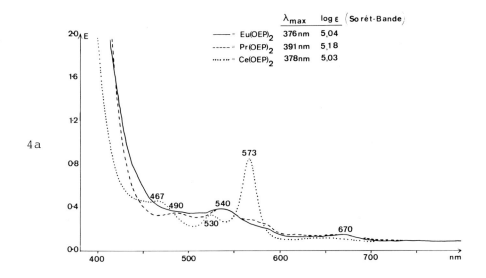

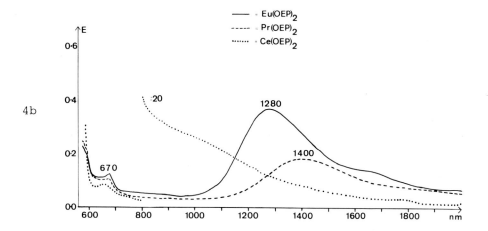

Figure 4: Electronic absorption spectra of the lanthanoid bis-porphyrinates Ln(OEP)$_2$ (Ln = Ce, Pr, Eu). a) visible region, b) near infrared region.

appears at rather high energy. - 2. There is an extra band in
Ce(OEP)$_2$ at 467 nm which was found in Ce(TTP)$_2$ at 485 nm. This
could be due to porphyrin to metal charge transfer transitions
which have been similarly observed in cerium(IV) acetylacetonate, Ce(acac)$_4$, at about 400 - 500 nm as very broad features
(16). - 3. The absorption in the visible does not fall to zero.

A noteworthy difference between the spectrum of Ce(OEP)$_2$ and
the spectra of both Pr(OEP)$_2$ and Eu(OEP)$_2$ is seen in the 500 -
700 nm region: the former has a typical Q(0,0) band at 573 nm,
the latter show rather broad features at 540 and 670 nm; absorption at 575 and 655 nm is characteristic for porphyrin π
radicals, e.g. Zn(OEP)Br (17).

The near infrared spectra (Fig. 4b) allow a novel, clear distinction between those species that contain only the normal
dinegative porphyrin ligands and those with one uninegative,
i.e. oxidized, porphyrin ligand: Only Pr(OEP)$_2$ and Eu(OEP)$_2$
show strong bands ($\varepsilon > 10^3$) at 1400 and 1280 nm, respectively,
both in the solid state (KBr pellets) and in solution (ε independent of concentration). These bands seem typical of a π
radical coupled with a second porphyrin disk; similar low
energy transitions at ∼900 nm have been found by Fuhrhop et
al. (17) with π-dimers, e.g. [Zn(OEP)Br]$_2$, in which two porphyrin radicals are combined. These bands are absent in all the
species $\underline{A}$ and Ce(OEP)$_2$ as well as all the derivatives of Ce^{IV}
and Pr^{III} investigated earlier (5). It has to be tested whether these bands can be regarded as "intervalence" charge
transfer bands between the two adjacent porphyrin disks
existing in different oxidation states.

Further studies are in progress in order to elucidate the
structures as well as the redox and optical emission behaviour
of these complexes.

Acknowledgment

Financial assistance of the Deutsche Forschungsgmeinschaft,
the Centre National de la Recherche Scientifique de France,
the Fonds der Chemischen Industrie and the Vereinigung von
Freunden der Technischen Hochschule Darmstadt is gratefully
acknowledged.

We are very much indebted to Prof. Dr. B. Elschner and Dr.
H. Schäfer (Darmstadt) for magnetic measurements, Prof. Dr.
E. Wölfel, Dr. H. Paulus (Darmstadt), Prof. Dr. J. Fischer
and Dr. A. Mitschler (Strasbourg) for x-ray diffraction ana-
lyses, Priv.-Doz. Dr. J. Veith for mass spectra, and Prof.
Dr. M. Wicholas and Dr. S. Braun (Darmstadt) for NMR spectra.

References and Notes

1.   Paper XXXVI: Botulinski, A., Buchler, J.W., Tonn, B.
     Wicholas, M., submitted to Inorg. Chem.

2.   a) Gmelin, Handbuch der Anorganischen Chemie, 8. Aufl.,
     Syst. No. 39, Rare Earth Elements, Part D 1, p. 96-101.
     Springer, Berlin 1980;
     b) Kasuga, K., Tsutsui, M., Coordin. Chem. Rev. $\underline{32}$,
     67-95 (1980).

3.   Chang, A.T., Marchon, J.C., Inorg. Chim. Acta $\underline{53}$, L 241 -
     L 243 (1981).

4.   a) Moskalev, P.N., Kirin, I.S., Zh. Fiz. Khim. $\underline{46}$ (1972);
     Russ. J. Phys. Chem. $\underline{46}$, 1019-1022 (1972);
     b) Walton, D., Ely, B., and Elliot, G., J. Electrochem.
     Soc. $\underline{128}$, 2479-2482 (1981).

5.   Buchler, J.W., Kapellmann, H.-G., Knoff, M., Lay, K.-L.,
     Pfeifer, S., Z. Naturforsch. $\underline{38b}$, 1339-1345 (1983).

6.   Buchler, J.W., Knoff, M., unpublished observations.

7.   Srivastava, T.S., Bioinorg. Chem. $\underline{8}$, 61-76 (1978).

8.   Kirin, I.S., Moskalew, P.N., Ivannikova, N.V., Russ.
     J. Inorg. Chem. $\underline{12}$, 497-498 (1967).

9.   Fischer, J., Mitschler, A., Paulus, H., unpublished ex-
     periments; d(Ce-N/endo) $\sim$2.76 A, d(Ce-N/exo) $\sim$2.50 A,
     d(Ce-Ce) = 3.75 A.

10. Buchler, J.W., Folz, M., Habets, H., van Kaam, J.,
    Rohbock, K., Chem. Ber. <u>109</u>, 1477-1485 (1976).

11. Buchler, J.W., Eikelmann, G., Puppe, L., Rohbock, K.,
    Schneehage, H.H., Weck, D., Liebigs Ann. Chem. <u>745</u>,
    135-151 (1971).

12. Spitsyn, V.I., Kiselev, Y.M., Martinenko, L.I.,
    Prusakov, V.N., Sokolov, V.B., Dokl. Akad. Nauk <u>219</u>,
    621-624 (1974).

13. Shimomura, E.T., Philippi, M.A., Goff, H.M., Scholz, W.F.,
    Reed, C.A., J. Am. Chem. Soc. <u>103</u>, 6778-6780 (1981).

14. Weiss, A., Witte, H. Magnetochemie, p. 174. Verlag Chemie,
    Weinheim 1973.

15. Gouterman, M., in D. Dolphin (Ed.): The Porphyrins,
    Vol. III, p. 1-165. Academic Press, New York 1978.

16. Ciampolini, M., Mani, F., Nardi, N., J.C.S. Dalton <u>1977</u>,
    1325-1328.

17. Fuhrhop, J.H., Wasser, P., Riesner, D., Mauzerall, D.,
    J. Am. Chem. Soc. <u>94</u>, 7996-8001 (1972).

18. Preliminary x-ray data show that $Eu(OEP)_2$ is indeed a
    doubledecker molecule: Fischer, J., Mitschler, A.,
    experiments in progress.

Received June 29, 1984

Discussion

Myer: In case of $M_2(P_4)_3$ systems, the outer rings are deformed. What is
the nature of the deformation and how much is its magnitude?

Buchler: The external rings are bent away from the metal and are saucer-
like deformed; the deviation of the plane of the 4 nitrogen atoms from the
mean plane of all porphyrin ring atoms is just being calculated in Strasbourg.

Myer: Do you have any evidence for the differences in the force constant
of vibrational modes of the rings on the outside relative to the inner ring?

Buchler: Apart from a band at about 830-840 cm^{-1} which is split in the
tripledeckers, the IR spectra of the doubledeckers and the tripledeckers
show practically the same bands.

Gouterman: Could the rings be tilted in $Ce(OEP)_2$?

Buchler: Inasmuch the structure of $Ce(OEP)_2$ can be deduced from NMR, and
the structure of $Eu(OEP)_2$ from X-ray crystallographic data that are known

right now, the rings are parallel to each other.

Gouterman: Has anyone examined the solid state conductivity of $Pr(OEP)_2$?

Buchler: No. But as soon as we shall have more information on structure
and redox properties, we certainly will pick out a suitable candidate for
such measurements.

Fuhrhop: How does the electronic spectrum of the one-electron reduction
product of $Ce_2(OEP)_3$ look? Is there evidence for a mixed valence state?

Buchler: We were not yet able to do oxidation-reduction chemistry because
the pure $Ce_2(OEP)_3$ is practically insoluble after crystallization. However,
we could use the solutions that come right off the chromatography column.

EXCITED STATE ELECTRON TRANSFER OF CARBONYL[MESOTETRA(P-TOLYL)
PORPHYRINATO]OSMIUM(II) COMPLEXES. ONE AND TWO ELECTRON PHOTO-
REDOX PROCESSES.

Arnd Vogler, Josef Kisslinger
Institut für Anorganische Chemie, Universität Regensburg
D-8400 Regensburg

Johann Walter Buchler
Institut für Anorganische Chemie, Technische Hochschule
Darmstadt, Hochschulstr. 4, D-6100 Darmstadt

Introduction

Naturally occuring iron porphyrin complexes (hemochromes) are
well known to be photosensitive or to participate in light-
driven reactions. The photodissociation of CO from carbonyl
hemoglobin and related compounds has been studied extensively.
In order to identify the reactive excited state we investigated an analogous ruthenium complex.(1) The substitution of
Fe(II) by the homologous Ru(II) modifies the electronic structure of the metalloporphyrin in a well-defined way. The study
of the photochemical and photophysical properties of the ruthenium compound contributed thus to a better understanding of
the photochemistry of carbonyl hemoglobin. The present work is
an extension of this investigation. Iron or ruthenium is replaced by osmium the heaviest member of the homologous group
of metals.

Other hemochromes such as the cytochromes participate in light-
induced electron transfer reactions of the photosynthesis. Although the photoredox behavior of some iron porphyrins was stu-
died(2-5), the investigation of these compounds is generally

hampered by their kinetic lability. For this reason the corresponding osmium porphyrins ("osmochromes"), which are substitu-
tionally inert, were suggested to serve as models for hemochromes.(6) Hence we expected that the study of the photoredox behavior of osmochromes would shed some light on corresponding
reactions of cytochromes. In this regard it is important that
many Fe(II) and Os(II) complexes may be oxidized at comparable
potentials while Ru(II) requires higher potentials.(6,7) Excited state electron transfer reactions of ruthenium porphyrins
were already investigated.(8,9)

A further important aspect of the present work is related to
the growing interest in metalloporphyrins as attractive candidates for photosensitizers in artifical photosynthesis.(10) In
such systems excited metalloporphyrins undergo one-electron
transfer as primary photoreaction. In subsequent processes
these one-electron redox reactions must be coupled by catalysts
such as metallic platinum because photochemical water splitting
or other useful photoreactions require multi-electron transfer
steps. These systems could possibly be simplified if the excited metalloporphyrin would itself participate in multi-electron
redox reactions. For this purpose osmium porphyrins are a good
choice. Not only the oxidation states II and III but also IV
and VI of osmium in porphyrin complexes are stable and spectroscopically well characterized.(6,11-13) While the photophysics
of some osmium porphyrins was investigated extensively(12,13),
only a few, mostly qualitative observations of the photochemistry of these compounds were reported.(14-17) The present
study adds new important data and thus gives a rather complete
picture of the photoreactivity of these compounds.

Results

Electronic Spectra

The electronic spectra of [Os(TTP)(CO)L] with TTP = mesotetra-(p-tolyl)porphyrin and L = methanol or pyridine are very simi-lar.(17,18) In the absorption spectrum (Fig. 1) of [Os(TTP)(CO)(CH$_3$OH)] (1) the Soret band appears at λ_{max} = 411 nm (log ε = 5.35) while the β band occurs at λ_{max} = 523 nm (log ε = 4.21). The α band shows up as a shoulder at λ = 555 nm (log ε = 3.55). Two low-intensity bands appear at λ_{max} = 593 nm (log ε = 3.33) and 645 nm (log ε = 2.72; shoul-der). The corresponding absorption bands of [Os(TTP)(CO)(pyridine)] (2) in CH$_2$Cl$_2$ occur at λ_{max} = 412 nm (log ε = 5.48), 522 nm (log ε = 4.25), 560 nm (log ε = 3.69), 591 nm (log ε = 3.53) and 651 nm (log ε = 3.02).

Both compounds show a photoluminescence at λ_{max} = 653 nm for 1

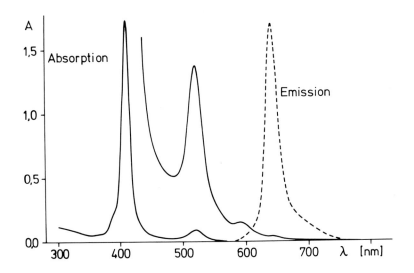

Fig. 1: Electronic spectra of 1 in CH$_2$Cl$_2$. Absorption (———): 7.8×10^{-6} M (8.5×10^{-5} M for long-wavelength region) solution, 1-cm cell. Emission (----): Intensity in arbitrary units, not corrected; λ_{exc} = 546 nm.

(Fig. 1) and 651 nm for *2* (not corrected for photomultiplier
response) at room temperature and in a variety of solvents
such as CH_2Cl_2 and at different exciting wavelengths (365, 405
and 546 nm). The emission of *1* decays with a first-order rate
constant of k = 1.1×10^8 s^{-1}. By comparison with the integrated
emission intensity of europium   thenoyltrifluoroacetonate (ϕ =
0.56 in acetone)(19) at $\lambda_{exc.}$ = 366 nm the emission quantum
yield of *1* was estimated to be ϕ = 4×10^{-3}. The emission was
hardly quenched by oxygen.

Photosubstitution. The complexes *1* and *2* do not undergo any
efficient photosubstitution of the axial carbonyl ligand upon
long-wavelength irradiation (λ_{irr} > 390 nm). The spectrum of
2 dissolved in pyridine changed to that of $Os(TTP)(pyridine)_2$
(16,17) upon irradiation of the β or Soret band of *2* only after
very long irradiation times. The quantum yield of CO substi-
tution was estimated to be ϕ < 3×10^{-7} (λ_{irr} = 405 nm).

Luminescence Quenching. Several electron donors and acceptors
were observed to quench the emission of *1* and *2* in solution
at room temperature. However, in some cases additon of the
quencher resulted in changes of the absorption spectra of *1*
and *2*, indicating ground state interaction or a chemical reaction. Two compounds, N,N-dimethylaniline (DMA) and p-nitro-
benzaldehyde (NBA), did not affect the absorption spectra of
the osmium porphyrins but quenched the emission according to
a Stern-Volmer kinetics. For *1* in CH_2Cl_2 (λ_{irr} = 546 nm) the
slopes of the Stern-Volmer plots were Ksv = 439 M^{-1} (DMA)
and Ksv = 378 M^{-1} (NBA). From the lifetime of *1* (τ = 9.4×
10^{-9} s^{-1}) the quenching constants k_q = 4.7×10^{10} $mol^{-1}s^{-1}$ (DMA)
and k_q = 4.0×10^{10} $mol^{-1}s^{-1}$ (NBA) were obtained. Continuous
irradiation of *1* in the presence of the quenchers did not lead
to any permanent chemical change. The quenching was thus associated with a reversible process.

Relative quantum yields of emission of *1* were determined in
two different solvents (CH_2Cl_2 and CCl_4) at two different exciting wavelengths (365 and 520 nm). At 520 nm (ß-band) the integrated emission intensities were equal in both solvents. At
365 nm (Soret band) the emission efficiency in CCl_4 was only
0.77 of that obtained in CH_2Cl_2.

Photooxidation. In solutions of chlorocarbons such as CH_2Cl_2,
$CHCl_3$ or CCl_4 the complexes *1* and *2* underwent an irreversible
photochemical reaction upon irradiation with light of $\lambda >$
300 nm. Typical spectral changes which accompanied the photolysis are shown for *2* in Fig. 2. Clear isosbestic points were
observed up to about 80 % conversion. Later, secondary photolysis obscured the spectral changes. The nature of the photoproduct did apparently not depend on the ligand L. The product
is characterized by the following absorption bands: λ_{max} =
395, log ε = 4,70; λ_{max} = 442 nm (shoulder); λ_{max} = 503, log ε =
4,02; λ_{max} = 584, log ε = 3.85; λ_{max} = 610 nm (shoulder). Although the photoproduct could not be identified definitely, it

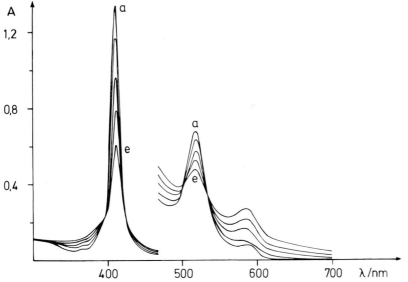

Fig. 2: Spectral changes during the photolysis of 4.4×10^{-6} M *2*
in CCl_4 at (a) 0 and (e) 13-min irradiation time, $\lambda_{irr} > 420$ nm,
1-cm cell; absorbance × 0.1 for long-wavelength region.

is assumed to be [Os(IV)(TTP)Cl$_2$]. This is supported by several
arguments. Many transition metal complexes were observed to be
photooxidized in chlorocarbon solvents.(20,21) Most compounds
undergo one-electron oxidations. Two-electron oxidations occur
if the metal exists in a stable oxidation state which is higher
by two units.(22) This is certainly true for Os(II) which may
not only be oxidized to Os(III) but also to Os(IV). Indeed,
Os(II)(octaethylporphyrin)[P(OCH$_3$)$_3$]$_2$ was observed to be photooxidized to Os(IV)(octaethylporphyrin)Cl$_2$.(15) Nevertheless,
an Os(III) porphyrin must be considered to be a reasonable alternative as final photoproduct. However, the oxidation of
Os(II) to Os(III) is usually not associated with large spectral changes in contrast to those observed during the photooxidation. The absorption spectra of Os porphyrins with the
metal in higher oxidation states (IV and VI) are characterized
by the appearance of longer-wavelength ($\lambda > 600$ nm) absorption
bands,which are due to the occurance of low-energy charge
transfer (CT) transitions (12), in agreement with our observation. If the photooxidation produces Os(III) porphyrins the
complexes [Os(TTP)(CO)L]$^+$ or [Os(TTP)LCl] would be reasonable
candidates. [Os(TTP)(CO)L]$^+$ can be excluded because it should
easily be reduced back to [Os(TTP)(CO)L]. This was not observed. The spectrum of [Os(TTP)ClL] should depend on L to some
degree. Again, the photoproduct was apparently the same regardless wether L was methanol or pyridine.

Quantum yields of photooxidation were dependent on the irradiating wavelength and the chlorocarbon solvent. For *1* the
following efficiencies were obtained: λ_{irr} = 333 nm, ϕ = 0.35
in CCl$_4$ (0.034 in CH$_2$Cl$_2$); λ_{irr} = 365 nm, ϕ = 0.045 (0.005);
λ_{irr} = 405, ϕ = 0.039 ($\sim$ 0); λ_{irr} = 546 nm, $\phi < 0.001$ ($\sim$ 0).

Discussion

The electronic spectra of *1* and *2* show only $\pi\pi^*$ bands of the
porphyrin ligand (Fig. 1). The B(0,0) or Soret bands appear at

24 330 cm^{-1} (*1*) and 24 270 cm^{-1} (*2*). Q bands occur at
19 120 cm^{-1} (0,1 or β) and 18 020 cm^{-1} (0,0 or α) for *1* and
19 160 cm^{-1} (0,1 or β) and 17 860 cm^{-1} (0,0 or α) for *2*. Most
remarkable are the longer-wavelength absorption bands of lower
intensities which we assign to vibrational levels of the T$_1$
(ππ*) state. These bands appear at 16 860 cm^{-1} (0,1) and
15 500 cm^{-1} (0,0) for *1* and 16920 cm^{-1} (0,1) and 15 360 (0,0)
for *2*. To our knowledge this seems to be the first case where
spin-forbidden ππ* transition of a porphyrin are identified
in absorption. The third-row transition metal osmium induces
apparently enough singlet-triplet mixing by spin-orbit coupling ("heavy atom effect") in the porphyrin. As a consequence
the spin selection rule breaks down partially. Several argue-
ments support this assumption.

The emission of *1* (Fig 1) at 15 310 cm^{-1} and *2* at 15 630 cm^{-1}
is certainly the ππ* phosphorescence from the T$_1$ (0,0) state
of the porphyrin. The energy separation between the Q (0,0)
and the T$_1$ (0,0) absorption band (ΔE ∿ 2500 cm^{-1}) is similar
to that observed for other metalloporphyrins.(12,23) In agreement with a strong spin-orbit coupling the lifetime of the T$_1$
state is extremely short (τ ∿ 10^{-8} s for *1*). For this reason
1 and *2* show a relatively intense phosphorescence even at room
temperature and in solution which is hardly quenched by oxygen.

The emission maxima of *1* and *2* are almost resonant with the
longest energy absorption maxima (the luminescence spectra are
not corrected; the real emission maxima should appear at lower
energies). In absorption and emission these maxima are thus
assigned to the singlet-triplet ππ* (0,0) transitions. Light
absorption into this long-wavelength band by the 632 nm line
of a He/Ne laser is indeed associated with this emission. This
observation excludes the possibility that this absorption is
due to an impurity or is caused by another electronic transition which occurs at lower energies than the ππ* transitions.

The assignment of the $\pi\pi^*$ T_1 (0,1)bands of 1 and 2 is certainly also reasonable. The energy gap between the (0,0) and (0,1) Q (ΔE = 1100 cm^{-1}) and T_1 (ΔE = 1360 cm^{-1}) bands of 1 are comparable. The corresponding values of 2 are ΔE = 1300 cm^{-1} and 1560 cm^{-1}. (The Q (0,0) band was not exactly located, see Fig. 1.) In this context it is remarkable that in contrast to our observation the lowest excited state of the corresponding octaethylporphyrin (OEP) complex [Os(OEP)(CO)(pyridine)] is not the T_1 ($\pi\pi^*$) but a CT (Os to porphyrin) transition.(12)

Photochemistry

Photosubstitution. In analogy to similar M(CO)(porphyrin) complexes with M = Fe and Ru (1) the compounds 1 and 2 may be expected to undergo a photodissociation/substitution according to (S = coordinating solvent):

$$[Os(TTP)(CO)L] \xrightarrow{h\nu} [Os(TTP)L] + CO$$

$$[Os(TTP)L] + S \rightarrow [Os(TTP)LS]$$

It has been reported that this reaction indeed occurs.(14,16, 17). However, it is rather inefficient at least for long-wavelength irradiation (λ > 390 nm). The photolysis of 2 in pyridine produced Os(TTP)(pyridine)$_2$ with a quantum yield ϕ < 3×10^{-7}. This observation is not surprising. In the case of iron carbonyl porphyrins such as carbonyl hemoglobin an efficient photolysis occurs ($\phi \sim 1$) because the reactive ligand field (LF) excited state, which is the lowest-energy excited state of these iron porphyrins, is effectively populated from $\pi\pi^*$ states.(1) If Fe is replaced by Ru the energy of the lowest LF state is raised slightly above the lowest $\pi\pi^*$ state of the porphyrin ligand. This leads to a considerable drop of the photosubstitution quantum yield because the most efficient deactivation path is now the radiative (phosphorescence) and radiationless transition from $T_1(\pi\pi^*)$ to the ground state.(1)

Since osmium as a third-row transition metal has the largest
LF splitting of the homologous metals the energy of the reactive LF state increases again. Excited state deactivation
occurs now almost exclusively via the T_1 ($\pi\pi^*$) state by phosphorescence and radiationless transition to the ground state.

Phosphorescence Quenching by Electron Transfer. The first reduction potentials of 1 and 2 were reported to be $E_{1/2}$ =
+0.71 V and +0.68 V vs. SCE.(6,17) Taking into account the
energy of the emitting T_1 ($\pi\pi^*$) states (0,0 band of 1 at
15 500 cm^{-1} or 1.92 V and of 2 at 15 360 cm^{-1} or 1.90 V) they
should be oxidized at $E_{1/2}$ = -1.21 V and -1.22 V. Accordingly
the T_1 ($\pi\pi^*$) state should be susceptible to a variety of
electron transfer reactions in solution. However, only very
fast, essentially diffusion-controlled redox reactions can
occur due to the very short life time of T_1 ($\tau \sim 10^{-8}$ s for 1).
This expectation was confirmed. Obeying a Stern-Volmer kinetics the phosphorescence of 1 in CH_2Cl_2 was quenched oxidati-
vely by NBA ($E_{1/2}$ = -0.86 V vs. SCE) with a second order rate
constant of k_q = 4.0×10^{10} $mol^{-1}s^{-1}$. Reductive quenching of 1
(k_q = 3.7×10^{10} $mol^{-1}s^{-1}$) was observed with DMA ($E_{1/2}$ =
+0.78 V). Although the first oxidation potential of 1 (or 2)
is not known it should occur at potentials less negative than
approximately -1.1 V according to this observation. Oxidative
and reductive quenching takes apparently place by reversible
electron transfer, since it was not associated with any permanent chemical change. This quenching can not occur by an energy
transfer mechanism because these quenchers do not possess low-
energy excited states.

Photooxidation

Many transition metal complexes are photooxidized in halocar-
bon solvents.(20,21) In most cases the metal complexes undergo
one-electron oxidations. Two-electron oxidations of the metals

are favored if higher oxidation states are stable and accessible
at moderate potentials. This was observed for Pt(II)(bipyridyl)
Cl_2 which undergoes a light induced oxidative addition in chlor-
carbon solvents such as $CHCl_3$. Pt(IV)(bipyridyl)Cl_4 is obtained
as stable product.(22) Higher oxidation states such as IV and
VI are also stable and well known for osmium porphyrins.(6,
12,17) It is thus suggested that the photolysis of *1* or *2* in
chlorocarbon solvents e.g. CCl_4 proceeds according to the
following reaction sequence:

$$[Os(TTP)(CO)L] + CCl_4 \xrightarrow{h\nu} [Os(TTP)(CO)L]^+ Cl^- + \cdot CCl_3$$

$$[Os(TTP)(CO)L]^+ Cl^- + \cdot CCl_3 \longrightarrow [Os(TTP)Cl_2] + CCl_2 + CO + L$$

An internally excited state of *1* or *2* undergoes an one-electron
transfer to the solvent in the primary photochemical step. The
chloromethyl radicals such as $\cdot CCl_3$ are strong one-electron
oxidants. Consequently, the oxidized osmium porphyrin obtained
in the primary step is further oxidized in a subsequent second
electron transfer to yield $Os(TTP)Cl_2$ as final product. The
two-electron reduction of CCl_4 leads to the release of two
chloride ions, which are attached to the osmium as ligands,
and the generation of the carbene CCl_2.

Generally, excited state electron transfer requires a long-
lived excited state because the excited molecule has to wait
for a diffusional encounter with an electron donor or acceptor.(9) Usually only the lowest-energy excited state meets
this requirement. This applies also to the phosphorescence
quenching of *1* by DMA and NBA. However,the photooxidation of
1 or *2* in chlorocarbon solvents is efficient only upon irradiation of the shorter-wavelength absorption bands. This means
that higher excited states of the complexes are involved in
the initial electron transfer step. This seems to be surprising because of the very short lifetime of higher excited
states. However, long lifetimes are apparently not required

because the complex is already in close contact with the solvent molecules as acceptors prior to excitation.

The formation of the primary one-electron redox products is
assumed to be preceded by the population of a charge-transfer-
to-solvent (CTTS) excited state from an internal excited
state of the complex.(20) The energy of this CTTS state depends
on the reduction potential of the complex and the oxidation
potential of the chlorocarbon solvent. If CH_2Cl_2 is replaced
by CCl_4 the CTTS state is shifted to lower energies because
CCl_4 ($E_{1/2}$ = -0.78 V) is a stronger oxidant than CH_2Cl_2
($E_{1/2}$ = -2.33 V).(21) Consequently, higher internal excited
states of the complex are required to populate the reactive
CTTS state if CCl_4 is replaced by CH_2Cl_2 in agreement with
our observation. For the photooxidation of 1 the light-sen-
sitive wavelength region is blue-shifted when CH_2Cl_2 is used
as solvent instead of CCl_4. These considerations apply also
to the wavelength dependence of the quantum yield of phosphorescence of 1. Since the energy of the Q ($\pi\pi^*$) state
($\lambda_{exc.}$ = 546 nm) is not sufficient to populate the CTTS state
in CH_2Cl_2 or CCl_4, the phosphorescence intensity is the same
in both solvents. However, the emission yield decreases in
CCl_4 compared with CH_2Cl_2 if the B ($\pi\pi^*$) is excited ($\lambda_{exc.}$ =
405 nm). In CCl_4 the radiationless deactivation from the B
state to the emitting $\pi\pi^*$ triplet competes apparently with
the population of the reactive CTTS state. In CH_2Cl_2 the CTTS
state seems to be at much higher energies. It can not be populated any more efficiently from the B state of 1.

Acknowledgment. Financial support for this research by the
Deutsche Forschungsgemeinschaft and the Fonds der Chemischen
Industrie is gratefully acknowledged. We are grateful to Dr.
K. Oesten for providing samples of the osmium porphyrin
complexes.

References

1. Vogler, A., Kunkely, H.: Ber. Bunsenges. Phys. Chem. <u>80</u> 425-429 (1976).

2. Bartocci, C., Scandola, F., Ferri, A., Carassiti, V.: J. Am. Chem. Soc. <u>102</u>, 7067-7072 (1980).

3. Bizet, C., Morliere, P., Brandt, D., Delgado, O., Bazin, M., Santus, R.: Photochem. Photobiol. <u>34</u>, 315-321 (1981).

4. Bartocci, C., Maldotti, A., Traverso, O., Bignozzi, C.A., Carassiti, V.: Polyhedron <u>2</u>, 97-102 (1983).

5. Maldotti, A., Bartocci, C., Traverso, O., Carassiti, V.: Inorg. Chim. Acta <u>79</u>, 174-175 (1983).

6. Buchler, J.W., Kokisch, W., Smith, P.D.: Struct.Bonding (Berlin) <u>34</u>, 79-134 (1978).

7. Brown, G. M., Hopf, F. R., Meyer, T. J., Whitten, D. G.: J. Am. Chem. Soc. <u>97</u>, 5385-5390 (1975).

8. Young, R. C., Nagle, J. K., Meyer, T. J., Whitten, D. G.: J. Am. Chem. Soc. <u>100</u>, 4773-4778 (1978).

9. Rillema, D. P., Nagle, J. K., Barringer Jr., L. F., Meyer, T. J.: J. Am. Chem. Soc. <u>103</u>, 56-62 (1981).

10. Darwent, J. R., Douglas, P., Harriman, A., Porter, G., Richoux, M.-C.: Coord. Chem. Rev. <u>44</u>, 83-126 (1982).

11. Buchler, J. W.: The Porphyrins, Dolphin, D., Ed., Academic Press, New York 1978, Vol. I, Part A, Chapter 10, p. 389-483.

12. Antipas, A., Buchler, J. W., Gouterman, M., Smith, P. D.: J. Am. Chem. Soc. <u>100</u>, 3015-3024 (1978) and <u>102</u>, 198-207 (1980).

13. Serpone, N., Netzel, T. L. Gouterman, M.: J. Am. Chem. Soc. <u>104</u>, 246-252 (1982).

14. Hopf, F. R., Whitten, D. G.: The Porphyrins, Dolphin, D., Ed., Academic Press, New York,1978, Vol. II, Part B, Chapter 6, p. 161-195.

15. Serpone, N., Jamieson, M. A., Netzel, T. L.: J. Photochem. <u>15</u>, 295-301 (1981).

16. Collman, J. P., Barnes, C. E., Woo, L. K.: Proc. Natl. Acad. Sci. USA <u>80</u>, 7684-7688 (1983).

17. Oesten, K.: Doctoral Thesis, Technische Hochschule Darmstadt, 1984.

18. Buchler, J. W., Herget, G., Oesten, K.: Liebigs Ann. Chem. 2164-2172 (1983).

19. Winston, H., Marsh, O. J., Suzuki, C. K. Telk, C. L.: J. Chem. Phys. <u>39</u>, 267-271 (1963).

20. Vogler, A., Losse, W., Kunkely, H.: J. Chem. Soc., Chem.
    Comm. 187-188 (1979).

21. Vogler, A., Kunkely, H.: Inorg. Chem. $\underline{21}$, 1172-1175 (1982)
    and references cited therein.

22. Vogler, A., Kunkely, H.: Angew. Chem. Int. Ed. Engl. $\underline{21}$,
    209-210 (1982).

23. Gouterman, M.: The Porphyrins, Dolphin, D., Ed., Acadmic
    Press, New York 1978, Vol. III, Chapter 1, p. 1-165.

Received July 13, 1984

Discussion

Fuhrhop: How do you explain the formation of a molecular complex between a positively charged central ion in porphyrins and a positively
charged pyrazinium ligand? Are there other examples?

Vogler: The coordination of a positively charged ligand to a metal cation is by no means rare in coordination chemistry. Ligands such as the
nitrosyl cation (NO^+), diazonium cations ($N \equiv N^+$-R), and carbyne cations
(C^+-R) are well-documented examples.

Gouterman: Do you get CO loss from Os(TTP)CO(L) when you pump into the
Soret band? The S_2 state may live some picoseconds and can then lead to
ligand field states.

Vogler: This is exactly the process which in our opinion is responsible
for the photosubstitution of CO upon irradiation into the Soret band.

Eilfeld: Were you able to trap the dichlorocarbene formed in the course
of photolysis of [Os(TTP)(Co)L], e.g. by cyclo-additions?

Vogler: We did not do this in this work. However, chlorocarbenes are
well known to insert readily in carbon-halogen bonds. In a related case
(photooxidation of Pt(dipyridyl)Cl_2 in $CHCl_3$; see ref. 22 of our paper)
we were able to detect the insertion product ($CHCl + CHCl_3 \longrightarrow CHCl_2$-$CHCl_2$).

Buchler: You mentioned that in the oxidized iron tetrabenzporphyrin carbonyl pyridine radical cation [Fe^{II}(TBP)COPy]$^+$ the CO stretching frequency is 2020 cm^{-1} while it appears at 1975 cm^{-1} in the neutral Fe^{II}(TBP)
COPy. This increase in frequency indicates a destabilisation of the CO
bond to Fe^{II} only if the correction for the increase in charge according
the rules of Ibers et al. has already been applied.

Vogler: You are correct. But this problem does not interfere with my
argument. We only wanted to show that the removal of a porphyrin π-elec-
tron does not lead to a loss of the CO ligand.

Myer: Have you studied the dissociation profiles? The Soret band has shoulders on both the high and low wavelength side. The dissociation profile will help in determining what transition is involved.

Vogler: The efficiency of dissociation was so low that we could only determine an upper limit of the quantum yield in the Soret region. For this reason, wavelength-dependent measurements were not carried out.

PHOTOPHYSICAL PROPERTIES OF HEMATOPORPHYRIN AND ITS DERIVATIVES AND THEIR USE IN TUMOR PHOTOTHERAPY

A. Andreoni and R. Cubeddu
Centro di Elettronica Quantistica e Strumentazione Elettronica del C.N.R.,
Istituto di Fisica del Politecnico, Milano, Italy

Introduction

Hematoporphyrin Derivative (HpD), prepared from Hematoporphyrin by chemical modification (1, 2) is increasingly used as a photosensitizer in the phototherapy of several kinds of neoplasias. Recent studies (3, 4) indicate that endogenous oxygen plays a major role in the photoinduced killing of tumor cells possibly involving the intermediate $^1\Delta_g$ O_2 generated by electronic energy transfer from the porphyrin triplet state. The reported greater phototherapeutic activity of HpD as compared with that of Hp (5) is likely to arise from a higher affinity of the former substance for tumor tissues, resulting in a specific greater accumulation and/or in a slower release of the drug (6). HpD is known to be a mixture of porphyrins which, in the injectable solution, are present both as monomers and aggregates, being the amount of the aggregates considerably greater than in Hp. The monomers were identified as different porphyrins (e.g. Hematoporphyrin, Hydroxyethylvinyldeuteroporphyrin, etc), while the aggregates were found to be either of hydrophobic origin or complexes stabilized by a chemical bond (7, 8). Actually these complexes resulted to be stable in water also at concentration values where the other aggregates dissolve (9). Recently it has been suggested that the stable aggregates originate from a dimeric structure with an ether linkage (Di-hematoporphyrin-ether, DHE) (8, 10). This fraction, which can be separated from the HpD solution by gel filtration (8, 11), is also marketed by Oncology Research and Development (Cheektowaga, N.Y., USA) under the trade name of Photofrin II (PII) and was found to be the component most active for the "in vivo" photosensitization of tumors (12). The components of HpD can also be separated by high performance liquid chroma-

Optical Properties and Structure of Tetrapyrroles
© 1985 by Walter de Gruyter & Co., Berlin · New York

tography. However, it was reported that the active component can be eluted
from the reverse phase column only by using a mixture of tetrahydrofuran
(THF) and water (9:1 by volume) but not with a solvent consisting of tetrahydrofuran, water and methanol (MeOH) in equal volumes (8).

In this paper we report on the photophysical properties of PII as compared
to those of Hp in different solvent systems. In particular, PII was investigated in THF, MeOH and water. Time resolved fluorescence measurements
revealed the presence of a new fluorescence decay component of ∿ 1.86 ns
that has been attributed to DHE. Results on a new blue-shifted emission
peaked at 580 nm, which is formed more favourably in PII in water, are also
reported. This band was first observed in a solid tumor "in vivo" (13) and
seems to be important in the mechanisms of biological uptake of porphyrins
(14). Moreover, in order to evidentiate the properties of the HpD in cell
cultures, we describe some experiments performed on normal and transformed
cell lines that demonstrate a major photosensitizing activity on tumor
cells. These data suggest that the tumor response "in vivo" originates also
at cellular level.

Material and Methods

Hematoporphyrin, free base, from Porphyrin Products (Logan, Utah, USA), was
dissolved and characterized as described elsewhere (15). HpD is the substance marketed by Oncology Research and Development (Cheektowaga, N.Y.,
USA) as Photofrin. Photofrin II was obtained from Oncology Research and
Development. Being the chemical composition of most substances undetermin-
ed, all solutions to be compared were made at the same w/v concentration.

The organic solvents used were spectroscopic grade samples obtained from
Merck. All other chemicals were commercially available reagent-grade products.

The absorption spectra were taken by a Perkin-Elmer 554 UV-VIS spectrophotometer with 2 nm-slits using matched quartz cuvettes of 1 cm optical pathway. The emission and excitation spectra were measured by a Perkin-Elmer

650-40 spectrofluorometer with 5 nm-slits in both the excitation and observation monochromators. The spectra were not corrected for either the
lamp spectrum, nor the monochromator and photomultiplier response.

The light pulses to be used in the experiments as the excitation pulses are
provided either by an Ar-ion mode-locked laser or by a dye laser (16) pum-
ed by an atmospheric-pressure N_2 laser. The Ar laser output consists of a
train of pulses with duration 100-200 ps depending on the wavelength at the
repetition rate of $\sim$ 70 MHz and could be tuned to the wavelength of 514,
364 and 351 nm with average power of 1, .200 and .150 W, respectively. The
514 nm emission is also used to pump synchronously a Rhodamine 6G dye laser
that gives pulses of $\sim$ 2 ps duration with tunable wavelength. The dye laser
pulses can be frequency doubled to obtain tunable UV picosecond pulses. In
all cases the pulse repetition rate is reduced to $\sim$ 350 kHz using an acous-
to-optic dumper. The N_2-pumped dye laser gives pulses of $\sim$ 150 ps duration
tunable all over the visible-near UV spectral range with peak power typically of $\sim$ 100 kW and repetition rate up to 100 Hz. These two laser sources
are used in the different experiments depending on the excitation wavelength needed. The solutions to be measured are contained in a 1 cm-path-
way quartz cell put in front of detector. The fluorescence is detected at
90° through suitable filters. The single-photon timing technique is used.
The light detector is a single-photon avalanche photodiode (SPAD) that is
biased above the break down voltage in which a non-proportional self-sus-
taining avalanche current can be triggered by the absorption of a single
photon (7). A home-made time-to-amplitude converter (TAC) and a multichannel analyzer are used to measure the delay distribution of the detected
single photons with respect to the reference start pulses synchronized with
the excitation light pulses. To increase the data acquisition rate, the excitation pulse and the fluorescence photons are made to provide the stop
and the start signals respectively to the TAC. The time resolution of this
apparatus is $\sim$ 350 ps using the Ar-laser pulses and $\sim$ 100 ps using the syn-
chronously-pumped dye-laser pulses. The averaged fluorescence decay curves
are in all cases transferred to a Tektronix 4051 graphic system for processing and plotting.

The irradiation source for the cell culture experiments was a continuous-wave

dye laser (Rhodamine B) tuned at 631 nm, pumped by an Argon laser. The
wavelength was measured by a Jarrel-Ash Model 82-410 monochromator. The irradiation average power as measured by a calibrated thermopile, was adjusted to 100 mW/cm^2.

Two different strains of culture cells have been studied, one epithelial
(undifferentiated) and one of mesenchimal origin (fibroblast) from rat
thyroid, each studied in parallel with its corresponding malignant coun-
ter part previously produced "in vitro" by retrovirus transformation. FRT
Cl 1 is a strain of epithelial, non-tumorigenic and undifferentiated cells
derived from adult normal rat thyroids and established as a permanently-
-growing strain. FRT Cl 1 KiKi transformed strain has been obtained from
the previous one by infection with an RNA tumor virus (Ki-MSV(Ki-MuLV)).
Viral transformation of the epithelial cells has been thus achieved using
the sarcoma virus, and transformation has been demonstrated, as already
reported (18), by morphological changes in individual cells and colonies,
by the presence of reverse transcriptase in the culture medium of infected
cells, by the acquired ability to clone in semi-solid media and by growth
in syngeneic animals forming solid, carcinoma-like tumors. FRT FIB is a
fibroblastic (as defined by morphological appearance of single cells and
colonies) strain also derived from adult rat thyroids. They are capable of
apparently indefinite growth "in vitro". This appears to occur very easily
in rat cells and - unlike human fibroblasts - without any sign of crisis or
spontaneous transformation. FRT FIB are unable to grow in semisolid media
or to form tumors in syngeneic animals. FRT FIB KiKi were obtained from the
previous normal cells by infection with the same RNA virus (Ki-MSV(Ki-MuLV)).
True transformation has been obtained, as demonstrated by the same criteria
already mentioned for the FRT Cl 1 KiKi cells.

Incubation and irradiation were performed as previously described (18). Suspensions of 10^7 trypsinized cells in 10 ml of medium lacking serum were
treated with 0 (controls) or 20 μg/ml HpD for 2 hours in a CO_2 incubator at
37°C with occasional mixing. After incubation, if laser irradiation was to
follow, 200 μl aliquots of washed cell suspensions were put in 5 mm optical
path quartz cuvettes and uniformly irradiated (mixing every min.). At the end
of experiments, cell viability was evaluated counting the cells in a haemo-

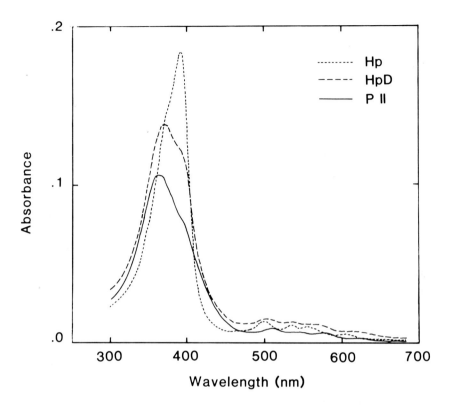

Fig. 1. Absorption spectra of fresh solutions of Hp, HpD and PII (1.25 µg/ml in water).

cytometer before and after irradiation using trypan blue (1%) in isotonic (0.9 NaCl) solution. All experimental points represent the average of several experiments.

To measure the cellular content in HpD after incubation, the cells (10^6 per sample) were washed and pelletted. The pellet was then dissolved in a mixture of H_2O (0.5 ml) and Protosol (NEN, 2 ml). The whole material was transferred in a quartz cuvette (1 cm light path) and the absorption were then measured.

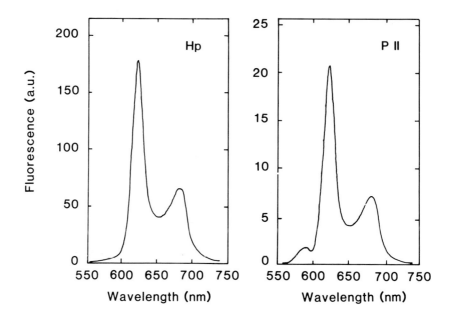

Fig. 2. Emission spectra of 1.25 μg/ml water solution of Hp and PII under excitation at 405 nm (a.u. = arbitrary units).

Results

Photophysical Properties

The absorption spectra of freshly-prepared solutions (1.25 μg/ml) of Hp, HpD and PII in water are shown in Fig. 1. The spectra indicate a different degree of aggregation. In fact, while for Hp the spectrum is peaked at 395 nm (i.e. the monomer absorption peak (19)), for HpD and PII the peak in shifted to 365 nm (i.e. the aggregate absorption peak). These results are confirmed by the emission spectra (see Fig. 2) obtained under excitation at 405 nm, where the absorbances of the solutions are very similar. The difference in fluorescence intensity that was observed between Hp and PII (a factor of ∿ 9) accounts for the larger amount of non-fluorescent aggregated species present in the latter substance (8, 9, 19). All the spectra exhibit peaks at 615 and 675 nm with similar shapes. However, a further peak at 580 nm appears in the spectrum of PII.

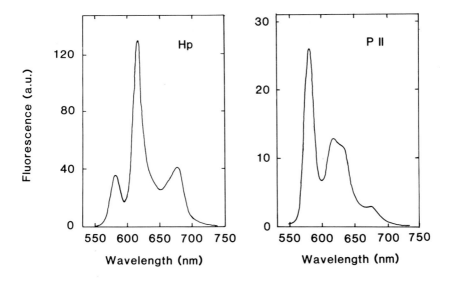

Fig. 3. Emission spectra of 1.25 μg/ml water solutions of Hp and PII under excitation at 405 nm after 48 h incubation at 37°C (a.u. = arbitrary units).

The time-resolved fluorescence of the two solutions was measured through cut-off filters (λ > 565 nm) under pulsed excitation at 405 nm. Both exper- imental decay curves could be fitted by two exponential components with time constant ∿ 14.2 ns and ∿ 3 ns and initial amplitudes 88.7% and 11.3%, respectively, for Hp and 64.1% and 25.9% for PII, indicating an increase in the fast decay component.

In order to simulate the temperature conditions in which the 580 nm band was reported to occur "in vivo" or "in vitro" (3, 14), the solutions were kept at 37°C in the dark for 48 h. The resulting emission spectra, measur- ed at room temperature after the incubation period and excited at 405 nm, showed a relevant increase of the 580 nm band in both solutions (see Fig. 3). For PII, this peak is greater than the one at 615 nm and the appearan- ce of another band at 635 nm is noted. These modifications in the emission spectra were found to be irreversible. The excitation spectra of the 580 nm and 615 nm emission were found to be peaked at 405 nm and 395 nm, re- spectively (see Fig. 4).

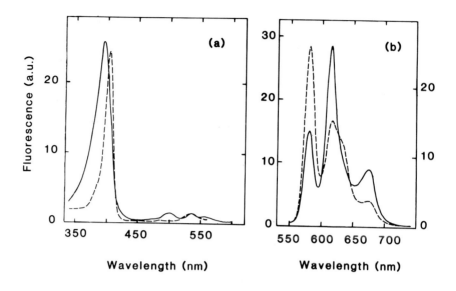

Fig. 4. Excitation (a) and emission (b) spectra of 1.25 μg/ml water solution of PII after 48 h incubation at 37°C. (a): observation at 580 nm
(----) and 615 (——); (b): excitation at 405 nm (----, left vertical scale)
and 395 nm (—— , right vertical scale) (a.u. = arbitrary units).

Excitation	Observation	τ_1	A_1	τ_2	A_2
405 nm	$\lambda > 565$ nm	14.71	11.3	2.58	88.7
	IF 580	-	-	2.51	100.0
	IF 615	14.56	17.6	2.66	82.4
395 nm	$\lambda > 565$ nm	14.86	28.7	2.75	71.3
	IF 580	-	-	2.48	100.0
	IF 615	14.80	37.9	2.74	62.1

TABLE 1: Fluorescence-decay time constants (τ_1 and τ_2) and relative amplitudes (A_1 and A_2) for a 1.25 μg/ml water solution of PII after 48 h incubation at 37°C.

These findings are in agreement with the time resolved measurements on

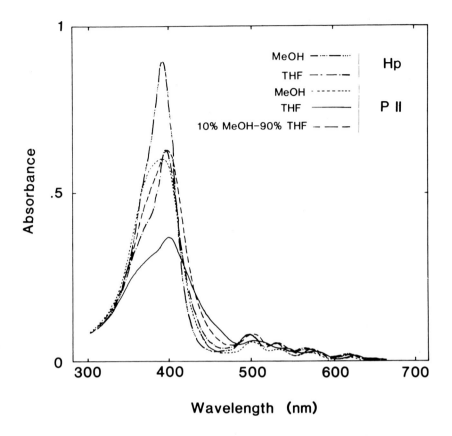

Fig. 5. Absorption spectra of Hp and PII solutions (5 µg/ml) in organic solvents.

the PII solution under excitation at 405 nm and 395 nm. The fitting parameters of the decay curves are reported in Table 1. The data indicate that: (i) the relative amplitudes of the two components depend on the excitation wavelength when the fluorescence is observed at λ > 565 nm. (ii) The amplitude of the fast component is always higher for excitation at 405 nm. (iii) Only this component is present for observation through an interference filter at λ = 580 nm. These results indicate the presence of two molecular species and it is possible, therefore, to attribute the fluorescence time constant of ∿ 2.6 ns to the molecular species with emission peaks at 580 nm and 635 nm and the time constant of ∿ 14.2 nm to the mono-

meric form of Hp, with emission peaks at 615 and 670 nm, in agreement with
our previous results. Moreover, for the PII solution, the 580 nm and 635 nm
bands increase with time even at room temperature and, in a few days, the
615 nm and 670 nm bands disappear. Only the fast component could be detected in this case in agreement with our attribution. The formation of this
new molecular species seems to require the presence of both stable aggregates and monomers. Actually the 580 nm band develops faster in PII than
in Hp and at low concentration values. According to this behaviour, we suggest that this species originates from monomers chemically bound to polymeric structures of porphyrins. Similar findings were reported for uroporphyrin when covalently bound to amino terminal agarose gel (20). On the basis of the above considerations, the appearance of the 580 nm band "in vivo"
and "in vitro" could arise from the binding of monomers to aggregates or
other polymeric structures.

The strong aggregation effect for PII in water did not allow us, even with
our time resolution ($\sim$ 70 ps for excitation at 300 nm), to detect any fluorescence from aggregated components. Since the aggregation is mainly due to
hydrophobicity, we have studied the same substances in organic solvents.

The absorption spectra of 5 µg/ml Hp and PII solutions in MeOH and THF are
reported in Fig. 5, together with that of PII solution in a mixture of 10%
MeOH and 90% THF. In all organic solvents the PII spectra are peaked at
395 nm which is indicative of a monomerization effect. However, while the
spectrum in MeOH is rather similar to that of Hp, a shoulder around 365 nm
is left in THF. The shape of the PII spectra were found to be indipendent
of concentration in the range 0.3 to 20 µg/ml, thus indicating the presence
of a stable equilibrium. In agreement with the absorption data an increase
in emission intensity for PII was observed in organic solvents respect to
water. Time-resolved fluorescence measurements were performed on Hp and PII
solutions in the spectral range λ > 565 nm under excitation at 364 nm. The
decay of Hp could be fitted by a single exponential component of $\sim$ 11.9 ns.
This is the monomer decay (19) which is shortened, as compared with that in
water, due to the more efficient singlet-state quenching by oxygen. A similar decay was found for PII in MeOH with time constant $\sim$ 10.7 ns, while
in THF the fluorescence decay could be fitted by two components with time

constants of 11.47 and 1.86 ns and relative amplitudes of 63% and 37% respectively. In agreement with the results of the absorption spectra the fluoresce decay resulted to be independent of concentration in the range 0.3 to 20 µg/ml. On the basis of these findings and of the data in ref. (8), we are led to attribute the 1.86 ns decay component to DHE. This dimer therefore seems to be peculiar of the PII chemical composition as a result of the preparation procedure of this drug from Hp.

It must be noted that the decay component attributed DHE was not detected in MeOH. To clarify this aspect, we have measured the time-resolved fluorescence of PII in mixtures of MeOH and THF. The addition of 10% MeOH is found to cause the disappearance of this component. Correspondingly, the absorption spectrum (see Fig. 5) becomes very similar to that measured in pure MeOH. Thus MeOH seems to be able to break the ether linkage in DHE.

Photosensitization of cell cultures

To investigate the existence of tissue or cellular properties responsible for the greater sensitivity of tumors to HpD phototherapy, as compared with that of the normal surrounding tissues, normal and transformed cell strains were irradiated after incubation with HpD. Normal and transformed epithelial cells (FRT Cl 1) responded to the treatment with a patterns qualitatively similar but quantitatively different. Treatment resulted in a cytocidal effect which increased with time of irradiation. The decrease in cell viability was linear when plotted on a semi-logarithmic scale from 0 to 10 min. The selectivity of the effect (i.e. preferential damage of the tumor cells) also increased with time of irradiation. Fibroblastic cells (FRT FIB) responded in a similar fashion but both cytocidal effect and selectivity were greater, to the point that no significant survival was found in the case of malignant cells after 3 min. of irradiation.

Comparative analysis of the results obtained with the two cell systems, epithelial and fibroblastic, reveals that the ratio in sensitivity between transformed and normal cells, measured as percentage of cells killed per unit time of irradiation, is significantly greater in the case of fibro-

blastic cells (ratio: 4.46) than in the case of the epithelial cells (ratio: 1.25). The difference in sensitivity was found to be related to the
larger amounts of HpD present in the tumor cell lines as compared with the
corresponding normal ones. In fact, following the procedure previously described (see Materials and Methods) the cellular HpD content was evaluated
by measuring the optical density of the solubilized samples at 400 nm. The
ratio of HpD content between transformed and normal strains resulted to be
1.44 for the FRT Cl 1 and 2.51 for the FRT FIB lines in agreement with the
survival data. The differential sensitivity and uptake seem to be related
to the cell transformation rather than to the faster duplication or metabolic rate of tumor cells. In fact these experiments performed on the same
cell line (FRT FIB) in completely different metabolic conditions (i.e. con-
fluence arrested or log phase growing) showed identical HpD uptake and survival rate.

Acknowledgements

The authors wish to acknowledge that the experiments on cells were performed in cooperation with Dr. S.F. Ambesi-Impiombato, Dr. D.Tramontano, Mr.
M.Esposito and Mr. M.Mastrocinque from CEOS (Naples, Italy).

References

1. Dougherty, T.J., Kaufman, J.E., Weishaupt, K.R., Boyle, D., Mittleman,
   A.: Cancer Res. 38, 2628 (1978).

2. Dougherty, T.J., Lawrence, G., Kaufman, J., Boyle, D.G., Weishaupt,
   K.R., Goldfarb, A.: J. Nat. Cancer Inst. 62, 231 (1979).

3. Dougherty, T.J., Gomer, C.J., Weishaupt, K.R.: Cancer Res. 36, 2330
   (1976).

4. Dougherty, T.J., Thoma, R.E., Boyle, D.G., Weishaupt, K.R.: Cancer Res.
   41, 401 (1981).

5. Dougherty, T.J., Weishaupt, K.R., Boyle, D.G., Kaufman, J., Johnson, R.:
   In: Progress in Photobiology (Castellani, A., ed.) Pergamon Press, London, 1977, p. 435.

6. Lipson, R.L., Baldes, E.J., Olsen, A.M.: J. Natl. Cancer Inst. 26, 1
   (1961).

7. Berenbaum, M.C., Bonnett, R., Scourides, P.O.: Br. J. Cancer 45, 571 (1982).

8. Dougherty, T.J., Potter, W.R., Weishaupt, K.R.: In: Porphyrins in Tumor Phototherapy (Andreoni, A., Cubeddu, R., eds.) Plenum Publishing Co., New York, 1984.

9. Andreoni, A., Cubeddu, R., De Silvestri, S., Laporta, P.: Chem. Phys. Letters 88, 33 (1982).

10. Bonnett, R., Berenbaum, M.C., Kaur, H.: In: Porphyrin in Tumor Phototherapy (Andreoni, A., Cubeddu, R., eds.) Plenum Publishing Co., New York, 1984, p. 67.

11. Andreoni, A., Cubeddu, R.: Chem. Phys. Letters 100, 503 (1983).

12. Dougherty, T.J., Boyle, D.G., Weishaupt, K.R., Henderson, B.A., Potter, W.R., Bellnier, D.A., Wityk, K.E.: In: Porphyrins Photosensitization, (Kessel, D., Dougherty, T.J., eds.) Plenum Publishing Co., New York, 1983, p. 3.

13. van der Putten, A.J.M., van Gemert, M.J.C.: In: Proceedings of Laser '81 Opto-Electronik, München, West Germany, 1981.

14. Berns, M.W., Wilson, M., Rentzepis, P., Burns, R., Wile, W.: Lasers in Surgery and Medicine 2, 261 (1983).

15. Cannistraro, S., Jori, G., Van de Vorst, A.: Photochem. Photobiol. 27, 517 (1978).

16. Cubeddu, R., De Silvestri, S., Svelto, O.: Opt. Commun. 34, 460 (1980).

17. Cova, S., Longoni, A., Andreoni, A., Cubeddu, R.: IEEE J. Quantum Electr. QE-19, 630 (1983).

18. Andreoni, A., Cubeddu, R., De Silvestri, S., Laporta, P., Ambesi-Impiom bato, F.S., Esposito, E., Mastrocinque, M., Tramontano, D.: Cancer Res. 43, 2076 (1983).

19. Andreoni, A., Cubeddu, R., De Silvestri, S., Jori, G., Laporta, P., Reddi, E.: Z. Naturforsch. 38c, 83 (1983).

20. Spikes, J.D., Burnham, B.F., Bonner, J.C.: In: Porphyrins in Tumor Phototherapy (Andreoni, A., Cubeddu, R., eds.) Plenum Publishing Co., New York, London, 1984, p. 61.

Received July 25, 1984

Discussion

Lamola: When porphyrins chelate Mg^{++} or Zn^{++} there is a characteristic red shift of the Soret band and blue shift of the fluorescence emission. It is also well known that porphyrins scavenge zinc in aqueous solution very efficiently. Have you discounted zinc complexes as being responsible for the "new" porphyrin that you have observed?

Cubeddu: We did not. Our assumption was based on a similar behaviour reported by J. Spikes et al. in uroporphyrin I where a similar shift of the
absorption spectrum was observed when monomeric uroporphyrin I is bound to
agarose. Moreover, the formation of zinc complexes should be more favour-
able in the presence of a large amount of monomers, that is, for hematoporphyrin free base solution rather than for Photofrin II solution, while
our observation did show the opposite. In any case, we will consider your
arguments and will verify the possible presence of zinc.

Hoff: I wonder whether it would be worthwhile to use chlorin derivatives
instead of hematoporphyrin, since chlorins have a much higher absorbance
in the Q_y region.

Cubeddu: There are several substances that are known to be photosensitizers. However, before they can be used for photochemotherapy of tumors
they must be tested to verify if they satisfy several requirements as, for
example, preferential accumulation in tumors with respect to the surrounding tissue, toxicity, time of clearance, cytocidal activity, etc. To my
knowledge, Hematoporphyrin Derivative is the only drug that seems at present to satisfy all requirements. Any new drug that can be proved to be an
alternative will be of very valuable interest. In particular I agree that
chlorin derivatives, due to their absorption properties, could be interesting candidates to start a new research.

McDonagh: With respect to the question of Dr. Hoff, I might point out
that chlorins are known to cause photosensitivity in mammals and might,
therefore, in principle be used for photochemotherapy. For example, accumulation of phylloerythrin (derived from metabolism of ingested chlorophyll)
in sheep and cattle can lead to acute, tissue-damaging photosensitivity.

I do not think that it is accurate to say that hematoporphyrin itself does
not accumulate in tumors. In fact, hematoporphyrin was used in the very
first study by Diamond and coworkers which showed that porphyrins can be
used to photosensitise tumors in vivo, a study that led to the present interest in photodynamic therapy of cancer. Those studies have subsequently
been confirmed by others using different tumor models in vivo.

It is also premature to imply that di-hematoporphyrin ether is the principle component of the crude mixture of substances marketed under the trade
name of Photofrin II. In fact, the ether which would itself be a mixture of
stereoisomers, has not yet been characterised and there is yet no reliable
published evidence that it is a major constituent of the Photofrin II mixture.

One advantage of the monophotonic process, which would not be shared by the
biphotonic process, is that it leads to a phototoxic product that is freely
diffusible and has a relatively long lifetime.

Lastly, I would question the scientific value of carrying out sophisticated
photophysical measurements on complex ill-defined mixtures of uncertain,
and possibly non-reproducible, composition.

Cubbedu: I think one must be very careful before referring to pure hematoporphyrin free base. Actually, the final real composition of the hematoporphyrin solution depends very strongly on the manipulation performed to dis-

solve the powder in water. As a matter of fact, observing the absorption spectrum, we have found that depending on NaOH concentration and on the time the hematoporphyrin powder is left in the NaOH solution before neutralization, the hematoporphyrin solution absorption spectrum resembles more or less that of Hematoporphyrin Derivative. No phototherapeutic effect was observed by us in vivo injecting to mice a solution of hematoporphyrin free base. Similar results have been reported by several authors as T.T. Dougherty et al. and R. Bonnet et al. Referring to the composition of Photofrin II we have reported the data of other authors. T.T. Dougherty and coworkers have shown that Photofrin II is a mixture of several monomeric porphyrins plus a heavier component that was described to be a covalent hematoporphyrin dimer and, more precisely, an ether. [These results are published in the book "Porphyrins in Tumor Phototherapy", edited by A. Andreoni and R. Cubeddu, Plenum Publishing Co., New York, London (1984)]. This assumption has been supported also by other groups, as R. Bonnet et al. and D. Kessel et al.

Our results, however, simply show the existence of a new molecular species with a time decay constant of ~1.8 ns that is present only in Photofrin II solution (and to a minor extent in Hematoporphyrin Derivative solution) and is not present in hematoporphyrin free base solution. In agreement with the mentioned authors we have attributed this time decay to dihematoporphyrin ether.

I think we have shown with our cell irradiation experiment that pulsed irradiation produces a more efficient killing of cells treated by Hematoporphyrin Derivative. This clearly indicates that the photoproducts obtained from a two-step process are more cytotoxic than singlet oxygen.

I do not believe that the final description you have made does fit HpD solution. As I have already mentioned, several papers in the literature describe the HpD composition, and the reproducibility of the drug is evidenced by the very encouraging results that have been obtained on more than 1500 patients treated all over the world.

Finally, I would like to point out that HpD photochemotherapy is all based on the interaction of light with porphyrin molecules. I therefore think that it is very important to study the optical properties of these molecules as absorption, emission and fluorescence lifetimes. In particular, these last parameters have been shown by us to be of importance to provide evidence for new molecular species that cannot be optically revealed by other means. The relevance of these results is related also to their use in detecting the presence of these molecular species in cells and tissues.

Holzwarth: Would you not expect any problems in living tissue with the two-photon reaction scheme? Any other compound present with small absorption at this wavelength might also give rise to two-photon processes which may cause toxic side effects.

Cubeddu: We have already tested a pulsed irradiation up to 30 minutes with an average power of 400 mW at the wavelengths of 631 nm (20 Hz repetition rate) on both cells cultures and experimental solid tumors in mice. By irradiating an area of ~1 cm^2 we have not observed any effect in the absence of Hematoporphyrin Derivative.

Ketterer: You have a marked difference between the uptake of porphyrin by
transformed thyroid cells versus fibroblasts. The latter are much more effective. Do you know where the porphyrin is stored etc. in the transformed
fibroblasts? This is an excellent system for study. Such differentials in
uptake as you describe could have very important implications for therapy
of mixed tumors.

Cubeddu: We do not know at the moment the porphyrin localization in our
cell systems. I agree that it will be of great interest to find the porphyrin localization in the cells and to understand the difference in uptake between FRT-Fibro and FRT-Clone 1.

Song: In those animal experiments you described, you used a CW laser as
continuous actinic source. Is it possible that two-photon processes are
also present in tumor phototherapy?

Cubeddu: The intensity values used in continuous wave irradiation therapy
are too low to allow any efficient two-step process.

Song: You have attributed a shorter τ_F of PII in methanol to oxygen. Did
you not deoxygenate the solution?

Cubeddu: We have observed a shorter time decay of monomeric porphyrins
in organic solvents with respect to that found in water. This is due to
quenching of the singlet excited state by oxygen that is present in larger
amounts in the organic solvents.

OPTICAL PROPERTIES OF FERRIHEME-QUINIDINE COMPLEXES

G. Blauer

Department of Biological Chemistry, The Hebrew University of Jerusalem,
91904 Jerusalem, Israel

Summary

Circular dichroism (CD) and light-absorption spectra were measured for
ferriheme (ferriprotoporphyrin IX) in the presence of excess (+)-quinidine
and in dilute aqueous solutions (26 to 27°C) of both pH 7.4 and 11.5.
Unique CD spectra were observed in the range of 360 to 650 nm, particularly at pH 11.5, where very large ellipticities were recorded at 420 nm.
Ferriheme or (+)-quinidine alone were optically inactive under the above
conditions of measurement, indicating highly specific complex formation in
the presence of both components. By preliminary ellipticity measurements
at both pH 7.4 and 11.5, a 1:1 mole ratio between ferriheme and quinidine
in the complexes appeared to be predominant. Ultracentrifugation revealed
a slow formation of higher aggregates in the presence of both ferriheme and
quinidine, depending also on pH. Time-dependent changes of the CD spectra,
extending over several days, were also observed, particularly at pH 11.5.
As in the case of ferriheme-quinine complexes investigated previously, at
least part of the optical activity observed may originate from interactions among ferriheme molecules arrayed chirally within specific aggregates formed from ferriheme and (+)-quinidine. The slow increase in the
absolute ellipticity values observed may reflect dynamic changes in the
relative orientation of the complex components within the aggregates.

Introduction

Interactions between FP* and quinine which is a stereoisomer of Qd (Scheme 1)
have been described (1,2). Complexes in benzene between FP and quinine or

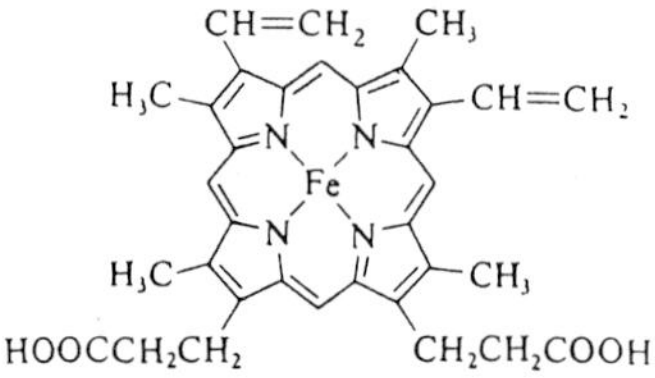

Scheme 1. Structures of iron-protoporphyrin IX (I), quinidine (II), and quinine (III).

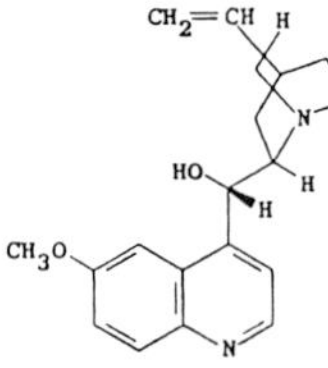

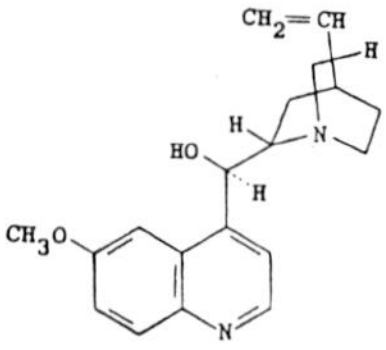

related antimalarial drugs were investigated (2). More recently, FP was considered to act as a receptor under physiological conditions for anti-malarial drugs based on quinoline, such as chloroquine [(3) and refs. cited therein]. Complex formation in aqueous solution between chloroquine and FP was characterized by light absorption (4). In addition, complexes between FP and (—)-quinine in aqueous solution, as measured by CD and light absorption, were reported (5). Fairly large and time-dependent CD bands were observed near 400 nm at pH 7.4. This complex showed charac-teristic aggregates in the analytical ultracentrifuge. It has been tenta-tively suggested that the optical activity observed in the 400 nm region ($[\theta]^{410} \simeq$ -2 x 10^5 deg·cm²·decimole⁻¹ FP) is due, at least in part, to interacting molecules of FP arranged chirally in specific complex aggre-gates (5). Very recently, Panijpan et al. (6) confirmed the large nega-tive CD band in an FP-quinine complex at pH 7.4, while no adjacent positive band was reported.

The present publication deals with similar investigations on FP-(+)-Qd com-plexes formed at both neutral and alkaline pH values. Under certain condi-tions (pH 11.5) a similar CD band near 400 nm and the corresponding rota-tory strength were larger by an order of magnitude than analogous values measured previously in FP-quinine complexes at pH 7.4 (5). (The latter com-plex at pH 11.5 also showed a much smaller and different CD spectrum near 400 nm.) Possible structures of these unique FP-Qd complexes are discussed below. FP or Qd alone had no measurable ellipticities between 360-650 nm.
*Abbreviations used: FP, ferriprotoporphyrin IX; Qd, quinidine; CD, cir-cular dichroism.

Materials and Methods

Hemin (chloro-ferriprotoporphyrin IX) was obtained from Sigma (Type I).
Stock solutions usually contained about 10 mg hemin in 25 ml of 0.02 N
aqueous NaOH. These solutions were stored in plastic vessels at 0 to 4°C
and in the dark for periods of up to about 10 days. The light-absorption
spectra of the stock solutions were checked regularly. Solutions which
showed significant changes in the 380 to 550 nm region were discarded.

(+)-Quinidine sulfate (trihydrate) was obtained from Sigma and (−)-quinine
sulfate (dihydrate) from Aldrich. Stock solutions of about 5×10^{-3} M were
prepared by addition of aqueous HCl (pH ∿ 4). These solutions were kept
in plastic bottles and were stored in the dark at 0 to 4°C. The ultraviolet
absorption spectra were checked frequently at pH 7.4. No significant
spectral changes were observed in the stock solution during storage of up
to three months. It was assumed that no significant structural changes in
the Qd and quinine occur under the above conditions of storage.

Poly-α,L-lysine-HBr was obtained from Miles-Yeda (Rehovot). The parent
carbobenzoxy derivative had a molecular weight of about 90,000. Buffers
and other salts used were of analytical grade.

Light-absorption spectra were measured on a Cary Model 14 spectrophotometer. The cell compartment was kept at 25 to 26°C. Distilled water was
used as reference solvent. In the wavelength range given, contributions
of quinidine to the absorbance were negligible.

Circular dichroism was measured on a Cary Model 60 recording spectropolarimeter equipped with a Model 6002 accessory. Calibration of the instrument
was checked daily with (+)-10-camphor sulfonic acid (Eastman Kodak, highest purity). Both the complexes and the reference solvent (distilled water)
were measured in the same cell. The contribution of the quinidine to the
measured ellipticity was usually not significant under the conditions given
(see Results). The optical path of the cells was chosen to give absorbance values below about 1.2 for most experiments. The temperature in the
cell compartment was usually 27.5 ± 1°C.

Molar ellipticities $[\theta]_\lambda$ given were calculated according to: $[\theta]_\lambda = \dfrac{10\theta_\lambda}{c \cdot d}$

where θ_λ is the observed ellipticity at wavelength λ in degrees; c, the concentration in moles$\cdot$l^{-1} total FP; and d, the optical path length in decimeters.

Ultracentrifugation. A Beckman Model E analytical ultracentrifuge was used at 50,740 rev./min with Schlieren optics. Sedimentation constants were corrected to 20°C.

Preparation of the complexes. Usually an alkaline stock solution of FP was added to aqueous solutions of buffer and salt, and the pH was adjusted to the required value by concentrated acid or base. [By this procedure significant amounts of both sodium or chloride ions will be added to the sodium chloride included originally (see legends to Figures), e.g., about 10 mM of extra chloride will be present in the case of 4 x 10^{-4} M FP at neutral pH.] Required amounts of the slightly acid quinidine stock solution were then added to a given volume (usually 3 ml). Plastic vessels were used in order to minimize adsorption of the quinidine to the walls of the container.

pH values were checked at the end of an experiment. Solutions were protected from light.

Results

Time dependence

Under the conditions given for Fig. 1, an FP-Qd complex at pH 7.4 reaches constant and relatively small ellipticity values in the Soret region near 400 nm, after about 2-3 h from the time of its preparation (curves A and B). In sharp contrast, an FP-Qd complex at pH 11.5 shows very little optical activity during the first hours, but later the absolute ellipticity values at 389-390 nm, and in particular at 419-420 nm, increase considerably for several days, reaching very large and fairly constant values when kept for over 100 h at 26°C (Curve E, Fig. 1). On the other hand, FP-quinine com-

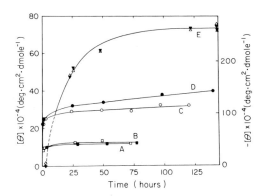

Fig. 1. *Time course of the appearance of ellipticity extrema in FP-Qd and FP-quinine systems in aqueous solution.* Temp. of measurements, 27.5 ± 0.5°C. In all experiments at pH 7.4, 0.01 M sodium phosphate buffer and 0.05 M sodium chloride were included. Except for the time of measurements, all samples were kept at 26.0 ± 0.5°C. Left-hand ordinate refers to all curves, except E at 420 nm. For comparison of the kinetic curves, the observed absolute θ-values are in each case given as molar ellipticities based on total FP. Curve A: FP, 1.0 x 10⁻⁴ M; Qd, 2.4 x 10⁻⁴ M; pH 7.40 ± 0.05. θ-Values refer to 396-397 nm. Curve B: FP, 4.0 x 10⁻⁴ M; Qd, 8.0 x 10⁻⁴ M; pH 7.40 ± 0.05. θ-Values refer to 396-397 nm. Curve C: FP, 1.0 x 10⁻⁴ M; quinine, 2.0 x 10⁻⁴ M; pH 7.40 ± 0.05. θ-Values refer to 410 nm. Curve D: FP, 4.0 x 10⁻⁴ M; quinine, 8.0 x 10⁻⁴ M; pH 7.40 ± 0.05. θ-Values refer to 410 nm. Curve E: FP, 1.0 x 10⁻⁴ M; Qd, 2.0 x 10⁻⁴ M; NaCl, 0.05 M; pH 11.5 ± 0.1. No phosphate buffer included. θ-Values refer to: Δ, 389-390 nm; ▼, 419-420 nm (right-hand ordinate). The solid curve is calculated for a first-order conversion to an optically active complex. After about 120 h at 26°C, some turbidity was noticed.

plexes at pH 7.4, included for comparison in Fig. 1 (Curves C and D), show much smaller but slowly increasing ellipticity values, which do not seem to level off even after about 100 h. All complexes at pH 7.4, given in Fig. 1 on a large time scale, show an initial rapid increase in ellipticities within the first hour from their preparation, indicating two or more different kinetic processes. In contrast, the FP-Qd complex at pH 11.5 (Curve E), indicates an induction period with regard to the development of ellipticity (particularly at 420 nm) during which a relatively small optical activity only is measured. All these observations demonstrate the uniqueness of the FP-Qd system at high pH. The solid curve (E) in Fig. 1 represents a first-order process for both CD bands at

389-390 nm and 419-420 nm, respectively. A corresponding rate constant
has a value of 0.045 h^{-1} under the conditions given (calculated on a Wang
2200 desk computer system), corresponding to a half-time of 15.4 h for
conversion to an optically active complex. Additional experiments will
be required in order to substantiate these kinetics which appear to be
applicable at higher degrees of conversion.

Light absorption

Figs. 2 and 3 include the light-absorption spectra of FP in the presence
of excess Qd and of electrolytes at pH 7.4 and 11.5, respectively. Because
of the time dependence of the ellipticity observed in these cases, the
light-absorption spectra were also measured both after a few hours and
after several days from the time of preparation. At pH 7.4 (Fig. 2), the
spectra obtained after about 4 h and 8 days from preparation of the solution were practically the same (Curve A). Pronounced maxima were recorded
at 610-612 nm, 495 and 397 nm, respectively. In the absence of Qd, a diffuse spectrum was obtained in the visible region exhibiting some shoulders
of bands between 540 and 610 nm. A broad and apparently composite band
was measured around 360-370 nm with lower absorbance than that around 397
nm obtained in the presence of Qd (Curve B, Fig. 2). This spectrum changed
little with time after 8 days (Curve C, Fig. 2). Thus, there is evidence
from spectral data alone for the formation of a complex between FP and Qd
at neutral pH. The spectrum of the complex (Curve A, Fig. 2) resembles
that of alkaline FP (see Curve C, Fig. 3). A similar absorption spectrum
was reported for the FP-quinine complex at pH 7.4, but the molar absorbance values of the maxima were lower in the latter case (5). At pH 11.5,
the spectrum between 460 and 650 nm in the presence of excess Qd (Curve A,
Fig. 3) is very similar to that obtained in its absence (Curve C, Fig. 3),
while the Soret band maximum in the presence of Qd is somewhat lower and
shifted by about 7 nm to longer wavelengths. After 5 days at 26°C, the
total absorbance in the presence of Qd has increased considerably between
about 450 and 650 nm (Curve B, Fig. 3), while the Soret band maximum is
slightly lower than that measured 1 h after preparation of the solution
and is shifted to the red by about 8 nm relative to Curve A. In the absence
of Qd, only minor spectral changes are observed between 1 h and 8 days

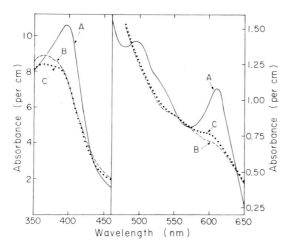

Fig. 2. *Light-absorption spectra of FP in the presence and absence of Qd at pH 7.40 ± 0.05.* FP, 2.0 x 10^{-4} M; Qd, 4.0 x 10^{-4} M; sodium phosphate, 0.01 M; NaCl, 0.05 M; temp., 25-26°C. Curve A: measurements were started 235 min after preparation of the final solution; after 8 days (191 h) from preparation, a similar spectrum was obtained. FP in the absence of Qd (other conditions as given above): Curve B: measured 210 min from time of preparation. Curve C: measured after 195 h at 26.0 ± 0.5°C.

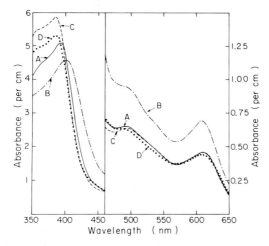

Fig. 3. *Light-absorption spectra of FP in the presence and absence of Qd at pH 11.55 ± 0.05.* FP, 1.0 x 10^{-4} M; Qd, 2.0 x 10^{-4} M; NaCl, 0.05 M; temp., 25-26°C. Curve A: measurements were started 60 min after preparation of the final solution. Curve B: measured after 5 days at 26.0 ± 0.5°C; this solution was slightly turbid. For comparison, FP in the absence of Qd (other conditions as given above): Curve C: measurements were started 40 min from the time of preparation. Curve D: measured after 5 days at 26.0 ± 0.5°C.

(Curves C and D, Fig. 3). Thus, there is also some spectral evidence at pH 11.5 for complex formation between FP and Qd after about 1 h. In addition, the large spectral differences observed in the complex after 5 days indicate the occurrence of slow changes in the complex, in contrast to the FP-Qd complex at pH 7.4 (a difference in the total concentration of the complexes at the two pH values is not considered to affect the comparison). These time-dependent changes observed in the complex are strikingly accentuated in the CD data (see Curve E, Fig. 1 and below, Fig. 5).

CD spectra

In Fig. 4 the CD spectrum of an FP-Qd complex at pH 7.4 in the range of 330 to 650 nm is shown. There is a positive band centered at 397-398 nm and smaller bands also appear near the light-absorption maxima or shoulders in the 460-650 nm region (see Curve A, Fig. 2). There are only minor changes with time of the CD spectra between 3.5 h (Curve A) and 8 days (Curve B), except for the 580 to 650 nm range, where the ellipticity increases with time. Fig. 5 (Curve A) includes the CD spectrum of an FP-Qd complex obtained at pH 11.5, 5 days after its preparation (see also Curve E, Fig. 1). [After 155 min from the preparation of the complex only a small positive CD band centered at 395 nm ($\sim$ 14 millideg·cm^{-1}) was observed.] The main feature of this spectrum is the very large negative CD band centered at 420 nm with a relatively smaller and adjacent positive band around 390 nm. For comparison, the CD spectrum of an FP-poly-L-lysine complex at pH 11 is given in the range of 300 to 460 nm (Curve B, Fig. 5). Although the absolute values of the ellipticities are considerably smaller in the latter case, as compared to the FP-Qd complex, the similarity of the two CD spectra is obvious. Possible interpretations of these features will be discussed below. The reproducibility in the magnitude of the 420 nm band of the FP-Qd complex was not always satisfactory, and $\theta_{obs.}$ values were in the range of -2700 to -4400 millideg·cm^{-1} under otherwise identical conditions. The apparent rotatory strength R_{420} is calculated by the approximation (see Ref. 7):

$$R_k \doteq 1.23 \times 10^{-42} \frac{[\theta_k^\circ]\Delta_k}{\lambda_k^\circ} \qquad [1]$$

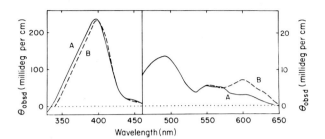

Fig. 4. *CD spectra of FP in the presence of Qd at pH 7.30 ± 0.05.* FP, 2.0
x 10^{-4} M; Qd, 4.0 x 10^{-4} M; sodium phosphate, 0.01 M; NaCl, 0.05 M; temp.,
27.5 ± 0.5°C. Curve A: measurements were started 3.5 h after preparation
of the final solution. Curve B: measured after 8 days at 26.0 ± 0.5°C.

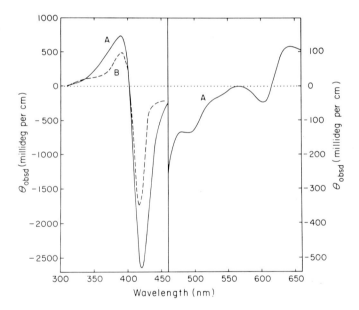

Fig. 5. *CD spectrum of FP in the presence of Qd at pH 11.45 ± 0.05 and of*
poly-L-lysine at pH 11.1 ± 0.1. Temp., 27.5 ± 0.5°C. Curve A: FP, 1.0 x
10^{-4} M; Qd, 2.0 x 10^{-4} M; NaCl, 0.05 M. This solution was kept at 26.0 ±
0.5°C and measured 5 days (122 h) after its preparation. Curve B: FP, 1.0
x 10^{-4} M; poly-L-lysine, 2.0 x 10^{-3} residue M. Measurements were started
1 h after preparation of the final solution. In the visible region, only
one measurable band, centered near 555 nm, was observed (-8 millideg·cm^{-1}).
Left-hand ordinate: FP-Qd in the range of 300 to 460 nm. Right-hand ordinate: FP-polylysine in the range of 300 to 460 nm and FP-Qd in the range
of 460 to 650 nm.

where $[\theta_k^\circ]$ is the molar ellipticity at the band extremum; λ_k° its wavelength position; and Δ_k the band half-width at $[\theta_k^\circ]/e$. R_{420} is about -13
Debye-Bohr magnetons for this extremely large band given by part of Curve
A, Fig. 5. The molar ellipticity is based on total FP, because the main
CD bands are in the Soret region and a 1:1 FP-Qd complex is formed (see
below). Its value is of the order of $(-3$ to $-4) \times 10^6$ deg·cm²·decimole⁻¹
under the conditions given. However, because of the heterogeneity of the
system (see below), the molar ellipticity based on total FP may represent
an apparent value only. The possibility of optical artifacts affecting
both the absorption and ellipticity of these aggregate systems should also
be considered, although such effects could not be substantiated so far.
Previously, a similar CD band centered at 410 nm was observed for an FP-
-quinine complex at pH 7.4; however, the extremum was much smaller (5).
Much lower ellipticities are also measured for the FP-Qd complex in the
visible region at 460 to 650 nm (Fig. 5) with extrema near the peaks of
the light-absorption bands (see above). In contrast to pH 7.4 (Fig. 4),
two proximate CD bands of opposite sign are observed not only in the
390-420 nm region but also near 600-650 nm.

In the absence of Qd, FP does not exhibit significant optical activity
when measured under the conditions given for Figs. 4 and 5. The ellipticities of the free Qd are practically zero above 360 nm and do not exceed
about +15 millideg·cm⁻¹ for 2×10^{-4} M solutions of Qd at pH 7.4 or +4
millideg·cm⁻¹ at pH 11.5, measured in the range of 300 to 360 nm.

It appears that CD is a convenient and sensitive tool for identification
and evaluation of FP-Qd complexes formed under different conditions.

Dependence of the ellipticity on the concentration of the complex components

In preliminary experiments, the mole ratios between the components of the
FP-Qd complexes were evaluated, using the large ellipticity extrema as a
direct measure for the formation of a complex. The total concentration of
both the complex components FP and Qd were kept constant (see, for example,
Ref. 8) at both pH 7.4 (Fig. 6a) and pH 11.5 (Fig. 6b). At both of these
pH values, the extrema obtained indicate primarily a mole ratio of 1:1

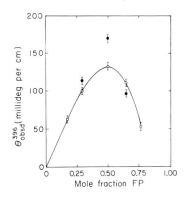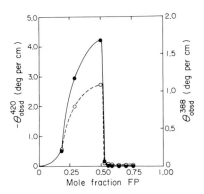

Fig. 6. *Ellipticity as a function of FP and Qd concentrations at different pH values.* Total constant concentration of FP and Qd: 3.4 x 10^{-4} M; NaCl, 0.05 M; temp., 27-28°C. (a) Left: pH 7.40 ± 0.05 (sodium phosphate, 0.01 M). The ellipticity at 396 nm was used as a measure for complex formation. ○, Measured 204 ± 5 min after preparation of complex. ●, Measured 52 ± 1 h after preparation. (b) Right: pH 11.55 ± 0.05. ●, Ellipticities measured at 419-420 nm (left-hand ordinate); o, measured at 388-390 nm (right-hand ordinate). Solutions were measured 46-48 h after preparation.
All solutions were kept at 26.0 ± 0.5°C before measurement.

between the complex components, using the ellipticities at 396 and 420 nm of the main CD band extrema at pH 7.4 and 11.5, respectively, and, in addition, at 388 nm at pH 11.5. However, there is a significant difference in the shape of the curves obtained at both pH values. Unlike pH 7.4, the high ellipticity at pH 11.5 drops abruptly to close to zero beyond a mole fraction of about 0.5 of FP. It should be noted that the ellipticity values given in Fig. 6b and obtained after 2 days from preparation of the solutions, do not represent final values at all mole ratios (see also Curve E, Fig. 1). As indicated in Fig. 6a, some ellipticity values were re-measured after an additional time (about 50 h as compared to about 200 min) and confirmed a 1:1 mole ratio for the optically active complex.

Several experiments carried out at an excess of Qd at pH 11.5 (constant FP, 1.0 x 10^{-4} M; NaCl, 0.05 M) showed that the ellipticity at 420 nm obtained after several days was practically independent in the concentration range of Qd of (2 to 10) x 10^{-4} M. At an excess of FP, the possibility cannot be excluded that additional complexes are being formed, as indicated by some shifts in the ellipticity extrema, particularly after 2 days. However, such complexes are not identifiable by the present data (see also Figs. 6a, b and Ref. 9). It should be noted that most of the present ellipticity data obtained at different concentrations and other conditions, are not amenable to a more quantitative analysis such as determination of equilibrium constants, mainly because of the dynamic effects described, the heterogeneity of the systems, difficulties in reproducibility and the uncertainties due to optical artifacts.

Effect of electrolytes

A number of preliminary experiments at pH 11.5 showed that FP (1×10^{-4} M) and Qd (2×10^{-4} M) produced precipitates after about 1 to 2 days at 26°C and at sodium chloride concentrations above about 0.25 M. On the other hand, at lower NaCl concentrations (0.02 M), only a very small optical activity could be observed near 400 nm, even when the solution was kept for about 200 h at 26°C. It appears that at pH 11.5 there is an optimum and defined concentration range of the electrolyte in which the large ellipticities can be observed.

Analytical ultracentrifugation

A number of sedimentation velocity experiments involving 4.0×10^{-4} M FP and 8.0×10^{-4} M Qd were carried out in order to obtain information on the state of aggregation and heterogeneity of the complexes. At pH 7.4 (0.05 M NaCl and 0.01 M sodium phosphate included), s_{20}-values of 7.3 S were obtained after about 1 h from the time of preparation of the complex, which is practically the sedimentation constant of FP in the absence of Qd (7.5 ± 0.5 S; average of 7 experiments obtained between 1 h and 8 days). After 5 days from preparation, an average of 17.6 S was obtained for the complex kept at 25°C. At pH 11.5 (0.05 M NaCl included), no significant sedimentation occurred 270 min from the time of preparation of the complex, while either after 7 h or after 5 days, very fast moving peaks were observed even at low centrifugation speeds. At the end of these experiments, about 25 to 30% of the original complex appeared to remain in the cell, as judged by the absorbance at 400 nm. In the absence of Qd and under otherwise similar conditions, FP did not show a measurable sedimentation pattern, although a small fraction of the total ferriheme may be present as aggregates sedimenting with about 3 S (10). Thus it appears that most of the complex at pH 11.5 is present as heterogeneous mixtures of lower molecular complexes and of high aggregates which are formed slowly during the first day after preparation of the complex. Hence the aggregate complexes formed at pH 7.4 and 11.5, respectively, differ also in their hydrodynamic behavior. Most of the above experiments were performed, for technical reasons, at two to four times higher concentrations of both complex components than in most of the corresponding optical measurements. However, analogous results were obtained in many of the latter experiments at different concentrations (see also Fig. 1).

Discussion

The present optical data require unique and different complex formation between FP and Qd at pH 7.4 and 11.5. The slow apparently first-order kinetics observed at pH 11.5, coupled with the hydrodynamic data, are consistent with a slow rearrangement in the structure of a complex between FP and Qd from an optically inactive to an highly active and thermodynamically more stable complex aggregate. This is in contrast to pH 7.4, where much smaller and stable ellipticities were recorded after about an hour from the time of preparation of the complex. On the other hand, FP-quinine complexes at pH 7.4 were found to exhibit CD spectra in the 400 nm region (5) similar to those observed in FP-Qd complexes at pH 11.5, although the ellipticities were one order of magnitude smaller in the quinine complex. The FP-quinine complexes were present as more uniform aggregates. They were also time dependent in their optical activity over periods of days. It appears that the previous tentative correlation of the CD spectra and their time dependence with a state of aggregation of the FP-quinine complexes (5) ["Aggregate-Induced Circular Dichroism (AGICD)"], may also be valid in the present case of FP-Qd complexes, where extremely large ellipticities and aggregation are recorded at high pH. However, smaller contributions to the optical activity observed under different conditions may originate from local perturbations in the complexes caused by the chiral centers of Qd, even in the absence of higher aggregated structures. Evidence with be sought to substantiate all these correlations.

The concept of AGICD gains some additional weight from the comparison of the CD spectrum near 400 nm of FP-Qd at pH 11.5 with that of an FP-poly-lysine complex (see also Ref. 11). The qualitative agreement between these two complexes is even better than in the case of an FP-quinine complex at pH 7.4 shown previously (5). ·On the basis of steric considerations and hyperchromic effects described previously on FP-poly-α,L-lysine complexes at pH 11, a structure was proposed for this hemichrome complex which included coordination of two ε-amino groups from different polylysine helices to the central FP iron (12, 13). Any neighboring FP molecules could then interact in a chiral and nearly co-linear arrangement of FP in the direction of the axes of the polypeptide helices. An analogous structure could be envisaged for the FP-Qd complex at pH 11.5, whereby the chiral Qd molecules bound each to an FP unit would produce a chiral (possibly helical) aggregated structure. Theoretical calculations are in progress to estimate the most likely relative orientations of molecules within such an aggregate structure, as well as the number of interacting units necessary to produce the large Cotton effects observed. In the present case, there is no evidence in the absorption spectra (Figs. 2 and 3) for hemichrome formation as in the case of the polylysine complex which shows a characteristic band at 540 nm (ε_{mM} = 13 mM^{-1}·cm^{-1}; see also Refs. 4 and 12). Therefore, octahedral coordination of two nitrogenous bases of the drugs to the FP iron should not occur in the present systems.

The sharp decrease in the optical activity observed beyond a mole ratio of 1:1 in the FP-Qd complex at pH 11.5 (Fig. 6b) deserves further study. A cooperative interference at a small excess of FP is indicated.

On the basis of the chemical structure of FP and Qd (Scheme 1), it is possible to postulate a variety of non-covalent interactions between the complex components and among themselves to account for the stability of the structures suggested above. Hydrophobic and electrostatic interactions are most likely, the latter including hydrogen bonding. As is evident in Scheme 1, there are many possible contacts between hydrophobic

regions in both FP and Qd, involving particularly the quinuclidine ring
of the latter. Electrostatic interactions could occur between FP carboxylate groups and protonated nitrogenous groups of Qd, depending on pH.
Hydrogen bonding could involve the Qd hydroxyl and FP carboxylate
groups. In addition, π-π electron interactions between close heterocyclic
ring structures of both FP and Qd may also contribute to the stability of
the aggregate complexes (see also Refs. 4 and 5). In a recent NMR study
of the interaction between FP and chloroquine in aqueous neutral medium,
intercalation of the quinoline residue of the latter between FP within a
polymer, has been postulated (14).

Acknowledgements

The skillful technical assistance of Mrs. J. Silfen and Mr. P. Yanai is
appreciated.

References

1. Phifer, K.O., Yielding, K.L., Cohen, S.N.: J. Exptl. Parasitol. 19,
   102-109 (1966).

2. Warhurst, D.C.: Biochem. Pharmacol. 30, 3323-3327 (1981).

3. Chou, A.C., Chevli, R., Fitch, C.D.: Biochemistry 19, 1543-1549 (1980).

4. Blauer, G., Ginsburg, H.: Biochem. Intl. 5, 519-523 (1982).

5. Blauer, G.: Biochem. Intl. 6, 777-782 (1983).

6. Panijpan, B., Mohan Rao, Ch., Balasubramanian, D.: Biosci. Rep. 3,
   1113-1117 (1983).

7. Moscowitz, A.: In: Optical Rotatory Dispersion (Djerassi, J., ed.)
   McGraw-Hill, New York, 1960, p. 150.

8. Brasted, R.C., Cooley, W.E.: In: The Chemistry of Coordination Compounds
   (Bailar, J.C., Jr., Busch, D.H., eds.) Reinhold, New York, 1956, pp.
   569-575.

9. Sillén, L.G.: In: Coordination Chemistry (Martell, A.E., ed.) Vol. 1,
   Van Nostrand-Reinhold, New York, 1971, p. 535.

10. Blauer, G., Zvilichovsky, B.: Arch. Biochem. Biophys. 127, 749-755
    (1968).

11. Yamamoto, S., Nozawa, T., Hatano, M.: Polymer 15, 330-334 (1974).

12. Blauer, G.: Biochim. Biophys. Acta 79, 547-562 (1964).

13. Yamagishi, A., Watanabe, F.: Biopolymers 21, 89-100 (1982).

14. Moreau, S., Perly, B., Biguet, J.: Biochimie 64, 1015-1025 (1982).

Received July 6, 1984

Discussion

Gouterman: Suppose you have high ellipticity with exactly 50:50 ratio of ferriheme : quinidine and then you add excess ferriheme. How long does it take for the ellipticity to fall off?

Blauer: This is one of our next experiments planned.

Song: Do you get an indication of a cooperative interaction in your complexes from the "melting curves" of temperature-dependent $[\theta]$?

Blauer: So far, we have not investigated the temperature-dependence of the ellipticity of the complexes.

Song: Is quinidine fluorescent? If so, does it get quenched by ferri-heme upon complexation?

Blauer: Quinidine is reportedly fluorescent in aqueous solution. There is a recent communication describing fluorescence quenching of an antimalarial quinoline derivative by ferriheme (see above, Ref. 6).

Beychok: Are these complexes formed with cyanoheme? With CO-heme?

Blauer: We have not investigated such complexes so far.

Woody: You mentioned that the large CD effects are observed only over a certain range of NaCl concentration. What is that range?

It is easy to understand a lower limit in ionic strength because at pH 11.5 there must be a very large negative charge on the aggregates. How do you explain the upper limit?

Blauer: Under conditions given, the optimum CD effects appear to be observed at 50 to 100 mM NaCl. Excess negative charges should indeed promote dissociation of the aggregates, while addition of a neutral electrolyte should facilitate association (see above, Ref. 10). Above about 0.25 M NaCl the aggregate precipitates, as mentioned in the lecture.

Woody: One can estimate an upper limit to the rotational strength due to the exciton effect between two hemes. For optimal geometry (heme planes parallel to each other and mutually perpendicular to the interheme vector), each exciton component has a magnitude of

$$|R_{\pm}| \;\; = \;\; \frac{\pi}{\lambda}\, r\mu^2$$

where λ = wavelength of transition, r = interheme distance, μ = transition moment magnitude.

The magnitude of the exciton rotational strength is proportional to the distance between the groups. Taking $r \sim 0.4$ nm, $\lambda \sim 400$ nm, $\mu \sim 2$ Debye, we obtain $|R_{\pm}| \sim 15 \times 10^{-38}$ cgs ≈ 15 Debye-Bohr magnetons. Since there are two Soret components, the upper limit for two hemes in van der Waals contact is ca. 30 DBM. Therefore, your estimate of the rotational strength of the negative lobe of the Soret band (ca. 13 DBM) can in principle be explained within the framework of the exciton model.

Buchler: My question refers to Fig. 6 (right part) of your paper. Do the optical absorption spectra or the turbidity/light refraction/light scattering of the solution also show this discontinuity as a function of the mole fraction FP which is observed with the ellipticity? To me the phenomenon looks like some colloidal chemistry effect - as if you change at this point the surface charge of a species which then suddenly aggregates (or disaggregates).

Blauer: It is difficult to observe a discontinuity in the light-absorption spectra because of the large contribution of free ferriheme and its similarity to the spectrum of the complex. There was no noticeable light-scattering effect under our conditions of measurements.

Gregory: High CD values are also found with chlorophyll in chloroplast grana and in aggregating light-harvesting complex (LHC). The CD is accompanied by (a) non-gaussian form (b) differential circular scattering resembling ORD. Do you find these? Is this a pattern for "AGICD"?

Blauer: The possibility of optical artifacts is considered in my paper. As already mentioned, so far some pertinent tests did not reveal such effects. It appears that "Aggregate-Induced Circular Dichroism" is not necessarily coupled with the phenomena mentioned by you.

Following an exchange with Dr. Myer whether or not CD in inhomogeneous systems is totally vitiated by internal scattering, given the usual Legrand/Grossean instrument, Dr. Woody comments as follows:

Woody: Observations reported some years ago on heme a might be related to your results. The failure of attempts to observe chiral phenomena in heme a were surprising in view of the generally accepted presence of a chiral center at the α-carbon of the side chain at position 2. In a symposium on cytochromes, King and coworkers (1) reported the observation of optical activity in heme a in aqueous solution. However, King also reported that the addition of pyridine, detergent, acid or base to the heme a sample led to disappearance of the observed optical activity. In addition, Lemberg (2) commented that he had observed similar phenomena, but with opposite signs! It is almost certain that these anomalies are due to the presence of large aggregates of extensively racemized heme a in aqueous solution. A small excess of one enantiomer can impose an overall chirality on the large aggregates and result in a detectable ORD/CD. However, when the aggregates are molecularly dispersed, the small enantiomeric excess is undetectable. The sign discrepancy is probably due to minor variations in the preparative procedure which led to a net retention in one case and a net inversion in the other. In your system, large aggregates also occur, but with the same chirality throughout the system, leading to much stronger effects.

1. Yong, F.C. et al. (1968) "Structure and Function of Cytochromes".
   Okunuki, K., Kamen, M.D. and Sekuzu, I., eds., University Park Press,
   Baltimore, pp. 196-203.

2. Lemberg, R. (1968) ibid., p. 215.

I would like to add a comment concerning the exchange between Dr. Myer and Dr. Gregory. The depolarization to which Dr. Myer referred does not affect the circularly polarized light which is used in CD measurements, as pointed out by Dr. Gregory. However, an analogous phenomenon, differential scattering of circularly polarized light does present serious problems in strongly scattering systems such as membrane suspensions. Furthermore, a common method for testing for such effects, a CD signal which is insensitive to the distance between the sample and the photomultiplier, may fail under certain circumstances. Tinoco, Maestre and Bustamante (1-3) have explored these questions theoretically and also experimentally, in the case of nucleic acid systems such as ψ-DNA and nucleoprotein complexes. They find that when chiral complexes have dimensions comparable to the wavelength of the light, differential scattering of circularly polarized light may be concentrated at large angles with respect to the direction of the beam. If this is the case, movement of the sample relative to the photomultiplier will not lead to substantial changes in the observed signal. Nevertheless, differential scattering would strongly affect the CD signal. Another criterion which should also be applied, but which is also not necessarily foolproof, is the absence of tails on the CD spectrum extending to wavelengths where the sample does not absorb. Of course, this may be more difficult in the case of heme proteins and especially chlorophyll-containing systems. Ultimately, as Tinoco, Maestre and Bustamante point out, one needs to study the differential scattering phenomenon directly. This can, in fact, provide additional information about large-scale chiral structures.

1. Tinoco, I., Jr., Bustamante, C. and Maestre, M.F. (1980) Ann.Rev.
   Biophys.Bioeng. 9, 107-141.

2. Bustamante, C., Tinoco, I., Jr. and Maestre, M.F. (1983) Proc.Natl.
   Acad.Sci.USA 80, 3568-3572.

3. Tinoco, I., Jr., Maestre, M.F. and Bustamante, C. (1983) Trends
   Biochem.Sci. 8, 41-44.

Scheer: My question refers to your last comments. Is there any good correlation between aggregation and the intense CD signal, e.g. in your breakdown region after addition of more than one mole of hemin to quinidine, or at about 50 % of the maximum CD signal?

Blauer: The problem of correlation may be due, at least in part, to the
difficulty in separation of the nonaggregated supernatant from the aggregate precipitate which may partly redissolve during the separation procedure.

Myer: Is the Job plot profile (Fig. 6) at pH 11.5 a constant feature at
all the time-spans of the incubation? The slide showed the profile after
50 h. How at about 20, 100 or say 150 hours?

Blauer: The time of close to 50 h was chosen for this plot because the
complex showed the highest molar ellipticity at excess quinidine and the
kinetics obtained under these conditions (Fig. 1) indicated close to 90 %
of the limiting ellipticity obtained after 150 h.

Sund: For the better understanding of Fig. 6, I would like to ask two
questions with respect to this Figure: Did you titrate the quinidine solutions or did you prepare separate solutions of different mole fractions?

Blauer: Each point in Fig. 6 corresponds to a separate experiment.

Sund: Since you got a strong decrease of the ellipticity at mole fractions higher than 0.5 only at pH 11.55 but not at pH 7.4, I would like to
know the pH dependence of the ellipticity and the sedimentation behaviour.

Blauer: According to our preliminary results, the type of complex formed
at pH 11.5 or 11.0 is not observed at pH 10, so that the transition to
this complex should occur between pH 10 to 11. As described in our paper,
most of the complex appears to sediment with very high velocity even at
low speeds of centrifugation. At pH 7.4, a more homogeneous sedimentation
pattern is obtained after some time with average values of 17 to 18 S at
20° C.

McDonagh: I have two questions and a comment.

Does hydroxylation of vinyl groups play a rôle in the long term changes
that you see, and are similar changes seen with meso-heme?

Blauer: Experiments involving meso-heme or other hemes with different
side chains are being planned.

McDonagh: Are μ-oxo dimers of heme involved in any of the aggregation
phenomena that you see?

I might also add that aggregation of heme in the absence of quinidine has
some clinical interest with respect to the treatment of acute porphyria.
Patients with acute attacks often respond favorably to intravenous infu-
sion of heme. However, there is considerable variation in the effectiveness of the treatment. One possible reason for the variability is variation in the state of aggregation of the infused heme solution.

Blauer: We have no evidence for a possible involvement of μ-oxo-dimers
of ferriheme in the ferriheme-quinidine aggregates. The mole ratio of 1:1
would preferably suggest alternating single ferriheme and quinidine molecules in an aggregate, however, a dimer-dimer structure may also be possible.

RESONANCE RAMAN STUDIES OF HAEMOGLOBIN

Kiyoshi Nagai

MRC Laboratory of Molecular Biology
Hills Road
Cambridge CB2 2QH
ENGLAND

Introduction

Haem groups can carry out a wide variety of biological reactions when incorporated into different proteins. Various derivatives of haem found in nature share a basic structure: a tetrapyrrole ring with variations in peripheral groups at the 2 and 4 positions. Futhermore, side chains of various amino acids including histidine, cysteine and methionine can coordinate to the haem iron. These variables determine primarily the electronic state of the haem and give rise to functional diversity.

Haemoglobin (Hb) provides a unique opportunity to study the effect of protein structure on the structure and function of the haem. The Hb molecule can exist in two alternative quaternary structures; namely, the T (Deoxy) structure with low oxygen affinity and the R (Oxy) structure with high oxygen affinity (1-4). The oxygen affinities of the haem in these two structures differ almost by three orders of magnitude (5,6,7). Using resonance Raman spectroscopy, we have tried to characterize the changes in the haem structure accompanying the R→T transition which results in a decrease in oxygen affinity.

Characterization of resonance Raman lines of haemoglobin

Resonance Raman spectra of haemoproteins have been extensively studied in the last decade, and the Raman lines in the high frequency regions have been well characterized (8). For instance, it has been shown that the prominent Raman line (ν_4) appearing between 1350 and 1380 cm^{-1} serves as an oxidation state marker (9) and the Raman lines around 1605-1645 cm^{-1} (ν_{10}) , 1535-1575 cm^{-1} (ν_{19}) and 1470-1505 cm^{-1} (ν_3) reflect

Table I Frequencies of Raman lines of deoxyhaemoglobin reconstituted with various deuterated haems (cm^{-1})

native deoxyHb	vibrational mode	2,4-CD=CH$_2$	2,4-CH=CD$_2$	1,3-CD$_3$	1,5-CD$_3$	Δ(CH$_3$ - CD$_3$)[a]	m-D$_4$[a]	Δx(m-H$_4$ - m-D$_4$)[b]
1623	vinyl	-	-	-	1620 (-3)	-	1621 (-2)	0
1608	ν$_{10}$	1609 (+1)	1603 (-5)	1605 (-3)	1608 (0)	-1	1595 (-13)	-11
1591	ν$_2$	1590 (-1)	1587 (-4)	1588 (-6)	1585 (-6)	-3	-	-2
1567	ν$_{11}$+ν$_{19}$	1561 (-5)	1564 (-3)	1661 (-6)	1565 (-2)	-3	1545 (-22)	-11
1529		1527 (-2)	-	1526 (-3)	1528 (-1)	-	-	-
1473	ν$_3$	1472 (-1)	1472 (-1)	1471 (-2)	1471 (-2)	-1	1469 (-4)	-4
1429		1416 (-13)	-	-	-	-	1424 (-5)	-
1357	ν$_4$	1355 (-2)	1357 (0)	1357 (0)	1356 (-1)	-1	1357 (0)	-5
1304	ν$_{21}$	1304 (0)	-	-	-	-	-	-426
1287		1284 (-3)	1283 (-4)	-	1284 (-3)	-	-	-
1218	ν$_{13}$	1219 (+1)	1213 (-5)	1211 (-7)	1216 (-2)	-1	-	-302
1176	ν$_{30}$	1176 (0)	1172 (-4)	1152 (-24)	1174 (-2)	-20	1174	-
1135		1133 (-2)	1130 (-5)	1129 (-6)	1125 (-11)	-32	1131	-
1120	ν$_{22}$	1095 (-25)	1116 (-4)	1112 (-8)	1113 (-7)	-10	1116	-
1090		-	1184 (-6)	1063 (-27)	-	-	-	-
793	ν$_6$	788 (-5)	778 (-15)	752 (-21)	786 (-7)	-14	-	-10
757		757 (0)	757 (0)	752 (-5)	756 (-1)	-	-	-
674	ν$_7$	672 (-2)	673 (-1)	662 (-12)	669 (-5)	-16	669 (-5)	-11
496		497 (+1)	-	-	-	-	-	-
407		406 (-1)	394 (-13)	-	408 (+1)	-	407 (0)	-
367		365 (-2)	367 (0)	363 (-4)	362 (-3)	-	366 (-1)	-
343		343 (0)	343 (0)	343 (0)	337 (-6)	-	339 (-4)	-
302		299 (-3)	300 (-2)	296 (-6)	299 (-3)	-	298 (-5)	-
216	Fe-His	216 (0)	215 (-1)	216 (0)	217 (+1)	0	216 (0)	-
157		155	155	157	-	-	-	-

(a) calculated frequency change upon deuteration of methyl grops for Ni(II)-octamethylporphyrin
(b) calculated frequency change upon deuteration of methine-proton for octaethylporphyrin
(c) numbers in parentheses are observed frequency shifts relative to native deoxyhaemoglobin

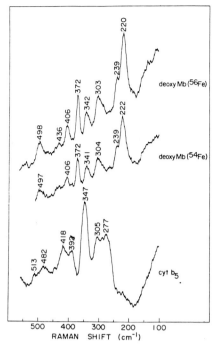

Figure 1. Resonance Raman spectra of native deoxy Mbs reconstituted with ^{56}Fe(II)-protoporphyrin-IX and ^{54}Fe(II)-protoporphyrin-IX and reduced cytochrome b$_5$. Taken from Kitagawa et al. (18).

the porphyrin core size (10,11) and can serve as spin-state markers. The assignment of Raman lines is essential for their interpretation. This can be done by specific isotopic labelling of the haem and normal coordinate analysis, but the low molecular symmetry of Fe(II)-protoporphyrin precludes a complete assignment of Raman lines. However, in the high frequency region, vibrations of the porphyrin macrocycle dominate its vibrational characteristics; consequently the resonance Raman spectrum of Fe(II)-protoporphyrin-IX is remarkably similar to that of Ni(II)-octaethylporphyrin. A close analogue of naturally occurring haem, Ni(II)-octaethylporphyrin has the highest possible symmetry (D$_{4h}$) and gives only 35 Raman active and 18 infrared active fundamental vibrational modes if the peripheral groups are regarded as single dynamical units (12). Abe et al. (13) and Kitagawa et al. (12) proposed a complete vibrational assignment of Ni(II)-octaethylporphyrin based on observed isotopic frequency shifts and normal coordinate analysis of the results.

In the low frequency region peripheral groups contribute significantly to the vibrational modes and thus the assigment for Ni(II)-octaethylporphyrin is no longer applicable to the Fe(II)-proto-porphyrin-IX system. Variations in the low frequency region of haemoproteins in the ferrous high-spin state (14,15) imply that these lines reflect interactions between the haem and the protein. It was important, therefore, to characterise these lines. Table I summarizes the frequencies of Raman lines of deoxy Hbs reconstituted with Fe(II)-proto-porphyrin-IX specifically deuterated at α-vinyl, β-vinyl, 1,3-methyl, 1,5-methyl (16) and meso-positions (17). The table also includes the frequency shift calculated for methyl-deuteration in Ni(II)-octamethyl-porphyrin (Abe, personal communication) and for meso-deuteration in Ni(II)-octaethylporphyrin (13). Vibrational mode numbers for Ni(II)-octaethylporphyrin (13) are indicated wherever the correspondence between Raman lines of deoxy Hb and those of Ni(II)-octaethylporphyrin is clear from the polarization property and isotopic frequency shift. The contribution of each peripheral group to these Raman lines can be estimated only qualitatively from these experiments.

The Fe(II)-His stretching Raman line

In many haemoproteins, the N_ε atom of histidine serves as an axial ligand. The mode of histidine coordination affects the electronic state of the haem and consequently its biological functions. To elucidate the role of the proximal histidine in such haemoproteins, it is vital to assign the Fe-His stretching Raman line. Fig 1 shows the resonance Raman spectra of native deoxy Mb and deoxy Mb reconstituted with protoporphyrin-IX-^{54}Fe(II). The 220 cm^{-1} line shows a significant frequency shift upon Fe isotopic substitution but other lines remain unshifted. The 220 cm^{-1} line is missing from the spectrum of reduced cytochrome b_5 in which two histidines coordinate to the haem iron. The corresponding Raman line of deoxy Hb at 216 cm^{-1} also undergoes a high frequency shift upon ^{54}Fe substitution (19). Subsequently, we studied the resonance Raman spectra of the picketfence-haem-(2-methylimidazole) complex which mimics the haem structure of deoxy Mb and deoxy Hb. The 209 cm^{-1} line exhibits a low frequency shift (3 cm^{-1}) upon perdeuteration of

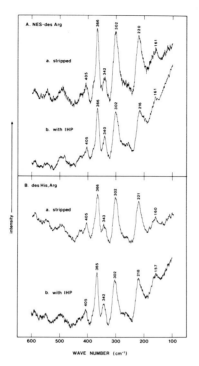

Figure 2. Comparison of the resonance Raman spectra of deoxyHbs in the T and R structures. (A) deoxy NES des-Arg(141α) Hb. (B) deoxy des-His(146β)-Arg(141α) Hb. In both A and B, (a) is without IHP (R structure) and (b) is with 2 mM IHP (T structure). Taken from Nagai et al. (19).

2-methylimidazole and a high frequency shift (3 cm^{-1}) upon substitution of ^{56}Fe with ^{54}Fe (20). Futhermore, as shown in Table I, the 216-cm^{-1} line is the only line which does not show a significant frequency shift upon any of the deuterations of the porphyrin. On this basis, we conclude that the 220 cm^{-1} line of deoxy Mb and the 216 cm^{-1} line of deoxy Hb are due to the stretching vibration of the Fe-His bond.

Effect of Quaternary Structure on the haem Structure in deoxy Hb

In the course of binding four molecules of oxygen, the haemoglobin molecule undergoes a change in quaternary structure (rearrangement of the subunits) from the T to the R structure (1,2,3). The oxygen affinity of the haem in the R structure is higher by a factor of several hundred than

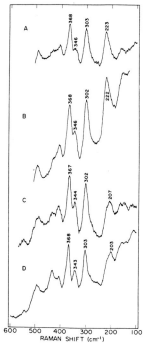

Figure 3. Resonance Raman spectra of the isolated α chain and valency hybrid Hbs. (A), the isolated α^{deoxy} chain at pH6.5; (B), stripped α_2^{deoxy} β_2^{+CN} at pH 9.0; (C) $\alpha_2^{deoxy}\beta_2^{+CN}$ with IHP at pH 6.5; (d) stripped deoxy Hb Milwaukee at pH 6.5. Taken from Nagai and Kitagawa (26).

that in the T structure (5,6,7). We have tried to characterise the change in the haem structure accompanying the R→T transition by resonance Raman spectroscopy. Sussner et al. (21) and Scholler and Hoffman (22) failed to observe any change in the Raman spectrum in the high frequency region upon R→T transition but Shelnutt et al. (23) reported that several lines in the high frequency region undergo small but significant frequency changes which are consistent with an increased electron density of the antibonding π^* orbitals of the porphyrin in the R structure. Our present work is mainly focused on the low frequency vibrational modes, especially the Fe-His(F8) stretching mode. Fig 2 shows the resonance Raman spectra of deoxy NES des-Arg(141α) Hb and deoxy des-His(146β)-Arg(141α) Hb. Both chemically modified Hbs are in the R structure even after deoxygenation but can be converted to the T structure by addition of inositol hexa-

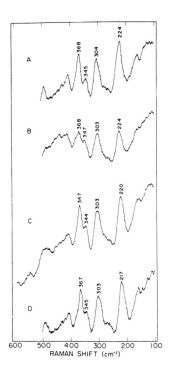

Figure 4. Resonance Raman spectra of the isolated β chain and valency hybrid Hbs. (A), the isolated β^{deoxy} chain at pH 6.5; (B), stripped $\alpha_2^{+CN}\beta_2^{deoxy}$ at pH 9.0; (C), $\alpha_2^{+CN}\beta_2^{deoxy}$ with IHP at pH 6.5; (D), stripped deoxy Hb M Boston at pH 6.5. Taken from Nagai and Kitagawa (26).

phosphate (IHP). For these deoxy Hbs, the Fe-His stretching line is observed at 221 cm^{-1} in the absence of IHP and at 216-218 cm^{-1} in the presence of IHP (19). The shifts of the Fe-His stretching to lower frequency upon R→T transition have since been observed for other mutant and modified Hbs (19,24,14).

Sugita (25) reported that the α subunit is predominantly responsible for the change in the Soret and visible absorption spectra of deoxy Hb observed on R→T transition. Therefore, in a second stage, we measured the Fe-His stretching frequency of the individual subunits to see if one of the subunits undergoes a larger change in the Fe-His stretching frequency (27). Fig 3 shows the resonance Raman spectra of the isolated α^{deoxy} chain and of valency hybrid Hbs in which the α subunits are in the

ferrous-deoxy and the β subunits in the ferric form. Upon excitation at
441.6 nm, the deoxy subunits dominates the Raman spectrum. All the
prominent Raman lines shown here are either greatly weakened or eliminated
on oxygenation of the ferrous subunit, which indicates that the ferric
subunit does not contribute significantly to the spectra of the
deoxygenated hybrids shown here. $\alpha_2^{deoxy}\beta_2^{+CN}$ is in the R structure in the
absence of IHP but can be converted to the T structure by addition of IHP.
The Fe-His stretching line of $\alpha_2^{deoxy}\beta_2^{+CN}$ is observed at 222 cm^{-1} in the
absence of IHP but at 207 cm^{-1} in the presence of IHP. Fig 3 also
includes the Raman spectrum of deoxy Hb M Milwaukee [Val(E11)67β→Glu]
which is in the T structure even in the absence of IHP. The α^{deoxy}
subunit of Hb M Milwaukee exhibit the Fe-His line at 203 cm^{-1} whereas the
frequencies of other Raman lines remain unchanged.

        Fig 4 shows the reciprocal experiments. $\alpha_2^{+CN}\beta_2^{deoxy}$ undergoes the
R→T transition on addition of IHP but the Fe-His stretching line shows
only a small frequency change. Deoxy Hb M Boston [His(E7)58α→Tyr] is in
the T structure even in the absence of IHP and shows the Fe-His line at
217 cm^{-1}. Therefore the shift of the Fe-His stretching line observed on
R→T transition is much larger in the α^{deoxy} subunit ($\Delta\nu = 20$ cm^{-1}) than in
the β ^{deoxy} subunit ($\Delta\nu = 7$ cm^{-1}).

Effect of quaternary structure on the state of the haem in nitrosyl Hb
        Unlike other ferrous liganded Hbs, nitrosyl Hb is readily converted
to the T structure by addition of IHP. This structural change is
accompanied by a large change in spectroscopic properties of the haem,
including electron spin resonance (27), visible (28,29) and infrared
absorption (30). Szabo and Barron (31) measured the resonance Raman
spectra of nitrosyl Hb in the absence and presence of IHP and showed that
the 1636 cm^{-1} line of nitrosyl Hb is split into two bands by addition of
IHP; one at 1636 cm^{-1} and the other at 1645 cm^{-1}. As shown in Fig 5, the
1636 cm^{-1} is shifted to 1626 cm^{-1} upon deuteration of meso-H. Several
other lines in the high frequency region undergo frequency shifts upon
meso-deuteration, as summerized in Table 1 and Table 2. By comparing the
Raman spectra of nitrosyl Hb in which haem in either the α or β subunit is
meso-deuterated, we can infer the spectral change in the individual
subunits (32). Fig 6 shows the Raman spectra of stripped $\alpha(H)_2^{NO}\beta(D)_2^{NO}$,

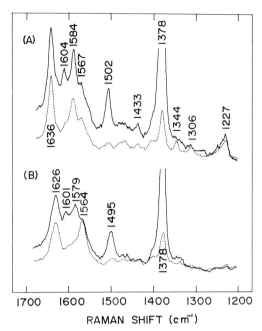

Figure 5. Polarized resonance Raman spectra of stripped $\alpha(H)_2^{NO}\beta(H)_2^{NO}$ (A) and $\alpha(D)_2^{NO}\beta(D)_2^{NO}$ (B). The solid line and broken lines indicate the electric vecror of the scattered radiation to be parallel and perpendicular to that of the incident radiation, respectively. Taken from Nagai et al. (32).

$\alpha(H)_2^{NO}\beta(H)_2^{NO}$, $\alpha(D)_2^{NO}\beta(H)_2^{NO}$, $\alpha(D)_2^{NO}\beta(D)_2^{NO}$ where H and D denote native and meso-deuterated haems. Fig 7 shows the Raman spectra of these hybrid Hbs in the presence of IHP. When the β haem is replaced by meso-deuterated haem, the 1637 cm^{-1} line of $\alpha(H)_2^{NO}\beta(H)_2^{NO}$ with IHP undergoes a low frequency shift whilst the 1645 cm^{-1} line remains unshifted (Fig 7). On the other hand, when the α haem is replaced by meso-deuterated haem, the 1645 cm^{-1} line shifts to 1637 cm^{-1} but the 1637 cm^{-1} line remains unshifted. Therefore, the 1645 cm^{-1} and 1637 cm^{-1} lines of $\alpha(H)_2^{NO}\beta(H)_2^{NO}$ with IHP belong to the α and β subunits, respectively. In this way, the frequencies of the Raman lines in the individual subunits were determined (Table II). The Raman spectrum of the β subunit remained unchanged but the ν_{10}, ν_{19} and ν_3 modes of the α subnunits showed large frequency shifts on addition of IHP. The comparison of these Raman spectra with those of

penta- and hexa-coordinate NO-haem complexes shows that in the absence of
IHP, haems in both subunits are hexa-coordinate but in the presence of IHP,
the α haem is penta-coordinate (32,34). This results is consitent with the
infrared (30), visible absorption (28,34) and EPR (35,36) studies of NO Hb
which showed that the Fe-His bond of the α subunit is cleaved or severely
stretched in the T structure.

Table II: Frequencies of Raman lines of NO-Hb and NO haem complexes (cm^{-1})

mode[a] ρ[b] ν[c]			stripped NO-Hb	NO-Hb with IHP subunits αNO	βNO	NO-heme complexes coordinateion penta[d]	hexa[e]
1	ν_{10}	dp	-10	1636	1645	1635	1647	1637
2	ν_2	p	-3	1604	1600	1602	shoulder	shoulder
3	ν_{19}	ap	-20	1584	1592	1584	1588	1585
4	ν_{11}	dp	-3	1567	1568	1567	1565	*
5	ν_3	p	-7	1502	1508	1501	1508	*
6		dp	0	1433	1435	1433	1434	*
7	ν_{20}	ap	0	1400	1402	1398	1403	*
8	ν_4	p	0	1378	1378	1378	1376	*
9		ap	0	1344	1346	1345	1342	*
10	ν_{21}	ap		1306	1306	1308	1300	*
11	ν_{13}	dp		1227	1225	1227	1227	*

[a]Mode numbers are taken from Abe et al. (1978)

[b]Depolarization ratio:

[c]Frequency shifts observed for stripped NO-Hb upon meso-deuteration.

[d]Protoporpyrin-IX-Fe(II)-NO in aqueous SDS solution.

[e]Protoporphyrin-IX-dimethylester-Fe(II)-NO-N-methylimidazole in N-methyl-
imidazole.

* Frequencies were not determined due to solvent peaks

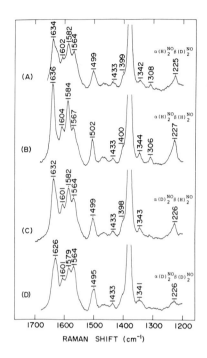

Figure 6. Resonance Raman spectra of stripped NO–Hb: (A) $\alpha(H)_2^{NO}\beta(H)_2^{NO}$; (B), $\alpha(H)_2^{NO}\beta(H)_2^{NO}$; (C), $\alpha(D)_2^{NO}\beta(H)_2^{NO}$; (D), $\alpha(D)_2^{NO}\beta(D)_2^{NO}$ in 0.05 M Bis-Tris, 0.05 M Tris, 0.1 M Cl⁻ pH 6.5. Taken from Nagai et al. (32).

Coordination of axial ligand in Hb M Boston and Hb M Iwate

Haemoglobins M are naturally occurring valency hybrids in which the haem irons in one pair of subunits are oxidized in vivo as a result of mutations in the vicinity of the haem. In Hb M Boston [His(E7)58α→Tyr], the distal and in Hb M Iwate [His(F8)87α→Tyr], the proximal histidine residues are replaced by tyrosine residues. The X-ray crystallographic study of deoxy Hb M Boston (37) has shown that in the abnormal α subunit, tyrosine (E7) coordinates to the haem iron and the Fe-His bond is disrupted. Unfortunately, no high resolution X-ray crystallographic study of Hb M Iwate has been reported. We have obtained the resonance Raman spectra of Hb M Boston and Hb M Iwate to compare their haem structures (38). Fig. 8 shows the resonance Raman spectra of Hb M Iwate and Hb M Boston in both the deoxy and oxy forms. These spectra are distinctly different from the spectrum of a equimolar mixture of met Hb and either

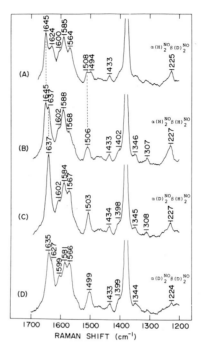

Figure 7. Resonance Raman spectra of stripped NO-Hb: (A) $\alpha(H)_2^{NO}\beta(H)_2^{NO}$; (B), $\alpha(H)_2^{NO}\beta(H)_2^{NO}$; (C), $\alpha(D)_2^{NO}\beta(H)_2^{NO}$; (D), $\alpha(D)_2^{NO}\beta(D)_2^{NO}$ in 0.05 M Bis-Tris, 0.05 M Tris, 0.1 M Cl$^-$ pH 6.5 with 2 mM IHP. Taken from Nagai et al. (32).

oxy or deoxy Hb. The depolarized Raman line of deoxy Hb M Boston at 1625 cm^{-1} is derived from the abnormal α subunit. It is therefore assignable to the ν_{10} mode of the abnormal α subunit. The frequency of the ν_{10} mode has been shown to be sensitive to the coordination number as well as the nature of the axial ligands (34). The observed frequency of this mode is characteristic of the five-coordinate ferric haem and this result is consistent with the X-ray crystallographic study. We have also examined the Raman spectrum of a model compound Fe(III)-protoporphyrin-IX-dibutyl-ester-p-nitrophenol synthesized by Dolphin. This compound mimics the haem structure of the abnormal α subunit in Hb M Boston and indeed shows the ν_{10} mode at 1628 cm^{-1} (Dolphin, Nagai and Kitagawa, unpublished results). Hb M Iwate also shows the ν_{10} mode at 1628 cm^{-1}, implying that the haem in the abnormal α subunit is also penta-coordinate with tyrosine as an axial ligand. Hb M Boston and Hb M Iwate show intense polarized lines at 1280

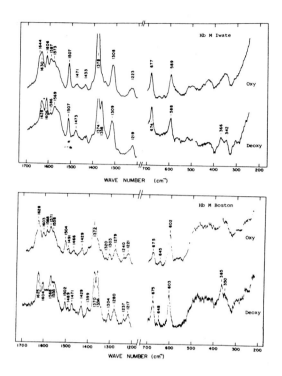

Figure 8. Resonance Raman spectra of the oxygenated and deoxygenated forms of Hb M Iwate and Hb M Boston. Taken from Nagai et al. (38).

cm^{-1} and 1308 cm^{-1}, respectively. Met Hb A derivatives do not show any corresponding Raman line in this region. Unlike other polarized porphyrin modes, these lines are intensified with excitation around 475 nm and above 500 nm. These Raman frequencies are close to the CO stretching frequency of phenolate observed for non-haem Fe(III)-tyrosine complexes such as transferrin (40) and dioxygenases (41). Accordingly, we assigned the 1278 cm^{-1} line of Hb M Boston and the 1308 cm^{-1} line of Hb M Iwate to the phenolate CO stretching mode of haem coordinated-tyrosine. Hb M Iwate and Hb M Boston show unusual Raman lines at 589 cm^{-1} and 603 cm^{-1}, respectively. The excitation profiles of these lines are similar to those of phenolate CO stretching lines and these frequencies are close to that of the Fe-O_2 stretching in oxy Hb (19,42). On this basis these lines are assigned to the Fe-O(phenolate) stretching lines. It is interesting that this line lies at lower frequency in Hb Iwate where the tyrosine lies proximal to the haem and the Fe-O bond is likely to be stretched by the T-structure, than in Hb Boston where the Fe-O bond is distal to the haem. Stretching of the Fe-O bond strengthens the CO bond, so that the CO stretching frequency is larger in Hb Iwate than in Hb Boston. In addition to these Raman lines Hb M Boston and Hb M Iwate show internal ring vibrational mode of tyrosine around 1600 cm^{-1} and 1500 cm^{-1}. Proto-porphyrin-IX-dibutylether-Fe(III)-p-nitrophenol exhibits neither the phenolate internal modes nor the Fe(III)-O(phenolate) stretching vibrational modes upon excitation between 441 and 515 nm. The coordination of phenol in the model complex is not restricted and it may bind to the haem iron perpendicular to the haem plane but in these abnormal Hbs, tyrosine binds to the haem only in the tilted geometry which allows direct interaction of the π orbitals of phenolate and that of porphyrin.

Discussion

X-ray crystallographic studies by Perutz and co-workers showed that the iron atom is displaced from the haem plane in deoxy Hb and the iron atom moves into the haem plane upon oxygenation. Perutz (1,2) proposed that the T structure is more stable when the iron atom is displaced from the haem plane and thus the T→R transition is caused by the movement of the iron atom accompanying the oxygen binding. Conversely, upon the R→T transition, globin pulls the proximal histidine away from the haem plane and reduces the oxygen affinity by restraining the iron atom from moving into the haem plane. Resonance Raman spectroscopy provides an ideal probe for the Fe-His bond if the Fe-His stretching Raman line is assigned. We have assigned the 216 cm^{-1} line of deoxy Hb to the Fe-His stretching mode based on the isotopic frequency shift on $^{56}Fe-^{54}Fe$ substitution and imidazole deuteration(18-20). If globin imposes a larger strain on the Fe-His bond in the T structure, it should be detected as a change in the Fe-His stretching frequency. Our resonance Raman study on deoxy Hb (19) shows that the R→T transition is indeed accompanied by a low frequency shift of the Fe-His stretching line. Furthermore, the experiment with valency hybrid Hbs (27) showed that the changes in the Fe-His stretching frequency upon R→T transition is larger in the α subunit ($\Delta\nu = 20$ cm^{-1}) than in the β subunit ($\Delta\nu = 7$ cm^{-1}). Therefore, the strain in the Fe-His bond imposed by globin is larger in the α subunit than in the β subunit. This is supported by the experiment with NO Hb (32) which showed that the Fe-His bond of the α^{NO} subunit is cleaved by the R→T transition. These experiments prove that the Fe-His bond plays an essential role in controlling the biological activities of the haem.

Acknowledgements
I thank Dr. M.F. Perutz for criticism, Drs. T. Kitagawa, H. Morimoto, G.N. LaMar, K.M. Smith, T. Jue for collaboration and Ben Luisi and Jennifer Cornwell for help in preparing the manuscirpt.

References
1. Perutz, M.F.: Nature 228, 726-734 (1970).

2. Perutz, M.F.: Ann. Rev. Biochem. 48, 327-386 (1979).

3. Baldwin, J.M.: Prog. Biophys. Mol. Biol. 29, 225-320 (1975).

4. Monod, J., Wyman, J., Changeaux, J.P.: J. Mol. Biol. 12, 88-118 (1965).

5. Imai, K.: Allosteric Effects in Haemoglobin, Cambridge University Press, Cambridge 1982.

6. Antonini, E., Brunori, M.: Hemoglobin and Myoglobin in Their Reactions with Ligands, North-Holland Publishers, Amsterdam 1971.

7. Roughton, F.J.W., Otis, A.B. Lyster, R.L.J.: Proc. R. Soc. London Ser. B 114, 29-54 (1955).

8. Asher, S.A.: Methods in Enzymology 76, 371-413 (1981).

9. Spiro, T.G., Burke, J.M.: J. Am. Chem. Soc. 98, 5482-5489 (1976).

10. Spaulding, L.D., Bhang, C.C., Yu, N.-T., Felton, R.H.: J. Am. Chem. Soc. 97, 2517-2525 (1975).

11. Spiro, T.G., Strong, J.D., Stein, P.: J. Am. Chem. Soc. 101, 2648-2655 (1979).

12. Kitagawa, T., Abe, M., Ogoshi, H.: J. Chem. Phys. 64, 4516-4525 (1978).

13. Abe, M., Kitagawa, T., Kyogoku, Y.: J. Chem. Phys. 64, 4526-4534 (1978).

14. Desbois, A., Lutz, M., Banerjee, R.: Biochim. Biophys. Acta 671, 177-183 (1984).

15. Desbois,A., Mazza, G., Stetzkowski, F., Lutz, M.: Bichim. Biophys. Acta 785, 161-176 (1984).

16. Nagai, K., Kitagawa, T., Jue, T., La Mar, G.N., Smith, K.M., Langrey, K.C.: manuscript in preparation.

17. Nagai, K., Kitagawa, T., Dolphin, D., Welborn, C.: manuscript in preparation.

18. Kitagawa, T., Nagai, K., Tsubaki, M.: FEBS Lett. 104, 376-378 (1979).

19. Nagai, K., Kitagawa, T., Morimoto, H.: J. Mol. Biol. 136, 271-287 (1980).

20. Hori, H., Kitagawa, T.: J. Am. Chem. Soc. 102, 3608-3613 (1980).

21. Sussner, H., Maycer, A., Brunner, H., Fasold, H.: Eur. J. Biochem. 41, 465-469 (1974).

22. Scholler, D.M., Hoffman, B.M.: J. Am. Chem. Soc. 101, 1655-1662 (1979).

23. Shelnutt, J.A., Rousseau, D.L., Friedman, J.H., Simon, S.R.: Proc. Natl. Acad. Sci. USA 76, 4409-4413 (1979).

24. Ondrias, M.R., Rousseau, D.L., Shelnutt, J.A., Simon, S.R.: Biochemistry 21, 3428-3437 (1982).

25. Sugita, Y.: J. Biol. Chem. 250, 1251-1256 (1975).

26. Nagai, K., Kitagawa, T.: Proc. Natl. Acad. Sci. USA 77, 2033-2037 (1980).

27. Rein, H., Ristau, O., Scheler, W.: FEBS Lett. 24, 24-26 (1972).

28. Perutz, M.F., Kilmartin, J.V., Nagai, K., Szabo, A., Simon, S.R.: Biochemistry 15, 378-387 (1976).

29. Salhany, J.H., Ogawa, S., Shulman, R.G.: Proc. Natl. Acad. Sci. USA 71, 3359-3362 (1974).

30. Maxwell, J.C., Caughey, W.S.: Biochemistry 15, 388-396 (1976).

31. Szabo, A., Barron, L.D.: J. Am. Chem. Soc. 97, 660-662 (1975).

32. Nagai., K., Welborn, C., Dolphin, D., Kitagawa, T.: Biochemistry 19, 4755-4761 (1980).

33. Tsubaki, M., Yu. N.-T.: Biochemistry 21, 1140-1144 (1982).

34. Nishikura, K., Sugita, Y.: J. Biochem (Tokyo) 80, 1439 (1976).

35. Nagai, K., Hori, H.,Yoshida, S.,Sakamoto, H., Morimoto, H.: Biochim. Biophys. Acta 532, 17-28 (1978).

36. Szabo, A., Perutz, M.F.: Biochemistry 15, 4427-4428 (1976).

37. Pulsinelli, D.D., Perutz, M.F., Nagel, R.L.: Proc. Natl. Acad. Sci. USA 70, 3870-3874 (1973).

38. Nagai, K., Kagimoto, T., Hayashi, A., Taketa, F., Kitagawa, T.: Biochemistry 22, 1305-1311 (1983).

39. Taraoka, J., Kitagawa, T.: J. Phys. Chem. 84, 1928-1935 (1980).

40. Tomimatsu, Y., Kiut, S., Scherer, J.R.: Biochemistry 15, 4918-4924 (1976).

41. Tatsuno, Y., Sueki, Y., Iwaki, M., Yagi, T., Nozaki, M, Kitagawa, T., Otsuka, S.: J. Am. Chem. Soc. 100, 4614-4615 (1978).

42. Brunner, H.: Naturwissenschaften 61, 129 (1974).

Received July 7, 1984

Discussion

Beychok: What is the distance between N_e of His-F8 and Fe in T state of α^{NO} subunit?

Nagai: It has not been determined.

Buchler: The rather low frequency of the Fe-N(His)-mode is rather surprising to the inorganic chemist. Could you comment on that, and could you explain why the frequencies in the T state are lower than in the R state?

Nagai: The Fe(II)-(2Me-Im) stretching band has been observed at 209 cm^{-1} for the picketfence haem-(2-methylimidazole) complex. This band exhibits a low frequency shift (3 cm^{-1}) upon perdeuteration of 2-methylimidazole. This frequency shift implies that the whole 2-methylimidazole is vibrating as a dynamical unit. If you calculate the force constant using the reduced mass

$$\mu = \frac{M_{Fe} \cdot M_{2-MeIm}}{M_{Fe} + M_{2-MeIm}}$$

where M_{Fe} and M_{2-MeIm} denote the mass units of the iron atom and 2-MeIm, respectively, the force constant appears to be comparable to those of Fe^{2+}-N bonds estimated for other coordination compounds. So I do not think that the frequency of the Fe-N(His) stretching band is particularly low.

X-ray crystallographic studies by Perutz and co-workers have shown that the iron atom is displaced from the haem plane in deoxyHb but the iron atom lies in the plane of porphyrin in oxyHb.

Perutz proposed that the T structure is more stable when the iron atom is displaced from the haem plane. If this is true, the R$\longrightarrow$T structural change should pull the proximal histidine further away from the haem plane and stretch the Fe-His bond. Our observation is consistent with the molecular mechanism of haem-haem interaction proposed by Perutz.

Myer: Penta-coordination means only 5 ligands. Is histidine displaced from coordination?

Nagai: The resonance Raman spectrum of the α^{NO} subunit in the T state NO Hb is identical to that of the penta-coordinate NO heme model complexes. The same results have been obtained with EPR, infra-red, and optical absorption spectroscopy. On this basis it is concluded that the Fe-His bond is cleaved in the T state. But we do not know yet the state of the histidine side chain.

Myer: The displacement of iron from heme is a result of multiple aspects:
a) the axial ligate, pulling etc.
b) the changes of core size
c) the alteration, if any, of M-N distances.

Nagai: If you compare the resonance Raman spectra of deoxy Hb between the

T and R states, you observe a large frequency difference of the Fe-His stretching line but the porphyrin modes do not show any significant frequency changes. Shelnutt et al. reported very small frequency changes of some porphyrin modes, but the result is not consistent with the change in the core size as far as I can remember correctly.

We can conclude from our experiment that the Fe-His bond is stretched in the T state, but we cannot draw any conclusion about the position of the iron atom with respect to the heme plane.

Myer: In the case of penta-coordinated hemes, it is expected that iron will be out of plane. Ligation on one side only most probably will pull the iron out of plane.

Nagai: Indeed, the iron atom of penta-coordinate high-spin haem complexes is displaced from the haem plane even in the absence of protein. This displacement is partly due to the large size of high-spin ferrous iron and partly due to the repulsion between haem and imidazole ring. In the T state the imidazole ring is tilted with respect to the haem plane so that either C_σ or C_ϵ is even closer to the haem plane. This may be why the Fe-N_2(His) bond is stretched in the T state.

Fuhrhop: Are pK_a values of the phenol ligands in Boston (distal) and Iwate (proximate) hemoglobins known?

Nagai: The Fe^{3+}-tyrosine bond is quite strong, unless you denature these Hbs. Therefore the Fe atom cannot be displaced with proton. However, if you add dithionite in the presence of carbon monoxide or nitric oxide, tyrosine is replaced by CO or NO and the His-Fe protoporphyrin-CO (or -NO) complex can be formed.

In the case of Hb M Iwate, the distal histidine (E7) serves as the 5th ligand and CO or NO binds from the other side of the heme (Nagai et al., Biochemistry, published between 1978-1981). But I have not measured the pK_a value of tyrosine under these conditions.

Holzwarth: Can any fluorescence be observed from the tyrosine sitting close to the iron in "Boston" hemoglobin?

Nagai: As far as I know, nobody has looked at fluorescence from Hb M Boston or Hb M Iwate, but I suspect that fluorescence from heme-coordinating tyrosine is quenched by the heme. There are some other tyrosines in the Hb molecule. I think that fluorescence from these tyrosines dominates in Hb M Boston and Hb M Iwate.

Blauer: Can the RR spectra also be used for diagnostic purposes, at least in the cases where you have spectral changes?

Nagai: The patient with M-type abnormal Hbs normally shows cyanosis due to brownish colour of the abnormal ferric subunits. The resonance Raman spectroscopy can distinguish between Hb M Iwate and Hb M Boston, therefore it can be used for diagnostic purposes. However, this can be done better with the EPR spectroscopy because all M-type Hbs so far reported show very characteristic EPR spectra.

OPTICAL PROPERTIES OF HEME BOUND TO THE CENTRAL EXON FRAGMENT OF GLOBINS AND TO CORRESPONDING FRAGMENTS IN CYTOCHROMES

Sherman Beychok

Departments of Biological Sciences and Chemistry
Columbia University
New York, New York

Introduction

In the hemoglobins, each of the four chains possesses an active site, and chain assembly serves to modulate the reactivity of the four sites. The fundamental event in evolution was the appearance of reversible oxygenation within the single chain, as in myoglobin and lamprey hemoglobin, followed much later by the tetrameric form exhibiting cooperative behavior (1-4). We have been investigating the effect of heme binding on the conformation and assembly of hemoglobin for many years (5-8). The investigations were prompted by the finding almost two decades ago that the secondary structure of the monomeric protein, myoglobin, differs in heme-free and heme-bound states. Subsequent experiments showed that apohemoglobin ($\alpha^\circ\beta^\circ$), the dimeric heme-free protein, similarly has a lower α-helix content than hemoglobin (6). The isolated heme-free chains, α° and β°, are unstable and more extensively disordered, but it is remarkable that while, at a given pH α° and β° are distinctly different from each other and from $\alpha^\circ\beta^\circ$, heme addition refolds them to α^h and β^h subunits with secondary structures very similar to each other and to hemoglobin. Moreover, either disordered globin chain can be refolded upon reaction with the complementary, highly ordered subunit yielding the half-filled molecule. This is very pronounced in the reaction at pH 6.7 between α° and β^h to form a half-filled molecule: the fraction of residues in helical segments in α° rises

Optical Properties and Structure of Tetrapyrroles
© 1985 by Walter de Gruyter & Co., Berlin · New York

from ~20% to > ~60% (a change in mean residue $[\theta]_{222}$ of
~11,200 deg-cm^2/decimol) in this so-called alloplex reaction
(8).

The binding of heme to the apohemoglobin dimer and to the separated globin chains thus confers not only reversible O_2 binding ability, but results in major conformational changes at all
levels of structure. The unstable heme-free proteins are
folded to stable structures and quaternary interactions are
established upon heme-binding. Heme is bound to a fragment of
the molecule encoded by the central exon of the globin chain
genes. This heme binding domain, its structural and functional potential, and its relation to similar domains in other
heme-binding proteins (its possible role in hemoglobin evolution) is the subject of a group of studies in our laboratory.
We wish to know 1) whether it is an independently folded domain, 2) how its folding pathway compares to that of the intact chains, 3) how its folding (both pathway and final state)
is affected by association with fragments encoded by other exons of the same gene and with the complementary chain, 4) what
its potentialities are as an isolated heme-binding domain with
respect to reversible O_2 binding and maintenance of a transient pentacoordinate ferrous state, and finally, 5) whether
there are groups of contact atoms, short sequences or structures shared by this domain and heme-binding fragments or domains in other proteins.

Results and Discussion

Effect of Heme Binding on Globin Chain Conformation. As noted
above it has been known for many years that removal of heme
from myoglobin (Mb) and hemoglobin (Hb) results in apoproteins
that are conformationally different from the native protein,
and that rebinding of heme restores the original conformation
to yield the fully active proteins. In both cases, there are

secondary structure changes suggesting that short helices or
parts of longer helical segments involving 10 or 15 residues
in each chain are unfolded in apomyoglobin and apohemoglobin.
These conformational changes are readily discerned by circular dichroism (CD) measurements in the peptide-absorbing, far-
ultraviolet spectral regions (7). On the basis of evidence
and arguments presented below, it is reasonable to suppose
that the unfolding is mainly confined to the region of the
heme pocket in both Mb and Hb. In the latter case, the conformation of these residues also affects the integrity of the
$\alpha_1 \beta_2$ interface that occurs between two symmetrical $\alpha\beta$ dimers.

Still more pronounced, even remarkable, changes in structure
accompany removal of heme from the separated α and β subunits
(α^h and β^h) (7-8). From the point of view of both secondary and
tertiary structure, these isolated subunits are conformationally very similar to their counterparts in intact tetrameric
Hb, except for reactivity with heme ligands and certain other
properties dependent on Hb quaternary structure. However, upon removal of heme there are dramatic structural changes, especially in the heme-free α globin chain (α°). Figures 1 and
2 demonstrate the reduction in α-helix content of both α and
β globin chains as compared to their heme-containing forms.
For both globins, the reduction in helicity is dependent on
pH. At pH 6.7, the residual helix content of α° is approximately 20%, compared to 70% for the heme-containing chain, and
the globin chain is thus highly disordered and unfolded. Intrinsic fluorescence measurements on this globin, which has a
single Trp residue, also reveal substantial exposure to solvent. Moreover, the molecule behaves as an aggregate rather
than as a monomer which is the predominant form of α^h at most
concentrations (8). These changes are completely reversible
with all structural and functional properties of the native
subunit entirely restored upon recombination with a stoichio-
metric amount of heme.

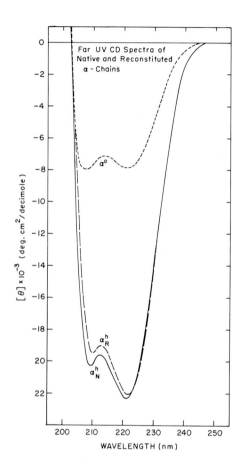

Figure 1.   Far-ultraviolet CD spectra of α-globin (-----) in
20 mM potassium phosphate, pH 6.7 at 4°C, and native (———)
and reconstituted (- - - -) α-cyanmet subunits in 0.1 M potassium phosphate, pH 7.5., at 20°C. From (8).

Such studies have both practical and theoretical importance
for understanding the mechanism of chain folding in this system.   From the practical point of view, the effects of heme
addition provide a model system for chain folding induced by
binding of a prosthetic group.   From a theoretical standpoint,
they suggest an important enlargement of the functional role
of heme.   Heme is now seen not only as the site of reversible

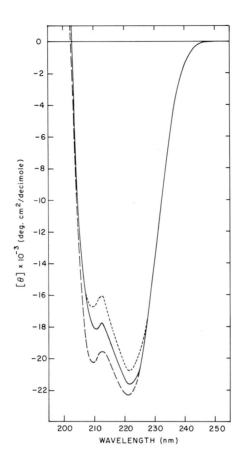

Figure 2. Far-ultraviolet CD spectra of β-globin (-----) in 0.1 M Tris-HCl, pH 8.0 at 4°C, and native (———) and reconstituted (- - -) β-cyanmet subunits in 0.1 M potassium phosphate, pH 7.5, at 20°C. From (8).

dioxygen binding, which can be regulated by allosteric effectors, but also as conferring stability and ordered structure on an otherwise unstable and highly disordered globin chain. Both the practical and theoretical aspects have been important in our studies of the relation of gene structure to protein structure, and in our attempts to define a genetic element that could be said to encode a heme-binding segment within diverse proteins.

The Heme-binding Domain. With the discovery of coding regions
(exons) separated by non-coding segments (introns) in eukaryotic genes, a series of new questions arose about the relation
between gene structure and gene product (9-14). Do exons code
for functional domains? Do exons encode peptide segments that
may fold independently? In an important proposal, Gilbert (10)
advanced the idea that the exon/intron structure could be a
mechanism for increasing the rate of evolution. If exons correspond to units of protein function, their recombination within introns could reassort protein functions to produce novel
proteins from pre-existing ones. Blake (11) then added, as a
corollary, that if exons encoded independently folded domains,
the probability that recombination would result in a folded,
stable structure would be increased.

These two hypotheses together generated what is now known as
'exon shuffling.' Clearly, the finding of homologous sequences or structures in different proteins corresponding to exon
units in different combinations (for example in hemoglobins
and cytochromes) would be very decisive, and the attributes of
different exon-encoded fragments - isolated, and in non-covalent association with other exon-encoded fragments - when compared to conformation and function within an intact native
protein, have to be characterized if we are to conclude that
exon shuffling has played a significant role in generating the
immense variety of proteins that now exists. Ideally, one
wants to establish not only the nature of finally folded states
but to see whether independent folding pathways (kinetics) exist.

Soon after the exon shuffling hypothesis was advanced, we
sought some experimental verification in the case of the β subunit of human hemoglobin, whose gene structure had been
solved. We isolated the central exon product and examined its

properties. Figure 3 presents a diagrammatic representation of
the β-globin intron/exon structure. With the single exception
of a plant globin, all globin genes consist of three exons,
separated by two introns, with boundaries at homologous positions. In terms of β-globin sequence, the exons code for
residues 1-30, 31-104 and 105-146; corresponding positions in
α-globins are 1-31, 32-99 and 100-141.

The junctions in the globins interrupt α-helical segments, residues 30 and 31 occurring in the B helix, and 104 and 105 in
the G helix. No clear separation exists at these points to
indicate a structural or independently folded domain. None-
theless, Gilbert (10), Blake (11) and Argos and Rossmann (16)
had all noted that the product of the central exon contains
the proximal heme-binding histidine and all but two of the residues that contact heme within a distance of 4 Å (22). Thus
it seemed worthwhile to attempt isolation of the central exon
product and to examine its properties. We took advantage of
the fact that the heme-free globins are generally susceptible
to proteolysis whereas the heme-containing subunits are resis-
tant.

Figure 4 shows arginine-specific clostripain cleavage sites in
both α and β human globins. In the case of β, arginine residues occur at the two boundaries corresponding to exon/intron
junctions. In addition, there is an arginine residue at position 40. This last site is considerably less susceptible to
clostripain, and digestion for 60 minutes gives an approximately equimolar mixture of the fragment 31-104 (the central fragment, $\beta°_{CF}$) and the slightly smaller fragment 41-104. These
can be separated by HPLC (17). Figure 5 shows Soret and visible absorption spectra resulting from combination of the central fragment with hemin dicyanide in comparison with intact,
native β°. Figure 6 demonstrates that the binding is stoich-
iometric with a sharp break at a ratio of one equivalent of
hemin dicyanide added to one mole of central fragment. These

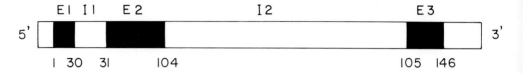

Figure 3. Diagrammatic representation of the β-globin intron/
exon structure. The black boxes E1, E2 and E3 represent expressed DNA sequences (exons); the white boxes l1 and l2 represent intervening sequences (introns). The open boxes at the
5' and 3' ends are untranslated regions of indeterminate
length. The numbers below the exons denote amino acid sequence positions of the β-globin. E2 represents the central
exon that codes for the heme-binding polypeptide consisting of
amino acids 31-104. The 5' → 3' orientation is indicated. A
similar arrangement exists for the α-globin gene. The small
intron (I1) for a α-globin occurs between codons 31 and 32
whereas the large intron (I2) interrupts the coding sequence at
99 and 100. When the amino acid sequences of α- and β-globin
are aligned to maximize functional homology, their intron/exon
junctions coincide; α-globin residue 31 is an arginine but residue 99 is a lysine, thus limited clostripain digestion yields
only two fragments: residues 1-31 and 32-141. From (15).

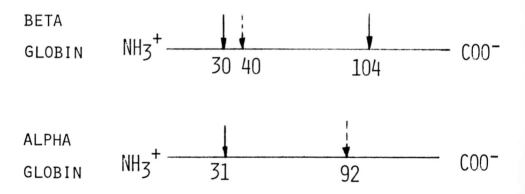

Figure 4. Clostripain cleavage sites of α and β globin. Solid
arrows indicate labile arginine sites. Dashed arrows denote
arginine sites less prone to cleavage. The numbers below
the line refer to the sequence number of the arginine residues.
For β globin, the solid arrow demarcate the exon-coded peptides. From (18).

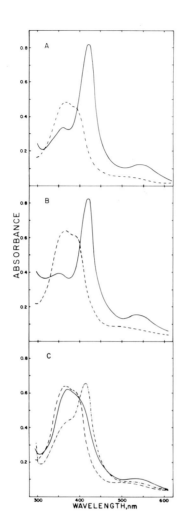

Figure 5.   Soret and visible absorbance spectra.   (A) Equimolar mixture of β° and hemin dicyanide (————————), 9.32 µM;
hemin dicyanide alone (------), 9.32 µM.   (B) Equimolar
mixture of β°$_{31-104}$, $_{41-104}$ and hemin dicyanide (————————), 12.7
µM; hemin dicyanide alone (-----), 12.7 µM.   (C)   Equimolar
mixture of β° peptic digest and hemin dicyanide (————————), 12.7
µM; equimolar mixture of bovine serum albunin and hemin dicyanide (-·-·-), 12.7 µM; hemin dicyanide alone (-----), 12.7
µM.   All solutions were at 4°C in 20 mM potassium phosphate/1
mM EDTA (pH 5.7).   From (18).

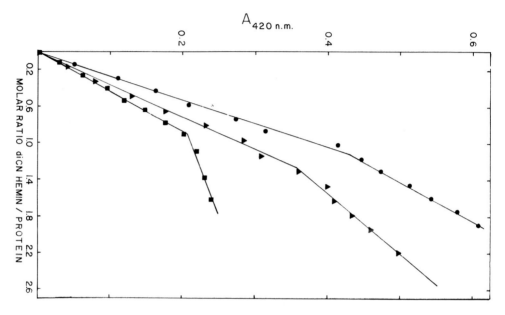

Figure 6.   Recovery of Soret absorbance intensity as a function
of added hemin dicyanide.   Protein solution (1.0 ml) was titrated by adding 2-µl increments of hemin dicyanide solution.
Concentration of β° (●), 6.17 µM; 60-minute β° digest (▲),
5.52 µM;   β° (■), 4.40 µM; concentration of hemin dicyanide
stock solutions: 900 µM, and 264 µM, respectively.   Titration
was carried out at 4°C in 20 mM potassium phosphate/1 mM EDTA
(pH 5.7).   From (9).

results demonstrated conclusively that the product of the central exon of the human β-globin gene is a functional domain
capable of binding heme tightly and specifically.

The question of whether the peptide segment can be considered
a structural domain, and whether it has similar conformations
as a fragment and as part of the intact protein chain is addressed in Table 1, which gives far UV circular dichroism parameters and secondary structure estimates of various proteins
and peptides under discussion.   The digest itself, that is β°
after clostripain hydrolysis but without fragment separation,
has a lower α-helix content than intact heme-free β°, so simply cleaving at the three sites already has a marked effect on

TABLE 1

Far Ultraviolet CD and Secondary Structure Charasteristics

Protein or peptide	$[\theta]_{222\,nm}$ (deg cm^2 per dmol)	Calculated % α helix*	Calculated no. of helical residues†
β^h	21,600	73	106
$\beta^0 + h$	21,600‡	73	106
$\beta^0_{digest} + \alpha^h + h$	18,200		
	(14,800)§	52§	76
$\beta^0_{rec} + \alpha^h + h$	17,500		
	(13,400)§	47§	69
β^0	10,600	39	57
$\beta^0_{digest} + h$	10,500	38	55
$\beta^0_{rec} + h$	9,500	35	51
β^0_{digest}	7,300	28	41
β^0_{rec}	7,500	29	42
$\beta^0_{cf} + \alpha^h + h$	13,700		
	(5,800)§	23§	17
β^0_{cf}	5,800	23	17
α^h	22,000	78	110
$\alpha^0 + h$	21,500‡	73	103
$\alpha^0_{digest} + \beta^h + h$	18,900		
	(16,200)§	57§	79
$\alpha^0_{digest} + h$	14,000	49	69
α^0	11,000	40	56
α^0_{digest}	7,200	28	39

+ Values in column 2 are means of at least three determinations, except for β^0_{rec} which is the mean of two determinations. The estimated uncertainty in $[\theta]_{222}$ is ± 5%. In the cases of the β^0_{rec} samples, the second digit in calculated α helix may not be significant due to uncertainty in concentration determination.

* Calculated according to Chen et al. (21).

‡ Calculated by multiplying the fraction of helical residues of the third column by 146 and 141 for α- and β-globins, respectively and 74 for β^0_{CF}.

† These values refer to the β components after subtracting the contribution made by α^h.

structure, although one cannot judge whether this refers to all components of the digest. Moreover, the isolated β^0_{CF} has a still lower value of helix content, again judging from the circular dichroism at 222 nm. It thus seems likely that the iso-

lated central fragment does not retain the conformation it has
in intact $\beta°$, which itself is less ordered than the heme-containing subunit.

More important still is that while heme is tightly and specifically bound to isolated $\beta°_{CF}$, its binding does not augment
the secondary structure, even in the presence of the complementary heme-containing subunit, α^h. In order to achieve substantial refolding, it is necessary that the side exon products be present ($\beta°_{rec}$) as well as α^h.

An analogous result is obtained with respect to the critical
property of reversible oxygen binding. When intact $\beta°$ globin
is reacted with hemin dicyanide, then reduced with sodium dithionite, the complex exhibits a typical deoxyhemoglobin β
spectrum in the Soret and visible spectral regions (Figure 7a).
Subsequent aeration yields the characteristic oxyhemoglobin β
spectrum. In contrast, when the same sequence of reactions is
carried out with $\beta°_{CF}$ or the $\beta°$ digest produced with clostripain, a series of hemo- and hemichromogen spectra results (Figure 7b). Here, again, it is only if the complementary α^h chain
is added along with the side exons (present in the $\beta°$ digest or
in $\beta°_{rec}$) that one obtains the set of deoxy- and oxyhemoglobin spectra (Figure 7c).

It is apparent, therefore, that reversible O_2 binding requires
the same complement of peptide fragments and intact α^h that was
necessary for refolding of the fragments produced by clostripain cleavage at positions 30 and 104. From the CD data of
Table 1, it can be calculated that $\beta°_{dig}$ or $\beta°_{rec}$ refolds
to an extent of 14% increase in α-helix content (20 residues)
when α^h is present before heme readdition to the non-covalently associated $\beta°$ peptide mixture. It is probable that this
refolding includes restructuring of the heme pocket in such a
way as to allow reversible O_2 binding. Craik et al. (15) interpreted these experiments to mean that the loss of structure

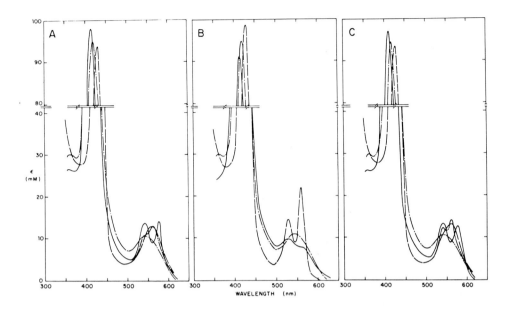

Figure 7. Spectral properties of cyanmet-, deoxy- and oxy-
derivatives of undigested and digested globins. (A) Cyanmet-
spectrum (– – –) of an equimolar mixture of β-globin and hemin
dicyanide; transient ferrocyanide spectrum (----) immediately
after sodium dithionite addition and the eventual deoxy-spectrum (- - -); oxy-spectrum (————) after dithionite removal.
The transient ferrocyanide spectrum and the deoxy-spectrum are
coincident below 500 nm. Solutions of the ferricyanide derivatives showed a visible absorption spectrum with a broad band
centered at 540 nm and a barely discernable shoulder at 560 nm.
The Soret band lies near 420 nm, and has an extinction coefficient about 10-fold greater than that at 540 nm. Ferricyanide derivatives of hemoglobin react directly with dithionite,
leading to reduction of the heme iron with ligand still bound,
the transient ferrocyanide derivatives then dissociating into
cyanide and deoxygenated ferrous heme-protein. The latter
transition from ferrocyanide to deoxyferroheme could be monitored for the separated α- and β-subunits, as the ferrocyanide
spectrum, with a maximum at 562 nm and a smaller peak at 541
nm, only gradually coalesced to the deoxyferroheme spectrum,
characterized by a single, broad asymmetric band in the visible
region with a maximum at 556 nm and a Soret band in the near UV
at ∿430 nm. After removal of dithionite by dilution, dialysis
or gel filtration, and exposure of the heme-protein solution to
oxygen, the characteristic oxyferroheme spectrum appears. The
two visible absorption bands are quite distinct from each
other, with one peaking at 577 nm and a broader one of slightly
lower intensity peaking at 541 nm. The Soret band is at 415
nm. (B) Cyanmet-spectrum (– – –) of an equimolar mixture of 60-

(Figure 7 cont.) min clostripain digested β-globin and hemin
dicyanide; hemochromogen spectrum (- - -) due to dithionite addition and a hemichromogen spectrum (———) after dithionite
removal. For α-globin, this spectrum also contains 9.3% oxy-
component presumably due to the small amount of undigested globin present.   (C) Cyanmet-spectrum (- - -) of an equimolar mixture of 60-min clostripain digested β-globin, the complementary
heme-containing α-subunit and hemin dicyanide; transient ferrocyanide spectrum (-----) immediately after dithionite addition
and the eventual deoxy-spectrum (———) after dithionite removal. The transient ferrocyanide spectrum and the deoxy-spectrum are coincident below 500 nm.

which accompanies clostripain cleavage at the two boundary arginine residues allows irreversible coordination of the distal

histidine to the heme iron, preventing dioxygen binding and

leading to hemochromogen (after dithionite) and eventual hemi-

chromogen formation (after aeration).

Coordination of the distal histidine is a feature of b-type

cytochromes, which undergo reduction-oxidation cycles upon ad-

dition of reductant, then O_2. Indeed, the spectra shown in

Figure 7b are similar to those obtained with the metcyano,

reduced and oxidized forms of b-type cytochromes.

Comparison of the Heme Binding Site of α^h and β^h With Those of

Cytochrome b_5, Cytochrome b_2 and Two Probable Sites in Mito-

chondrial Cytochrome b. These last results of the preceding

section led us to consider the possibility that a heme-binding

domain was encoded by an exon (or minigene) that has in the

course of evolution become associated with other gene elements,

and that such an exon might be recognized in several genes or

the corresponding protein products.

During the course of the investigations just described, Nobre-

ga and Tzagoloff determined the sequence of the cytochrome b

gene of yeast mitochondria (19). By comparing the base and

predicted amino acid sequences with homologous genes and gene

products from mitochondria of other species, they were able to

demonstrate that the cytochrome b gene of Saccharomyces cere-

visiae is split. In the strain they examined (referred to as
the short strain; see below), the gene is composed of two introns separating three coding regions. Moreover, these authors
have presented biochemical and genetic evidence that this single polypeptide chain protein binds two equivalents of heme
per mole (20) with the production of two functionalities, referred to as cytochrome b_t and cytochrome b_k. Since their
work, the corresponding mitochondrial gene has been sequenced
in other yeast strains and in different organisms, ranging all
the way to human. Since all these proteins are highly homologous, it is likely that any particulars about the heme binding
sites will be equally true of all mitochondrial cytochromes b.

Figure 8 shows portions of the amino acid sequence of human hemoglobin α and β chains, microsomal cytochrome b_5, yeast cytochrome b_2 and two stretches of yeast mitochondrial cytochrome
b. The segment of hemoglobin β sequence begins, arbitrarily,
at residue 20 and continues to 106; it includes all but one of
the heme contact residues (Leu 141). Residues β63 and β92 are
the distal and proximal histidines, respectively. Numbering of
the α chain is slightly different, but the sequences are shown
here aligned for maximum homology (22). Arrows mark the in-
tron-exon boundaries at positions 30-31 and 104-105 in the β-
globin chain; the corresponding positions in α-globin are 31-
32 and 99-100.

We have aligned the sequences of the cytochromes in Figure 8 to
maximize comparable positions for the known contacts and to facilitate comparison with the two suggested mitochondrial heme-
binding sites, but the positions of deletions are otherwise
arbitrary. Furthermore, as will be discussed below, the similarities of the mitochondrial sites with hemoglobin and cytochrome b_5 are more pronounced than those of the hemoglobins
with either cytochrome b_5 or yeast cytochrome b_2.

The two sections of mitochondrial sequences shown are both in

```
Hb  α     A G E Y G A E A L E R ‖ M F L S F P T T K T Ⓨ Ⓕ P H F - D L S H - - - - - G S A Q V K G
Hb  β     V D E V G G E A L G R L L V V Y P W T Q R Ⓕ Ⓕ E S F G D L S T P D A V M G N P K V K A

Mit b(1)  Y W W N M G S L L̲ G L C L̲ V I Q I V T G I Ⓕ M (              A,M                        )
Mit b(2)  G N D̲ I V S̲ W L W G G F S V S N P̲ T I Q R F F (              A,L                        )

Cyt b5    H N H S K S T W Ⓛ I Ⓛ H H K V Y D Ⓛ T K - Ⓕ L E (     ) E (                            )
Cyt b2    H N K P D D C W V V I N G Y V V Y D I T R - F L P (     ) N (                            )


Hb  α     Ⓗ* G K Ⓚ V A D A Ⓛ T D A V A H V D D M P N A L S A Ⓛ S E Ⓛ Ⓗ⁺ A D K Ⓛ R Ⓥ D P V Ⓝ Ⓕ K ‖ L Ⓛ
Hb  β     Ⓗ* G K K Ⓥ L G Ⓢ Ⓕ S D G L A H L D N L K G T Ⓕ A T Ⓛ S E Ⓛ Ⓗ⁺ C D K L H Ⓥ D P E Ⓝ Ⓕ R L Ⓛ

Mit b(1)  H̲* Y S S N I ELA F S S - V̲ E H̲ I I R̲ D V H N G Ÿ I - L̲ R Y L̲ H⁺ A N G A S F F F M V M F M H
Mit b(2)  H̲* Y L V P - - - F I I A̲ - A M - - V I M H L M - - A - - L̲ H⁺ I H G S S N̲ P L G I T G N L̲

Cyt b5    Ⓗ* - - - Ⓟ G G E E V L R E Q A G G Ⓓ Ⓐ T E N Ⓕ E D Ⓥ - - G Ⓗ⁺ S T D Ⓐ R E M Ⓢ K T Ⓕ I I G
Cyt b2    H* - - - P G G Q D V I K F N A G K D V T A I F E - P - - L H⁺ A P B V I B K Y I A P Q K L
```

Figure 8.   Comparison of the primary sequence of segments of Hb α, Hb β, yeast mitochondrial apocytochrome b, [Mit b (1) and Mit b(2)] cytochrome b_5 and cytochrome b_2. Superscripts * and + indicate axial ligands. In the case of Hb α and Hb β, these refer to distal and proximal histidines, respectively. Sequence letter code used in this figure and in the Tables is: A, Ala; C, Cys; D, Asp; E, Glu; F, Phe; G, Gly; H, His; I, Ile; K, Lys; L, Leu; M, Met; N, Asn; P, Pro; Q, Gln; R, Arg; S, Ser; T, Thr; V, Val; W, Trp; Y, Tyr.

the first coding sequences in this yeast strain, and run from residues 28 to 96 and 158 to 126, respectively, with the heme-binding sites demarcated by histidines 53 and 82, in site 1 (labelled Mit b[1]), and histidines 183 to 202 in the second site, Mit b(2). The number of residues between axial histidine ligands is accordingly 28 in Mit b(1), the same number as in the hemoglobin, and 18 in Mit b(2), five less than in the cytochrome b_5 site.

As noted above, Mit b(1), the first site in the linear sequence, has the same length as in normal hemoglobins. In this site, also, the distal histidine follows a phenylalanine by three residues. Table 2 compares the sequence immediately surrounding the distal and proximal histidine ligands in the two

TABLE 2

Comparison of Local Sequences
Surrounding Distal and Proximal Histidines

			Distal							Proximal		
Hb α	Q	V	K	A	H*	G		L	S	N	L	H+	A
Hb β	K	V	K	A	H*	G		L	S	E	L	H+	C
Mit b(1)	F	M	A	M	H*	Y		L	R	Y	L	H+	A
Mit b(2)	F	F	A	L	H*	Y		L	M	A	L	H+	I
Cyt b₅	F	L	E	E	H*	P		E	D	V	G	H+	S
Cyt b₂	F	L	P	N	H*	P		F	E	P	L	H+	A

The key for the one-letter code identifying the amino acid
residues is in the caption for Figure 8.

proposed mitochondrial sites and in the other four proteins.
The immediate sequence neighbors of two distal histidines in
the mitochondrial sites are very similar, those of the proximal site less so. In each of the pairs of related proteins,
however, both distal and proximal local sequences are more
closely related to one another than to any of the other local
sequences.

From the aligned sequences of Figure 8, and the known contact
residues of the hemoglobins and cytochrome b₅ (circled in Figure 8), we suggest a set of probable contacts for the two mitochondrial sites, and the cytochrome b₂ site shown in Table
3. Not surprisingly, the proportion of hydrophobic residues
is very high in all the sites, with Leu and Phe residues predominant. In these cases, at least, a fairly small number of

TABLE 3

Known Heme Contact Residues in Hb α, Hb β and Cytochrome b₅; (a)
Possible Heme Contact Residues in Mit b(1), Mit b(2)
and Cytochrome b₂ (b)

```
Hb α          Y F H* K      L    L L H† L V N F L L

Hb β            F F H*    V S F F L L H†    V N F L L


Mit b(1)    L L F M H*      D V Y    L H† A S    F

Mit b(2)      F F H* P      F M A L H† S S N


Cyt b₅      L L L F H* P    D A F    V H† A S    F

Cyt b₂      V I I F H* P    D V F    L H†
```

(a) Heme contact residues are those with any atom whose center
lies within 4.0 Å of an atom in the heme group. Contacts for
Hb α and Hb β are based on coordinates deposited by Joyce Baldwin with the Brookhaven Protein Data Bank. The contact residues in cytochrome b₅ are based on Argos and Rossmann (16); sequence data from Nobrega and Ozols (23).

(b) From the sequence of the heme binding dipeptide resulting
from tryptic digestion (24); see also Guiard and Lederer (25).

* and † indicate axial ligands. In the case of Hb α and Hb
β, these refer to distal and proximal histidines, respectively.

residues defines a heme pocket, the number varying from 12 to
15 in the six sequences and 14 or 15 in the three sites whose
structures are actually known. With Phe and Tyr taken as homologous residues, the proposed contact residues of Mit b(1) and
Mit b(2) are homologous to Hbβ at 43 and 50% of positions, respectively, according to the alignment given in Figure 8. Interestingly, the proposed contact Asp-Val-Tyr (DVY) can be a-
ligned with the corresponding Asp-Val-Phe and Asp-Ala-Phe in
cytochrome b₂ and cytochrome b₅, respectively. In addition,
although not used as close contacts, the positions in Hb α and
β corresponding to these Asp residues are Asp and Asn.

Additional observations about the sites are as follows: for

Mit b(1), 56 positions are compared to Hb α and β in Exon 2.
Of these, 16, or 29%, are homologous. Of the 30 residues from
distal to proximal histidines, 12, or 40% are homologous. Of
the 14 residues from the exon boundary to the distal histidine,
3, or 21%, are homologous and 2, or 14%, are homologous from
the proximal histidine to the other exon boundary. The site
itself, therefore, shows significantly greater homology than
sequences preceding or following, which do not show higher
homology than would be expected if two sequences chosen at
random were compared.

For Mit b(2), 48 positions are compared corresponding to a-
lignment with either Hb α or Hb β in Exon 2. Of these, 14, or
29%, are homologous, just as in Mit b(1). Of the 20 residues
from the distal to proximal histidines, 5, or 25%, are homologous, which is just slightly higher than homology in random
comparison. Of the 14 residues following the proximal histidine, only 1, or 7% is homologous. It is striking, however,
that of the 14 residues from the exon boundary to the distal
histidine, 9, or 64%, are homologous.

All this points very strongly to the idea that the Mit b(2)
heme binding site is an exon-encoded functional domain and that
the genetic element has, through recombination events, become
associated in split genes with different combinations of exons, while retaining the essential and recognizable heme-
binding function. We are unable to identify an exon-intron
boundary on the C-terminal side of the heme pocket corresponding to the boundary in the hemoglobin genes, but it is possible
that an exon fusion occurred in the cytochrome b gene. Certainly, the gene structure for cytochrome b shows evidence of
variability, given the different intron number in two yeast
strains, the reduced number in Aspergillus nidulans and the
absence of any introns in the human mitochondrial gene (26).
This variability may be contrasted with the conservation of
gene structure in hemoglobin chains. Those genes appear to

have conserved exon/intron structure that has persisted at
least since the time of appearance of separate α and β globin
genes (27).

Kinetics of Heme-Induced Refolding of α-Globin. Before proceeding to a concluding discussion of the domain-like characteristics of the heme-binding site and its relationship to a
particular genetic element, I wish to discuss recent results of
an investigation of the kinetics of refolding of human α-globin induced by heme binding (28). It was noted above that removal of heme from α^h and β^h results in gross conformational
changes, with extensive loss of secondary and tertiary structure and concomitant unfolding. These changes are fully reversible and we took advantage of this fact to characterize
the refolding kinetics when an equivalent amount of hemin dicyanide is added to disordered α globins (8). In this study,
conditions were selected to ensure that a large change in helicity (from ca. 30% or lower of residual helix in α° at pH
6.7, to ca. 75% of α helix in α^h) was due solely to the prosthetic group. α° **was** selected because, unlike β°, it refolds
completely upon hemin dicyanide addition, whereas complete β^h
refolding requires slow dialysis against buffer, precluding
stopped-flow studies. An important feature was the use of
fluorescence and CD, in addition to the Soret absorption, as
reaction probes.

As summarized in Table 4, the following sequence of events was
observed, based on the rates obtained with the three techniques
and calculated at a protein concentration of 24 μM. First,
heme enters the pocket-like site, with a half-time of 0.01 sec.
The bond to N_ε of His-F8 probably occurs simultaneously or soon
after, with a half-time of the order of hundredths of a second.
After an initial surge in Soret intensity (∿65%), a constant
rate of change is observed with a half-time of 40 sec, during
which the heme pocket assumes its final conformation. In con-

TABLE 4

Comparison of Rate Constants
Obtained from Different Measurements[a]

Stopped Flow Technique	Fast Phase	Slow Phase
Fluorescence	$3.3 \times 10^{-7} M^{-1} s^{-1}$	$2.3 \times 10^{-2} s^{-1}$
	$(80 - 85\%)^{b}$	$(15 - 20\%)$
Absorption	mixed	$1.7 \times 10^{-2} s^{-1}$
		$(35 - 45\%)$
Circular Dichroism	---	$2.6 \times 10^{-2} s^{-1}$, $6.0 \times 10^{-3} s^{-1}$
		$(30 - 35\%)$ $(65 - 70\%)$

[a] All measurements are for α°-globin at a concentration of 2.4 $\times 10^{-6} M$ and are $\pm \leq 15\%$.

[b] Numbers in parentheses give % of total parameter change. From (28).

trast, the recovery of ordered secondary structure in the molecule as a whole has no fast component, within 10% error. The
growth of α-helix has two phases with half-times of 27 and 116
sec. This sequence of events shows the heme pocket to be a
nucleus for the formation of the native conformation.

This work demonstrates that the residues of the heme pocket,
which are included within the central fragment, probably a-
chieve or closely approach their final three-dimensional structure well before the entire chain is folded, and suggests that
an element of tertiary structure originates within this peptide
fragment and serves as a nucleus for an independent folding
pathway.

It should be emphasized that the refolding occurs from a state

that is not entirely unfolded. This initial state may or may
not contain elements of the native structure. Moreover, since
the refolding is induced by heme addition, this process may
not represent a pathway taken by, say, a guanidine-unfolded
molecule that is restored to nondenaturing conditions. The
kinetics observed do, however, raise the possibility that the
heme pocket, or a segment of the entire central fragment, is a
kinetically independent unit in terms of folding.

Concluding Comments on Further Aspects of the Domain-Like Central Fragment. Figure 9 shows computer-generated drawings of
the backbone of the Hb central fragment as it exists in the
intact molecule, and of a corresponding segment in cytochrome
b_5, with the side chains of the contact residues of Table 3
included. The segment containing the heme contact residues
in Hb β is much larger than in cytochrome b_5, mainly due to
the presence in Hb β of two large segments that contain no
contact residues. The first of these encompasses residues 43-
62 and makes up part of the interhelical CD segment, all of
helix D, and the first six residues of the E helix. The second stretch includes the C-terminal residues of the E helix
and the EF interhelical residues 71-84. The cytochrome b_5
site is more economical, with no large loop between the first
contact residues and the distal histidine. At the base of the
site, it is as if the EF bend of Hb β had been pinched off.
As for the remainder of the site, Argos and Rossman have made
a detailed comparison of the secondary structure similarities
(16).

In view of the fact that the cytochrome b_5 site is so much
smaller, we were concerned about whether a site could be constructed for Mit b(2), since there are five fewer residues
still between distal and proximal histidines. The question
arises, whether both histidines could be ligands while reasonable ϕ, ψ angles were set for the intervening residues. Accordingly, after manipulating the cytochrome b structure on a

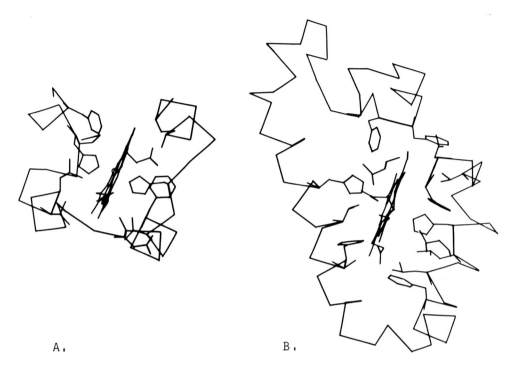

A.   B.

Figure 9. Computer generated drawings of the C_α backbone of
the residues forming the heme-binding site in cytochrome b_5
(A) and of the central β-globin exon product (B). Contact
residues (4 Å) are shown as is the heme group. Coordinates
deposited by Joyce Baldwin with the Brookhaven Protein Data
Bank. The contact residues in cytochrome b_5 are based on
Mathews et al. (41). Numbers in the figure refer to C_α residues of Hb β.

graphic display unit, we constructed a model of the Mit b(2)
site utilizing known ϕ,ψ angles in the cytochrome b_5 site for
the residues KFLEEHPGG__VL__QA__DATENF__VGH of this site (16).
This generated a site in which the distal and proximal histidines had the same spatial relationship to the enclosed heme
as in cytochrome b_5. This model was constructed merely to
demonstrate that a plausible site could be constructed, while
recognizing that the actual site may have quite a different
conformation.

Two final observations are of interest: The first is that both
apomyoglobin and apohemoglobin bind iron-free protoporphyrin
IX tightly (28). In the case of Hb β, the kinetics of this
combination are virtually identical to those of heme binding
(28). The other is that apomyoglobin binds bacterial chloro-
phyllides in the heme-binding site with generation of a green,
crystallizable holoprotein (30).

These observations lead us to surmise that the heme-binding
site may be a more general porphyrin-binding site which has
evolved to accomodate more than one kind of metalloporphyrin.
If this is the case, then the existence of heme-containing
and, possibly, chlorophyll-containing proteins may provide
powerful support for the hypothesis that exon-shuffling has
played a role in protein evolution. The evidence, however,
is far from conclusive and much remains to be done before such
a hypothesis is accepted as verified.

Acknowledgements

I am indebted to my associate, Charles S. Craik, who is responsible for the computer calculations and for preparing
Figure 9 as well as for constructing the feasibility model
of the Mit b(2) site. Research previously unpublished was
supported in part by grants from the National Science Foundation, NSF PCM 80-5614 and the National Institutes of Health,
NIH RO1 HL 16601.

References

1. Efstradiatis, A., et al.: Cell 21, 653-668.
2. Blanchetot, A., Wilson, V., Wood, D, Jeffreys, A.J.:
   Nature 301, 732 (1983).
3. Czelusniak, J., et al.: Nature 15, 297 (1982). Jeuseu, E.
   O., Paludau, K., Hyldig-Nielsen, Jorgensen, P., Marcker,
   K.A.: Nature 291 677 (1981).

4. Hunt, T.L., Hurst-Calderone, S., Dayhoff, M.O.: Atlas of Protein Sequence and Structure, (Ed. Dayhoff, M.O.) 229-251 Nat. Biomed. Res. Foundation · Washington, D.C. 1978.

5. Breslow, E., Beychok, S., Hardman, K.D., Gurd, F.R.N.: J. Biol. Chem. 240, 304 (1965).

6. Javaherian, K., Beychok, S.: J. Mol. Biol. 37, 1 (1968).

7. Yip, Y.K., Waks, M., Beychok, S.: J. Biol. Chem. 247, 7237 (1972).

8. Waks, M., Yip, Y.K., Beychok, S.: J. Biol. Chem. 248, 6462 (1973).

9. Darnell, J.E.: Science 202, 1257-1260 (1978).

10. Gilbert, W.: Nature 271, 501 (1978).

11. Blake, C.F.: Nature 277, 598 (1978).

12. Orgel, L.E., Crick, F.H.C.: Nature 284, 604 (1980).

13. Craik, C.S., Sprang, S., Fletterick, R., Rutter, W.J.: Nature 299, 5879 (1982).

14. Lewin, R.: Science 217, 921 (1982).

15. Craik, C.S., Buchman, S.R., Beychok, S.: Nature 291, 87 (1981).

16. Argos, P., Rossmann, M.G.: Biochemistry 22, 4951 (1972).

17. Craik, C.S.: Ph.D. Dissertation, Columbia University (1980).

18. Craik, C.S., Buchman, S.R., Beychok, S.: Proc. Natl. Acad. Sci. U.S.A. 77, 1384 (1980).

19. Nobrega, F.G., Tzagoloff, A.: J. Biol. Chem. 255, 9828 (1980).

20. Tzagoloff, A., Nobrega, F.G.: Biological Chemistry of Organelle Formation, ed. T. Bucher, W. Sebald, H. Weiss, Springer-Verlag, Heidelberg, 1 (1980).

21. Chen, Y.-H., Yang, J.T., Chau, K.H.: Biochemistry 13, 3350 (1974).

22. Dickerson, R.E.: The Proteins Vol. 2, ed. H. Neurath, Academic Press, New York, p. 603 (1964).

23. Nobrega, F.G., Ozols, J.: J. Biol. Chem. 246, 1706 (1971).

24. Labeyrie, F., Grodinsky, O., Jacquot-Armond, Y., Naslin, L.: Biochim. Biophys. Acta. 128, 492 (1966).

25. Guiard, B, Lederer, F.: J. Mol. Biol. 135, 639 (1979).

26. Anderson, S., Bankier, A.T., Barrell, B.G., de Bruijn, M.H.L., Coulson, A.R., Drouin, J., Eperon, I.C., Nierlich, D.P., Roe, B.A., Sanger, F., Schreier, P.H., Smith, A.J.H., Staden, R., Young, I.G. Nature 290, 457 (1981).

27. Kimura, M. Sci. Amer. 341, (5) 98 (1979).

28. Leutzinger, Y., Beychok, S.: Proc. Natl. Acad. Sci. U.S.A.
    78, 780 (1981).

29. Mathews, F.S., Levine, M., Argos, P.: J. Mol. Biol. 64,
    449 (1972).

30. Boxer, S.G., Wright, K.A.: J. Amer. Chem. Soc. 101, 6791
    (1979).

Received July 13, 1984

Discussion

Bode: Have you separately checked whether the fragments obtained by digestion are still associated in the digest?

Beychok: Yes. The fragments do remain strongly associated in the digest.
In fact, SDS gel electrophoresis does not result in fragment separation.
In order to separate, it is necessary to use Swank-Munkries gels, that is,
SDS-urea gels.

Blauer: First I would like to report that we have observed also a rapid
change in Soret absorbance upon recombination of myoglobin from its components apomyoglobin and ferriheme at neutral pH and 25 C within a few
minutes. This change amounts to about 90 % of the final value. In a second
phase, the absorbance increases slowly within about one hour to the final
value. The first phase is pH dependent. Could you comment on this?

Beychok: We have also observed this second phase, both with ferriheme and
with ferrihemecyanide. This second phase corresponds to dissociation of
aggregates of the hemin or cyanohemin. In the case of the cyanide form,
this exists as a monomer-dimer mixture at concentrations higher than about
5×10^{-6}M. We studied the rate of dissociation of the dimer under various
conditions and found this to be identical to the slow combination rate under the same conditions.

Blauer: What was the temperature relevant to your kinetic data given in
Table 4? Do you have a temperature dependence and what are the corresponding energies and entropies of activation?

Beychok: We certainly have a strong temperature dependence, but for all
of the results presented here, the temperature was 4°C. At higher temperatures, the globin chains are unstable even during the course of an experiment with continuous aggregation and precipitation.

Woody: Have you considered the possibility that the slow step in recovery
of secondary structure might be related to the phenomenon of heme scrambling studied by LaMar and coworkers? The bis-cyanomet heme will initially
combine in two nearly equivalent orientations and if the interchange of

methyl and vinyl groups interferes with proper folding, isomerization of about half the molecules will be required.

Beychok: We have not considered the possibility systematically, and it is certainly an interesting thought. Inasmuch as the slow step represents about two thirds of the amplitude of the CD change, this would require that the unfavorable orientation be preferred kinetically in the binding step. I do not believe our present data could distinguish this, but I have not really thought through a definitive way to test the possibility.

Myer: How would you apply the common gene model to cytochrome c?

Beychok: Well, as you know the heme in cytochrome c is bound into its site in an orientation rather different from that in the hemoglobins. I therefore do not know that we should expect a common ancestor or minigene for heme binding. However, it is interesting that in the one case of a split gene for cytochrome c, an intron divides the binding 'site' roughly in half. This occurs in a curious plant hemoglobin namely leghemoglobin. This raises the possibility that two-sided heme binding evolved after one-sided heme binding, but I think it much too premature to try to connect cytochrome c with the hemoglobins or the b-type cytochromes, in both of which cases histidine serves both axial ligands, there is no covalent linkage and the heme-orientations are similar.

HEME CONFIGURATION IN CYTOCHROME C

Yash P. Myer

Institute of Hemoproteins
Department of Chemistry
State University of New York at Albany
Albany, NY 12222 U.S.A.

Abstract

Resonance Raman spectroscopic measurements in the high-frequency region of
models with and without the unique structures of type c hemoproteins and
of cytochrome c and of low- and high-spin forms have been reported. Based
on the positions and nature of the porphyrin skeleton vibrational modes,
it has been concluded that in both spin states, the heme-protein linkage
has little or no effect on the core dimensions. The ligation of histidine
imidazole in low-spin forms causes a slight expansion of the tetrapyrrole
core, which is further loosened upon localization of heme in the protein
environment. In the high-spin form and in the absence of intrinsic
histidine imidazole, the iron configuration is in-plane and hexacoordinated with two water molecules at the two axial positions. The presence
of histidine imidazole, however, results in a pentacoordinated configuration with iron about 0.3 Å out of the porphyrin plane. This is
characteristic of type c high-spin systems and is possibly linked to the
unique location of the histidine residue providing the axial ligate.

Introduction

The hemoproteins are widely diverse in function, ranging from oxygen

carrier to oxygen storage to electron transport to ATP production to

degradation of physiological byproducts. The center of function in the

main is the iron tetrapyrrole structure, iron porphyrin, but it differs in

peripheral substitutions, coordination configuration, and protein

environment. The diversity of function of essentially the same group is

undoubtedly a reflection of the unique state of the reaction center, the

heme. A comprehension of the state of the reaction center and of its

Optical Properties and Structure of Tetrapyrroles
© 1985 by Walter de Gruyter & Co., Berlin · New York

interplay with structural and configurational variations and the unique
conformation of the environment is fundamental to the understanding of
structural-functional interrelationships and the reasons for the diversity
of function.

Of all the heme proteins, the cytochromes c are possibly the most widely
investigated family. Amino acid sequences of more than 90 cytochromes c
have been determined, the three-dimensional structures of more than 6
members of the family of eukaryotic cytochromes c and a number of other
cytochromes c have been elucidated, and the proteins have been subjected
to almost all the known physico-chemical studies (for reviews see 1-6).
These proteins contain iron protoporphyrin IX, heme, as do the hemoglobins
and myoglobins, but their function is mainly that of electron carrier
between ATP production sites. In addition to the heme-protein linkage
through the protein group/groups occupying one/both axial positions of
iron, a feature common to all heme proteins, the cytochromes c also
contain another protein-heme linkage at the vinyl side chains of the
tetrapyrrole structure. With the exception of only two members in the
family, the vinyl linkage is at both the chains forming thioether bonds to
the cysteine side chains. Another feature which makes the cytochromes c
unique is the structural location of one of the protein groups occupying
the axial position of iron, the imidazole of the histidine residue next to
the C-terminus cysteine-heme structure. In terms of the state of heme,
and within the limits of the resolution of the three-dimensional
structure, 1.5 - 1.8 $\overset{o}{A}$, depending upon the oxidation state of iron, it
appears the porphyrin macromolecule is planar and the iron is in the plane
of the tetrapyrrole structure (for reviews see 1-3; 7).

Developments in the field of resonance Raman (RR) spectroscopy since the
first studies of cytochrome c in 1972 (8) have resulted in a powerful
probe for the examination of the state of heme, including the ability to
examine the interplay between heme and its environment (for reviews see 9-
14). The technique allows selective examination of the porphyrin, the
metal-porphyrin, and the metal-ligand vibrational modes, a property
directly linked to the bonding energetics of the macromolecule. The probe
is sensitive to bond length changes of less than 0.001 $\overset{o}{A}$ and to

perturbations of the order of 0.1 electron in the electron density, limits
far lower than those of any other technique thus far available for
investigation of proteins. The vibrational modes contributing to RR bands
in the main have been assigned, the bands sensitive to the porphyrin
geometry have been identified, and correlations between the core
dimensions and the positions of a number of bands have been developed (for
reviews see 9, 11-13; 15-21). Metal-ligand RR bands, bands reflecting
sensitivity to the electron distribution of the macromolecule, to
interactions with the surroundings, and to peripheral substitutions have
been located (for reviews see 9, 12-14). During the last few years we
have used resonance Raman spectroscopy to probe and discern the nature and
the magnitude of the interdependence between the state of the iron-
porphyrin macromolecule and its environment in cytochrome c systems. This
report deals with some recent findings; a more detailed treatment of
aspects not dealt with in this report are to be found in a recent
publication from this laboratory (22), which also contains a description
of the instrumentation and the experimental approach.

Materials and Methods

Materials. Horse heart (HH) cytochrome c, types III and VI, was purchased
from Sigma Chemical Co., and used without further purification. Fragment
1 through 65 (H65) (sequence numbering of horse heart cytochrome c) (1)
with intact heme-protein structures was prepared using limited cyanogen
bromide cleavage and purification as detailed by Corradin and Harbury
(23), and the heme-containing fragment from residue 11 through 21, heme
undecapeptide (H11), by peptic hydrolysis according to the procedure of
Harbury and Loach (24). Further tryptic digestion of H11 yielded the
shortest of the heme c systems, the heme-protein fragment from residues 14
through 21, heme octapeptide (H8) (24). A further simplified model, H8nb,
heme octapeptide with blocked heme iron ligating groups, the α-amino group
of residue 14 and the imidazole of His-18, was prepared as described in an
earlier publication (25). Fig. 1 details the structural similarities and
differences of the model systems used. All other chemicals used were of
analytical grade.

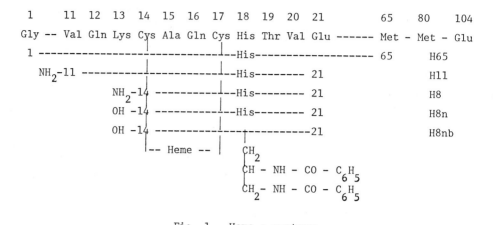

Fig. 1. Heme c systems.

Measurements. Resonance Raman spectroscopic measurements were performed at 30 ± 0.1 C using the setup described in earlier works (22). The spectrum was corrected for background contributions and smoothened where and when needed using the UNIVAC 1110/83 and the procedure described in the above publication.

Results and Discussion

RR spectra and analysis. The region of the electronic spectrum most easily accessible to RR investigations of hemes and hemoproteins is the visible region, containing the porphyrin π - π^*, plane-polarized transition, the Soret or the B band, and the α and β bands, i.e., the Qo and the Ql bands (for reviews see 26, 27). Excitation in this region of the spectrum results in intensity enhancement, primarily of the porphyrin, in-plane vibrational modes. Modes with contributions largely from vibrations with Alg and/or Eu symmetry are observed upon excitation in the B band region, while modes with A2g symmetry become dominant upon excitation in the Qo and Ql region (Table I of ref. 22). Almost all the observed RR bands of tetrapyrrole systems have been assigned, and the

Table I

Assignment, Potential Energy Distribution, Polarization and Position
of High Frequency RR Bands of Low-Spin Ferric
Protoporphyrin IX and Heme c Systems

Vib.	Mode	Assignment[a] Potential Energy Distribution (PED) (%)		Band Positions of (cm⁻¹)						
				Im₂-PPIX	Im₂-H8nb[b]	Im-H11	Im-H65	Im-Cyt	Cyt	Urea Cyt
$\nu10$	B1g	$C_\alpha C_m(49); C_\alpha C_\beta(17)$	dp	1640	1641	1638	1637	1637	1635	1640
$\nu37$	Eu	$C_\alpha C_m(36); *C_\alpha C_m(24)$	ip	1602	1599	1590	1590	1587	1585	1590
$\nu19$	A2g	$*C_\alpha C_m(67); *C_\alpha C_\beta(18)$		1586						
$\nu2$	A1g	$C_\beta C_\beta(60); C_\beta Et(19)$	p	1579	1587	1590	1589	1588	1585	1588
$\nu11$	B1g	$C_\beta C_\beta(57); C_\beta Et(11)$	dp	1562	1569	1567	1567	1564	1562	1568
$\nu38$	Eu	$C_\beta C_\beta(53); C_\beta Et(16)$	ip	1554						
$\nu3$	A1g	$C_\alpha C_m(41); C_\alpha C_\beta(35)$	p	1502	1503	1500	1503	1500	1503	1504
$\nu28$	B2g	$*C_\alpha C_m(52); *C_\alpha C_\beta(21)$		1469	1468w	1461	1468	1460	1467	1464
	A1g		p			1409		1404	1406	1414
$\nu29$	Eu, B2g		ip	1402				1411	1409	1411
$\nu20$	A2g	$*C_\alpha N(29); *C_\beta ET(24)$		1399						
$\nu4$	A1g	$C_\alpha N(53); C_\alpha C_m(21)$	p	1373	1373	1373	1373	1373	1372	1374
	Eu		ip			1318	1318	1316	1314	1319
$\nu21$	A2g	$\delta C_m H(53); *C_\alpha C_\beta(18)$	dp	1306	1317			1315	1316	
$\nu5+\nu9$	A1g		p	1260		1255	1249	1247	1248	1250
$\nu13$	B1g	$\delta C_m H(67); C_\alpha C_\beta(22)$	dp	1230		1230			1233	1234
	A1g		p			1170	1172	1177	1178	1171
$\nu30+\nu29$			dp	1167				1170	1171	1170
$\nu6+\nu8$				1130						
$\nu22$	A2g	$*C_\alpha N(37); *C_\beta Et(26)$		1125		1124	1129	1130	1126	1124

[a] Mode numbers, assignments and PED, Abe et al.(15); band correlation by Choi et al. (21). *, antisymmetric stretch; δ, deformation vibrations about the C2 axis of the pyrrole ring. Nature: dp, depolarized, $\rho = I_\perp/I_\parallel$ = 0.75 ± 0.05; p, polarized, $\rho \leq 0.70$; ip, inversely polarized, $\rho \geq 0.80$.

[b] Data from Choi et al. (21).

Table II

Characteristics of Porphyrin Core Size Related RR Bands of
Hemes and Hemoproteins

RR Band[a]	$\nu10$	$\nu37$	$\nu2$	$\nu11$	$\nu3$	$\nu28$
Earlier Labels[b]	V	IV	1590-70		II	
Polarization[c]	dp	ip	p	p	p	dp
Mode[c]	B1g	Eu	A1g	A1g	A1g	B2g
Potential Energy Distribution (PED) (%)[c]						
$C_\alpha C_m$ st	49	36,*24			*41	*52
$C_\alpha C_\beta$ st	17				35	*21
$C_\beta C_\beta$ st			60	57		
Core Size Correlation Parameters (Expression $\nu=K(A-d)$)[d]						
K (cm^{-1}/Å)	423.7	555.6	390.8	344.8	375.5	402.3
A (Å)	5.87	4.86	6.03	6.53	6.01	5.64
Configuration-Spin State Dependence[e]			Range (cm^{-1})			
LS:HC	1641-35	1590-85	1590-80	1565-60	1509-03	1469
HS:HC	1612-08	1562-55	1584-82		1483-75	
HS:PC	1633-28	1575-72	1576-74	1533	1499-95	1453

[a];Frequency numbering according to Abe et al. (15).   [b];Numbering according
to Spiro and Burke (19); 1590-70 labeled according to Myer et al. (22).
[c];See Table I for detailed description.[d];According to Spiro et al. (19)
and Choi et al. (17). Parameters for bands $\nu2$, $\nu11$ and $\nu28$ are derived
from consideration of only 5 iron porphyrin systems, whereas data for both
metalloporphyrins and hemoproteins are accounted for in the correlation of
other bands (18-20). e;Composite of RR data of hemes and hemoproteins with
known coordination configurations and spin states of iron (9, 10, 18-20,
22). LS:HC, low-spin hexacoordinated; HS:HC, high-spin hexacoordinated;
HS:PC, high-spin pentacoordinated.

contributing vibrations to each of the RR bands, especially those above
1000 cm^{-1}, have been discerned (17-21; for reviews see 9, 11, 13) (Table
I). It recently has been shown that not only a number of high-frequency
bands (18-20), but almost all the RR bands with contributions from
porphyrin skeletal modes above 1450 cm^{-1}, are relatable to the core
dimensions, the Ct-N distances, of the tetrapyrrole macromolecule through
linear, but distinct, relationships (21) (Table II). A composite of the

resonance Raman data of a variety of metalloporphyrins and hemoproteins
with known coordination configurations and spin states of iron
demonstrates that the positions of a number of the porphyrin-core-
sensitive bands fall into distinct and well-defined small frequency ranges
for given coordination-spin state combinations, e.g., bands $\nu 10$, $\nu 37$, $\nu 2$
and $\nu 3$ (Table II). A combination of the correlation of the observed
positions of the porphyrin-core-sensitive bands to the sets for various
permutations of the spin state of iron and its coordination configuration,
on the one hand, and on the other, analysis of band positions with respect
to core dimensions and to the contributing vibrational modes constitutes a
feasible approach to probe both the conformational and the configurational
state of heme in systems with unknown three-dimensional structures. With
the iron-pyrrole nitrogen distances available from the three-dimensional
structures of almost all the permutations of spin state-coordination
configurations of iron porphyrins (for reviews see 28-30), it is possible
to quantitate the geometry of heme, the iron-to-porphyrin plane
arrangement, by applying the simple triangulation principle using the
estimated Ct-N distances (for reviews see 9, 10, 13; 22). However, since
the resonance bands in general are a composite of multiple vibrations
(Table I), situations in which not all the structurally sensitive bands
conform to behavior strictly consistent with empirical correlations can
occur. This is reported for two systems differing in the state of the
vinyl chains, iron protoporphyrin IX vs. mesoporphyrin IX (17, 21). Given
that accurate information about the contributions to a certain vibrational
mode is known (Table I) (15, 16, 18, 21), such deviations provide a means
of determining the interplay between the energetics of porphyrin
structures and their environment, such as peripheral substitutions, heme-
protein interactions, etc. (for reviews see 12, 14; 31).

Low-spin heme c systems

Addition of extrinsic ligands with strong ligand-field strengths, e.g.,
imidazole, cyanide, ammonia, etc., to all heme c systems generates a low-
spin, hexa-coordinated configuration of iron with the characteristic
absorption spectrum shown in Figure 2 (for review see 4; 22-25).
Spectroscopically and potentiometrically, H8nb in the presence of

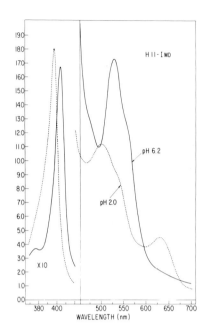

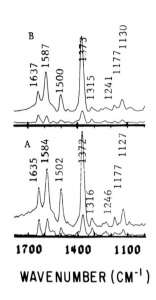

WAVENUMBER (CM⁻¹)

Fig. 2, Left.   Representative absorption spectra of low-spin, pH 6.2, and
                high-spin, pH 2.0, forms of heme c systems in the presence
                of imidazole. Spectra shown for H1̄1, 0.5 M imidazole.
                Ordinate, mM extinction coefficient.

Fig. 3, Right. Resonance Raman spectra of horse heart ferricytochrome c
                in the absence, A, and presence, B, of imidazole. Solution
                conditions: 0.05 M protein ,0.05 M phosphate + ligand
                (0.5 M), pH 6.8.   4579 Å excitation; 30 mW incident power;
                1 sec time constant.

imidazole and at neutral pH is characterized as containing imidazole

nitrogens as the axial ligates, Im/Im (25), whereas H8, H11, H65, and

cytochrome c under similar conditions have configurations with extrinsic

ligands at the sixth position of iron and imidazole nitrogen of His-18 at

the fifth position, Im/His (for review see 4; 23-25). Studies of the

reaction of the axial ligates to group-specific reagent, bromoacetic acid,

showed that the heme domain in the region of the extrinsic ligand only is

accessible to anionic reagents (4). Conformational studies using circular

dichroism spectroscopy showed that except for Im_2H8nb, all other systems

retain a finite degree of protein secondary structure which depends upon

the length of the polypeptide chain; Im-cytochrome c has a polypeptide

organization like that of the native protein (for review see 3).   In

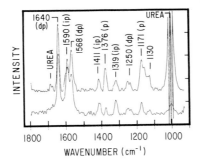

Fig. 4. Resonance Raman spectra of HH ferricytochrome c in the presence
of 4 M urea; pH 6.8, 0.05M phosphate; 5145 Å excitation (22).

contrast to the above, cytochrome c at neutral pH, form III, contains an
iron configuration of methionine-sulfur and imidazole of His-18, -S/His,
and the heme is almost completely (80%) shielded from solvent (1, 7). In
the presence of 4 M urea and at neutral pH, the low-spin -S/His configuration is maintained, but protein-heme interactions on both sides of the
heme are disrupted to the extent that the domain of both the axial ligates
becomes accessible (33). At alkaline pH's, near 11, cytochrome c becomes
a low-spin, hexacoordinated form, form IV, with an iron configuration
different from that of the neutral pH form, possibly ε-amino of a lysine
residue and imidazole of His-18, $-NH_2$/His (for reviews see 1, 4).

Resonance Raman spectra of several low-spin systems are shown in Figs. 3-
6a, and in Table I are listed the positions and the nature of the high-
frequency bands of all these systems, including the bands of Im -ferric
iron protoporphyrin IX, Im_2-Fe(III)PP IX (21), the simplest of the
systems. Alignment of the spectra is made using polarization nature and
enhancement as criteria. The positions of the porphyrin core-size-
sensitive bands of all the low-spin systems and of Im_2-PP IX are compared
in Table III along with calculated Ct-N distances.

Protein-heme covalent linkage. Diimidazole H8nb, Im_2-H8nb, differs from
Im_2-PP IX by the presence of the heme-protein covalent linkage intrinsic
to cytochromes of type c, the thioether linkages between the two cysteine
sulfurs and the β-carbon of the vinyl side chains (Fig. 1). With the
exception of the $\nu 2$ (Alg) and $\nu 11$ (Blg) bands, all other porphyrin-core-

Table III

Porphyrin Core-Size-Sensitive RR Bands of Low-Spin Ferric Protoporphyrin

IX and Heme c Systems and the Ct-N Distances

Band Polarization	$\nu 10$ (dp)	$\nu 37$ (ip)	$\nu 2$ (p)	$\nu 11$ (p) (cm^{-1})	$\nu 3$ (p)	$\nu 28$ (dp)	Coordin. Configu.	Ct-N Dist. (A) [a]
Systems								
$(Im)_2$ PP IX	1640	1601	1579	1562	1502	1469	Im/Im	1.994 ± 0.006
$(Im)_2$ H8nb	1640	1599	1587	1569	1502	1469	Im/Im	1.995 ± 0.004
Im H11	1638	1590	1590	1567	1505	1467	Im/His	2.001 ± 0.004
Im H65	1637	1590	1589	1567	1503	1468	Im/His	2.004 ± 0.006
Im Cyt c	1637	1587	1588	1565	1505	1467	Im/His	2.004 ± 0.005
CN' H11	1639	1590	1590	1565	1506	1459	CN'/His	2.003 ± 0.005
CN' H65	1638	1589	1589	1565	1507	1465	CN'/His	2.000 ± 0.002
CN' Cyt c	1638	1589	1590	1567	1507	1460	CN'/His	2.000 ± 0.003
Cyt c (pH 7)	1635	1585	1585	1562	1503	1467	-S/His	2.010 ± 0.003
4M Urea Cyt c	1640	1590	1588	1568	1504	1464	-S/His	2.000 ± 0.002

[a] Based upon the linear correlation $\nu=K(A-D)$, where ν is the observed position of the band and K and A the correlation parameters for a given band (Table 2). The Ct-N distance is the average value using bands $\nu 10$, $\nu 37$, $\nu 3$ and $\nu 28$ (see text).

sensitive bands for the two systems are at about identical positions (Tables I, III). Ignoring the deviation of the $\nu 2$ and $\nu 11$ bands, it appears that the presence of the covalent linkage to the vinyl side chains and the existence of the polypeptide moiety have little or no effect on the core dimensions of the porphyrin macromolecule. This confirms the idea that variations in the peripheral substitutions have no appreciable effect on the geometry of the tetrapyrrole skeleton (28).

The two bands exhibiting significant deviation between the non-c system, Im_2-PP IX, and the simplest c system, Im_2-H8nb, are the $\nu 2$ (Alg) and the $\nu 11$ (Blg) bands, which, according to normal coordinate analysis, are

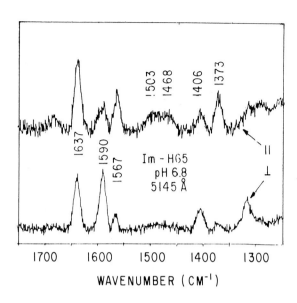

Fig. 5. Resonance Raman spectra of Im-ferric H65; 0.05M phosphate + 0.5 M
   imidazole. Conditions same as in Fig. 3.

predominantly C_β-C_β stretching modes (Table I, III). The sensitivity of
both of these bands to the nature of the group at the C carbon atoms, the
vinyl chains, has been demonstrated (17, 21, 34, 35). Differences in the
position of the $\nu 2$ band between the iron complexes of protoporphyrin IX
and its saturated analogs, etioporphyrin and octaethylporphyrin, have been
reported (35). A similar situation is reported for both the bands of the
two very close systems, Im_2-PP IX and Im_2-mesoporphyrin IX, which differ
only in the saturation of the two vinyl side chains; the high-frequency
shifts are of 13 and 9 cm^{-1} respectively (21). The observed difference of
about +8 and +7 cm^{-1} between Im_2-PP IX and Im_2-H8nb are the same in
direction, but of about 2/3rd the magnitude of change upon saturation of
the vinyl chains. Thus the formation of the thioether linkage at the β-
carbon of the vinyl chains has in part the same effect as the saturation
of the vinyl chain. The variation of the $\nu 2$ and $\nu 11$ positions between Im_2-
-PP IX and Im_2-H8nb, while all other porphyrin-core-sensitive bands retain
the same position, is consistent with the idea that the vibrational modes
involving the C_β-C_β stretch, but coupled with the vinyl modes, are in

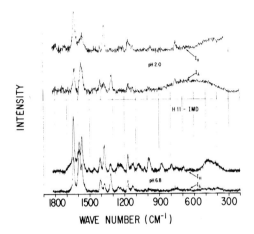

Fig. 6. Resonance Raman spectra of ferric Im-H11. A, Low-spin form, pH
    6.8. B, High-spin form, pH 2.0. Conditions same as in Fig. 3.

general sensitive to structural alterations involving the latter, but such
structural variations have little or no effect on the skeletal geometry of
the tetrapyrrole (17, 21, 28).

A number of explanations have been proffered to account for the selective
variations of the $\nu2$ and $\nu11$ modes (17, 18, 21). Electronic factors
associated with the conjugation of the vinyl chain, sensitivity to the
delocalization of the π electron of the porphyrin skeleton, and the
results of the distinct influence of the environment in heme proteins are
a few possibilities. At present it appears that a clear-cut explanation
to account for the positional variations of both of these bands among a
variety of iron-porphyrin systems and heme proteins is difficult.
However, it can be stated that the effect of the covalent linkage between
heme and the protein in type c hemoproteins is similar to the effect of
saturation of the vinyl chain in simple model systems, the slight
tightening of the pyrrole structures and/or the energetics of the modes of
the vinyl side chain, with little or no effect on the skeletal geometry of
the tetrapyrrole structure.

Recently, from a consideration of the complexes of iron protoporphyrin IX,
Choi et al. (21) have shown that both the $\nu2$ and the $\nu11$ bands and an
additional RR band, $\nu28$, exhibit empirical linear relationships to the
core dimensions of the tetrapyrrole structure.   The empirical
relationship of three other RR bands, $\nu10$ [formerly band V (19)], $\nu37$
[formerly band IV (19)], and $\nu3$ [formerly band II (19)] (Table II), on the
other hand, is based upon a consideration of not only a wide variety of
metalloporphyrins, but also a number of hemoproteins (18-20; for reviews
see 9, 10).   Findings that the frequencies of both $\nu2$ and $\nu11$ bands are
affected by the formation of the thioether linkages intrinsic to type c
heme proteins renders the applicability of core size relationships derived
from considerations exclusively of Fe-PP-IX complexes (21) inapplicable to
the analysis of data of type c sytems.   This however does not appear to be
the case for the $\nu28$ band.   The position of the band is the same for both
Im_2-PP IX and Im_2-H8nb (Table I, III); a similar situation is also
reported for NiPP IX and Ni-octaethyl porphyrin (21). For the analysis of
the RR data in terms of the core dimensions of type c systems, the only
reliable relationships are those of the four bands, $\nu10$, $\nu37$, $\nu3$ and $\nu28$.
Calculation of the Ct-N distances from the positions of these four bands
bands yields an average Ct-N distance of 1.995 $\pm$ 0.006 A for both Im_2-PP-
IX and Im_2-H8nb (Table III).   This is typical of low-spin, hexacoordinated
hemes with iron in the plane of the tetrapyrrole structure (28-30).

Coordination configuration and protein-heme interactions.   Another
structural feature typical of type c heme systems is a coordination
geometry of iron with the imidazole of the histidine residue adjacent to
one of the heme-attached cysteine residues occupying an axial position of
iron.   Comparison of the positions of the heme-core-sensitive bands of Im_2-
·H8nb with the Im/Im iron configuration to the positions for Im-H11, Im-
H65 and Im-cytochrome c, all with the Im/His configuration, reveals a
distinct pattern.   The $\nu10$, $\nu37$, $\nu28$ and $\nu11$ bands are shifted down-field,
and bands $\nu2$ and $\nu3$ appear to be shifted slightly up-field (Table III).
Estimates based upon the positions of $\nu10$, $\nu37$, and $\nu28$ yield an average
Ct-N distance of 2.003 $\pm$ 0.006 Å, which, though within the limits of the
Ct-N distance of Im_2-H8nb, 1.995 $\pm$ 0.006 A, is common to all three systems
(Table III).   Data in Table III further show that there are no notable

variations in the positions of the core-sensitive bands among Im-H11, Im-H65 and the Im-cytochrome c, which differ not only in the length of the polypeptide chain, 11 residues through 104 residues (Fig. 1), but also in the extent and the nature of the protein-organized structures (32). Im-H11 is shown to have minimal or no organization of the poly-peptide moiety, whereas Im-cytochrome c is more or less indistinguishable from the native protein. A similar situation is seen for these models in the presence of cyanide, conditions producing the CN^-/His iron configuration rather than Im/His. These findings establish that the coordination of the imidazole of the histidine residue in type c heme systems does alter, though only slightly, the core dimensions of the tetrapyrrole structure. In addition, there appear to be no noticeable porphyrin-core-size-linked implications resulting from either the increasing length of the polypeptide chain or the nature and the extent of protein organized structure. It should be noted that the above inference applies only to those systems whose iron configuration is extrinsic ligand/imidazole of iron and which contain heme groups exposed to solvent, i.e., those systems with minimal protein-heme interactions (32).

The neutral pH form of cytochrome c, form III, has a coordination configuration with methionine sulfur/histidine at the two axial positions of the iron and with the heme group more than 80% buried in a protein crevice (for reviews see 1, 4; 7). Almost all the structurally sensitive bands of ferric cytochrome c form III are further down-shifted by as much as 2 - 4 wave numbers from all other systems (Tables I and III). The core-sensitive bands of ferric cytochrome c fall on one end of the scale and those of Im_2-H8nb on the other; all the other systems fall in between. The heme in the native protein is definitaly more perturbed than in any other system. Estimation of the Ct-N distance yields an average length of 2.010 ± 0.003 Å, which is about 0.015 Å larger than the core dimensions of Im_2-PP_0 IX or of the simplest of the heme c systems, Im_2-H8nb, and about 0.005 Å larger than of systems containing Im/His or CN^-/His iron configurations (Table III). The porphyrin core in the native protein is thus larger than in any of the simpler systems, to the extent of about 0.015 Å.

The two features of ferric cytochrome c which may have heme configurational implications are the iron coordination configuration, methionine
sulfur/His, and the enclosure of more than 80% of the tetrapyrrole
structure in a protein crevice (1, 7). The effect of the addition of 4 M
urea to ferricytochrome c, in terms of the structurally sensitive RR
bands, is the high-frequency shift of almost all the bands; the final
positions are very similar to those of the simple heme c models (Fig. 4
vs. Fig. 6a; Table III), which lack both the coordination configuration
and the protein-localization aspects of the protein (24-26; for review see
4). The effect on the porphyrin core is to shrink it from about 2.01 to
2.00 $\overset{o}{A}$ (Table III). The addition of 4 M urea thus reverts the core size
of the tetrapyrrole structure to that of the simple systems. Since the
effect of the addition of 4 M urea has been shown to be the loosening of
the heme crevice, i.e., the disruption of protein-heme interactions,
without alteration of the coordination configuration of iron (33), the
normalization of the expanded porphyrin core in the native protein upon
addition of urea must be the result of the loosening of the heme crevice,
the elimination or the weakening of the protein-heme interactions. The
crevice location of heme in cytochrome c thus does affect its geometry,
enough to cause an expansion of 0.01 - 0.015 $\overset{o}{A}$ of the tetrapyrrole ring,
depending upon the point of reference (Table III).

High-spin heme c systems

The ferric heme c model systems (Fig. 1), both those with intact imidazole
of His-18 and with/without an additional imidazole, H8, Im-H8, H11, Im-
H11, H-65 and Im-H65, and those with a cleaved imidazole ring, H8nb and
Im -H8nb, at acidic pH's, pH 2 or below, generate forms which are spectroscopically typical of high-spin ferric hemes and hemoproteins: Soret peak
below 398 nm, visible peak at about 490 nm with inflections at about 540
and 570 nm, and the characteristic 620-nm peak (Fig. 2) (23-25; see review
4). Similarly, when horse heart ferricytochrome c is brought to a pH of
about 1, it also reverts to a typical high-spin form, form I (36, for
reviews see 1, 4).   Resonance Raman spectra of Im-H11 at pH 2.0 and of
cytochrome c at pH 1.0 (Figs. 6a and 7) represent in general the spectra
of the high-spin forms of the heme c systems containing intact imidazole

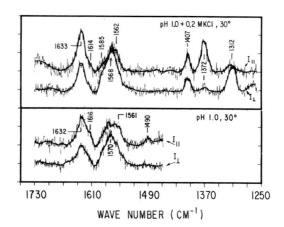

Fig. 7. Resonance Raman spectra of the pH 1 form of horse heart ferricytochrome c. Lower, in absence of KCl. Upper, in 0.2 M KCl.
Conditions same as in Fig. 3. Solid line, digitally smoothened.

of His-18. The high-frequency RR spectra of high-spin forms generated
from H8nb and Im-H8nb, models without the imidazole of the histidine
residue, are very similar to the spectrum shown in Fig. 6b, except for the
positional variations of a number of bands. In Table IV are listed the
positions of the structurally sensitive bands of all the high-spin forms
of type c systems and of a number of fer0ic porphyrin complexes and
hemoproteins with well-defined spin states, coordination configurations
and/or porphyrin dimensions (17, 20; for reviews see 9, 10, 29, 30).

The core-sensitive bands of high-spin heme c models clearly fall into two
distinct sets; the first is made up of the positions of H8nb and Im2-H8nb
(the models without the imidazole of His-18 available) with ν_{10}, ν_{37}, ν_2
and ν_3 at 1610-11, 1558-59, 1581-82 and 1479-76 cm^{-1} respectively, and the
second, with bands of H8, Im-H8, Im-H11 and Im-H65 (models with intact
imidazole of His-18, Fig. 1), at 1630-24, 1571-74, 1572-68 and 1500-1497
cm^{-1} respectively (Table IV). The RR spectrum of ferric cytochrome c,
however, contains features of both sets, but it is dominated by bands
corresponding to the second set (Fig. 7, Table IV). The concurrent
presence of bands conforming to both sets is an indication of the
simultaneous existence of the molecular species generating the two

Table IV

Porphyrin Core Size Sensitive RR Bands of High-Spin Ferric Heme Proteins, Iron Porphyrins and Heme c Systems and the Configurational Distances

Systems	[a]Bands and Positions				[c]Coordination		[b]Distances			Ref.
	$\nu10$ dp	$\nu37$ ip	$\nu2$ p (cm^{-1})	$\nu3$ p	Num.	Conf.	*Ct-N (Å)	*M-N	*M-Ct	
$(Me_2SO)_2$PPIX	1610	1559	1580	1480	6	-SO/OS-	*2.060	*2.065		17,20
$(Me_2SO)_2$OEP	1613	1563	1577	1481	6	-SO/OS-				20
$(H_2O)_2$TPP					6	H_2O/H_2O	*2.045	*2.045	*0.0	37
MetMb	1611	1562	1582	1482	6	H_2O/His				20
MetMbF	1609	1557	1583	1482	6	F^-/His				20,35
AquMetHb	1610	1561	1583	1481	6	H_2O/His				20
HRP(F^-)	1608	1555		1482	6	F^-/?				20,35
(Cl^-)TPP					5	/Cl^-	*2.023	*2.060	*0.39	38
(Cl^-)PPIX	1632	1572		1495	5	/Cl^-	*2.027	*2.062	*0.48	39
(Cl^-)PPIXDMeE	1633	1572	1573	1495	5	/Cl^-	*2.03		*0.30	18
$(OClO_3)$TPP					#5	/$OClO_3$	*1.981	*2.001	*0.28	38
RP Cyt c'(pH10)	1633	1578	1572	1500	5	/HHis	2.02		0.40	40
H8nb(8M Urea, (pH 2)	1611	1558		1499	6	H_2O/H_2O	2.06		0.0	TW
Im_2-H8nb (pH 2)	1610	1559	1581	1496	6	H_2O/H_2O	2.06		0.0	TW
Im-H8 (pH 1)	1630	1573	1572	1500	5	-/HHis	2.02		0.40	TW
Im-H11(pH 2)	1624	1571	1568	1497	5	-/HHis	2.03		0.39	TW
Im-H65(pH 1.5)	1628	1574	1564	1499	5	-/HHis	2.03		0.41	TW
Cyt c (pH 1) (30 C) (Maj)	1632	1570	1577	1490	5	-/HHis	2.03		0.41	TW
(Min)	1614	1552			6	H_2O/H_2O	2.07		0.00	TW
(0.2 M KCl)	1633	1568	1577	1493	5	-/HHis	2.03		0.41	TW

a: Bands and labeling according to Spiro and coworkers (17, 20).
b: Distances with * in front from X-ray data; other distances, the average value of distances calculated using the positions of $\nu10$ and $\nu37$ modes with the empirical relationships detailed in Table II. M-Ct distances calculated using triangulation with known M-N distances. TW, this work.
c: 6, hexacoordinated; 5, pentacoordinated; #, intermediate spin.

distinct RR sets. This is supported by the observation that the RR
spectrum of pH 1 cytochrome c upon addition of KCl becomes better defined,
with bands corresponding to the second set (Fig. 7, top). The precise
effect of the addition of KCl to acidic forms of cytochrome c is the
subject of future studies, but for the present it appears that the
apparent effect is merely the shifting of the distribution of the high-
spin forms in favor of the form generating the second set. Two distinct
high-spin molecular forms of the heme c systems are clearly evident. One
with RR bands of the first set is generated for model systems without
histidine imidazole, and the second is constituted of models with intact
histidine imidazole.

The positions of core-sensitive bands of set I are typical of the high-
spin hexacoordinated (HC) hemes and hemoproteins, while the bands of set
II are in excellent agreement with the bands observed for pentacoordinated
(PC) hemes and heme proteins (Table IV and references cited). The high-
spin forms of H8nb and of Im_2-H8nb thus appear to contain hexacoordinated
heme iron configurations, while all the other heme c systems, including
cytochrome c at pH 1 in the presence of KCl, have pentacoordinated heme
iron. The HC configuration of iron of acidic H8nb and Im_2-H8nb is
consistent with conclusions based on pH-linked spectroscopic and
potentiometric studies (25), which also suggest that the groups occupying
the two axial positions are two water molecules. The HS:PC iron
configuration of H8, Im-H8, Im-H11 and Im-H65 is, however, at variance to
the generally held view that these systems also contain hexacoordinated
heme iron (24, 25, 36; for review see 4). The fact that the RR spectra of
these systems are distinct and typical of pentacoordinated hemes and heme
proteins leads us to contend that the configuration of iron in all these
systems is indeed pentacoordinated. With regard to the configuration of
the porphyrin skeleton, the core-sensitive bands, ν10, ν37 and ν2, have
average Ct-N distances of 2.065 $\overset{o}{A}$ for H8nb and Im_2-H8nb. The Ct-N
distances of systems constituting the second set, Im-H8, Im-H11, Im-H65
and cytochrome c form I, however, fall into the range of 2.03 $\overset{o}{A}$. Analysis
of the RR data of $(Me_2SO)_2Fe(III)$-PP IX, a hexacoordinated high-spin heme
with two weak ligands at the two axial positions of iron (28-30), yields a
Ct-N distance of 2.06 $\overset{o}{A}$, which is identical to the core dimensions of

high-spin-hexa-coordinated heme c models, H8nb and acidic Im_2-H8nb (Table IV). The high-spin forms of c systems without the imidazole of His-18 are thus similar or identical in core size to HS:HC heme, while the high-spin forms of models and the protein with the imidazole of His-18 available have porphyrin cores about 0.03 $\overset{o}{A}$ smaller.

With regard to the iron-porphyrin geometry, whether one considers the traditional Fe-N bond length of 2.065 $\overset{o}{A}$ for high-spin ferric porphyrins (41) or a distance of 2.045 $\overset{o}{A}$, as in the case of ditetramethylene sulfoxide tetraphenylporphinato iron (III) perchlorate [(TMSO)$_2$Fe(III)TPP ClO_4] (42) and diaquo tetraphenylporphinato iron (III) perchlorate [(H$_2$0)$_2^-$Fe(III)TPP ClO_4] (for reviews see 29, 30), the estimated Ct-N distance of 2.065 $\overset{o}{A}$ of H8nb, Im_2-H8nb and the heme complex, $(Me_2SO)_2$Fe(III) PP IX, is enough to accommodate the iron in the center of the porphyrin skeleton. An in-center iron configuration is assigned to both (TMSO)$_2$Fe(III)TPP and (H$_2$0)$_2$Fe(III)TPP (42, 37), the HC:HS systems with identical weak ligand-field groups at the two axial positions of iron. It has also been acknowledged that the structural variations between Fe-TPP and Fe-PP IX can be expected to have little or no effect on the porphyrin core dimensions (17, 21), allowing the stipulation of a similar iron porphyrin configuration for the Me_2SO complex of Fe(III)-PP IX.

The available three-dimensional structures of pentacoordinated iron-porphyrins present two distinct situations: the high-spin complexes with an average Fe-N distance of 2.069 $\pm$ 0.008 $\overset{o}{A}$ and with iron out of the porphyrin plane by about 0.47 $\overset{o}{A}$, and the complexes exhibiting intermediate spin, S = 3/2, with an Fe-N distance of about 2.001 $\overset{o}{A}$ and iron displacement of about 0.28 $\overset{o}{A}$ (for review see 30). Spectroscopically, all the heme c models with RR structurally sensitive bands typical of pentacoordinated hemes and hemoproteins, e.g., Im-H8, Im-H11, Im-H65 and cytochrome c at pH 1, are typical of high-spin hemes and hemoproteins (Fig.2; 23-25), in particular with reference to the presence of a well-defined absorption maximum at about 620 nm (Fig. 2) which has been linked to the high-spin state of heme (43, for review see 26). Preliminary magnetic suscepti-bility measurements of Im-H11 at acidic pH's, μ = 5.46, are consistent with a typical high-spin ferric complex, S = 5/2 (unpublished data).

Similar studies of the pH 1 form of ferricytochrome c also showed it to be
high-spin, but the possibility of a thermal and/or quantum mechanical
equilibrium between a high-spin and a low-spin form of iron exists (for
reviews see 1, 4; 46). If it is acknowledged that the state of iron in
the pentacoordinated heme c models is typically high-spin, the estimated
Ct-N distance of 2.03 $\overset{o}{A}$ and an Fe-N distance of 2.069 $\overset{o}{A}$ generate a
geometry with iron out of the plane by as much as 0.4 $\overset{o}{A}$ (Table IV). If,
on the other hand, these systems turn out to have intermediate spins, S =
3/2, then a pentacoordinated configuration with iron in the plane of the
porphyrin structure would describe the state of heme, a situation yet to
be encountered. Studies presently in progress, magnetic susceptibility
measurements, should resolve the nature of the spin state of iron in these
complexes, and subsequently, the configuration of iron porphyrin.

The property of iron to attain different coordination configurations in
the case of the high-spin heme c systems appears to be linked to the
presence or the absence of the imidazole moiety of the His-18 residue.
Imidazole, though protonated under the solution conditions generating the
high-spin states, pH 2 or below, could provide the ligating group with
reduced ligand-field strength which generates a high-spin pentacoord-
ination configuration of iron. The pentacoordinated heme configuration
with the imidazole of the histidine residue as the axial ligate has been
shown to occur in another member of the type c heme protein family, R.
molischianum cytochrome c' (44, 45). It recently has been suggested that
the decreased ligand-field strength of the imidazole nitrogen upon
protonation at Nδ1 and/or the hydrogen bonding at Nδ1 results in
significant lowering of the ligand-field strength of this nitrogen, and
consequently, the generation of an HS or of an intermediate spin
pentacoordinated heme (46, 47). We, however, propose that although the
formation of the pentacoordinated iron configuration at acid pH's in heme
c models as well as in RM cytochrome c' may be due in part to the
availability of imidazole nitrogen with substantially lowered ligand-field
strength, it cannot be accounted for entirely on those grounds, but
rather, it is a composite of the above and the unique structural location
of the liganding group, the imidazole of His-18, a structure adjacent to
the C-terminal heme-protein linkage (Fig. 1). This proposal stems from

the following observation: Im_2-H8nb at acidic pH's with a substantial excess of imidazole, 0.5 M, and under conditions where imidazole is protonated, pKa 6.9, has a hexacoordinated high-spin iron configuration (see preceding sections), which clearly demonstrates that the availability of protonated imidazole is not sufficient to generate the pentacoordinated configuration. Since pentacoordinated iron appears to be a feature of the type c systems, the unique structural location of the histidine residue must provide the necessary energy to stabilize the coordination linkage between iron and imidazole nitrogen. It is noteworthy that this structure is indeed an invariant feature of all the cytochromes c thus far known (for reviews see 1, 4). It is not entirely unexpected to find a definite role associated with this unique structure of these systems.

References

1. Dickerson, R.E., Timkovich, R.: in The Enzymes, Academic Press, New York, Vol. III, pp. 398-547 (1975).

2. Salemme, F.R.: Ann. Rev. Biochem. 46, 299-329 (1977).

3. Timkovich, R.: in The Porphins (Dolphin, D., ed.), Academic Press, New York, Vol. VII, pp. 241-294 (1974).

4. Harbury, H.A., Marks, R.H.L.: in Inorganic Biochemistry (Eichorn, G.L. ed.), Elsevier, Amsterdam, pp. 902-956 (1973).

5. Brautigan, D.L., Ferguson-Miller, S., Margoliash, E.: in Methods in Enzymology (Fleischer, S., Packer, L., eds.), Academic Press, New York, Vol. 53, pp. 128-164 (1978).

6. Erecinska, M., Vanderkooi, J.M.: in Methods in Enzymology (Fleischer S., Packer, L., eds.) Academic Press, New York, Vol. 53, 165-191 (1978).

7. Takano, T., Dickerson, R.E.: J. Mol. Biology. 153, 79-94(1981).

8. Strekas, T.C., Spiro, T.G.: Biochim. Biophys. Acta 278, 188-192 (1972).

9. Felton, R.H., Yu, N.-T.: in The Porphyrins (Dolphin, D., ed.), Academic Press, New York, Vol. III, pp. 347-393 (1978).

10. Spiro, T.G.: in Methods in Enzymology Vol. 54, pp. 233-249 (1978).

11. Kitagawa, T., Ozaki, Y., Kyogoku, Y.: Advances in Biophys. 11, 153 (1978).

12. Asher, S.A.: in Methods in Engymology Vol. 76, 371-413 (1981).

13. Spiro, T.G.: in Iron Porphyrins (Lever, A.B.P., Gray, H.B., eds.),
    Addison-Wesley Publishing Co., London, Part II, pp. 89-159 (1983).

14. Rousseau, D.L., Ondrias, M.R.: Ann. Rev. Biophys. Bioeng. 12,
    357-380 (1983).

15. Abe, M., Kitagawa, T., Kyogoku, Y.: J. Chem. Phys. 69, 4526-
    4534 (1978).

16. Sunder, S., Bernstein, H.J.: J. Raman Spectroscopy 5, 351-371
    (1976).

17 Choi, S., Spiro, T.G., Langry, K.C., Smith, K.M.: J. Am. Chem. Soc.
    104, 4337-4344 (1982).

18. Spaulding, L.D., Chang, C.C., Yu, N.-T., Felton, R.H.: J. Am. Chem.
    Soc. 97, 2517-2525 (1975).

19. Spiro, T.G., Burke, J.M.: J. Am. Chem. Soc. 104, 5482-5489 (1976).

20. Spiro, T.G., Stong, J.D., Stein, P.J.: J. Am. Chem. Soc. 101, 2648-
    2655 (1979).

21. Choi, S., Spiro, T.G., Langry, K.C., Smith, K.M., Budd, D.L., LaMar,
    G.N.: J. Am. Chem. Soc. 104, 4345-4351 (1982).

22. Myer, Y.P., Srivastava, R.B., Kumar, S., Raghavendra, K.: J. Protein
    Chem. 2, 13-42 (1983).

23. Corradin, G., Harbury, H.A.: Biochim. Biophys. Acta 221, 489-496
    (1979).

24. Harbury, H.A., Loach, P.A.: J. Biol. Chem. 235, 3640-3645 and
    3636-3653 (1960).

25. Myer, Y.P., Harbury, H.A.: Annals New York Acad. Sci. 206,
    685-700 (1973).

26. Smith, D.W., Williams, R.J.P.: in Structure and Bonding (Hemmerich,
    P., Jorgensen, C.K., Nyholm, R.S., Williams, R.J.P., eds.), Springer-
    Verlag, New York, Vol. 7, pp. 1-45 (1970).

27. Makinen, M.W., Churg, A.K.: in Iron Porphyrins (Lever, A.B.P., Gray,
    H.B., eds.), Addison-Wesley Publishing Co., London, Part I, pp. 141-
    235 (1983).

28. Hoard, J.L.: in Porphyrins and Metalloporphyrins (Smith, K.M., ed.)
    American Elsevier, New York, pp. 317-376 (1975).

29. Scheidt, W.R.: Accounts of Chem. Res. 10, 339-345 (1977).

30. Scheidt, W.R., Gouterman, M.: in Iron Porphyrins (Lever, A.B.P.,
    Gray, H.B., eds.), Addison-Wesley Publishing Co., London, Part I,
    89-139 (1983).

31. Shelnutt, J.A., Rousseau, D.L., Dethmers, J.K., Margoliash, E.:
    Biochemistry 20, 6485-6497 (1981).

32. Myer, Y.P., Pande, A.: in The Porphyrins (Dolphin, D., ed.), Academic
    Press, New York, Vol. III, pp. 271-322 (1978).

33. Myer, Y.P, MacDonald, L.H., Verma, B.C., Pande, A.: Biochemistry 19,
    199-207 (1980).

34. Stong, J.D., Burke, J.M., Daly, P., Wright, P., Spiro, T.G.: J. Am. Chem. Soc. 102, 5815-5819 (1980).

35. Callahan, P.M., Babcock, G.T.: Biochemistry 20, 952-958 (1981).

36. Boeri, E., Ehrenberg, A., Paul, K.G., Theorell, H.: Biochim. Biophys. Acta 12, 273-282 (1953).

37. Scheidt, W.R., Cohen, I.A., Kastner, M.E.: Biochemistry 18, 3546-3552 (1979).

38. Hoard, J.L., Cohen, G.H., Glick, M.D.: J. Am. Chem. Soc. 89, 1992-1996 (1967).

39. Koenig, D.F.: Acta Crystallogr. 18, 663-673 (1965).

40. Strekas, T.C., Spiro, T.G.: Biochim. Biophys. Acta 351, 237-245 (1974).

41. Hoard, J.L., Harmor, M.J., Harmor, T.A., Caughey, W.: J. Am. Chem. Soc. 87, 2312-2319 (1965).

42. Mashiko, J.L., Kastner, M.E., Spartalian, K., Scheidt, W.R., Reed, C.J., J. Am. Chem. Soc. 100, 6354-6362 (1978).

43. Day, P., Smith, D.W., Williams, R.J.P.: Biochemistry 6, 1563-1566 (1967).

44. Weber, P.C., Bartsch, R.G., Cusanovich, M.A., Hamlin, R.C., Howard, A., Jordan, S.R., Kamen, M.D., Meyer, T.E., Weatherford, D.W., Xuong, Ng.H., Salemme, F.R.: Nature (London) 286, 302-304 (1980).

45. Weber, P.C., Howard, A., Xuong, Ng.H., Salemme, F.R.: J. Mol. Biol. 153, 399-424 (1981).

46. Weber, P.C.: Biochemistry 21, 5116-5119 (1982).

47. Landrum, J.T., Hatano, K., Scheidt, W.R., Rood, C.A.: J. Am. Chem. Soc. 102, 6729-6735 (1980).

Received July 4, 1984

Discussion

Buchler: On Table IV of your paper and at the end of your talk you derive from the RR spectra that Im-H8 (pH 1), Im-H11 (pH 2), and Im-H65 (pH 1.5) are penta-coordinated species with an imidazole coordianted to Fe^{III}. How can you rule out the occurrence of a penta-coordinated, Fe^{III} aqua or Fe^{III} hydroxo moiety, and the imidazole moiety being protonated and hence not coordinated?

Myer: Only circumstantial considerations have led us to suggest a configuration with protonated imidazole of histidine residue 18 as the ligate in penta-coordinated configurations of these heme c systems. The only difference between Im_2H8nb and Im-H8 is the availability of imidazole nitrogen of histidine, yet under almost identical solution conditions, pH 1.0-1.5, the former develops a hexa-coordinated bis aquo heme, while the latter yields a penta-coordinated configuration. The possibility of OH-coordination at pH values as low as 1 is most unlikely. A penta-coordinated heme with water is yet to be observed, while a penta-coordinated heme with imidazole of the histidine residue is an observed fact in the case of a bacterial cytochrome c.

Woody: Based on your results or on those of other workers, do you have any information concerning the effects of distortion of the porphyrin from planarity upon the vibrational frequencies you are studying?

Myer: Positions of a number of high-frequency RR bands are dependent upon the state of the porphyrin macrostructure. Spiro and coworkers did try to develop correlations between the positions of some of these bands and the deformation of the tetrapyrrole structure. It now appears that although a number of bands are sensitive to the geometry of the porphyrin ring, the core size of the ring relates rather satisfactorily to all the bands above 1450 cm^{-1}. This view is now generally accepted.

Rapoport: Could the question of hydrogen bonding of the imidazole NH be addressed by preparing the $N-CH_3$ derivative of cytochrome c?

Myer: Yes, methylation at $N\delta_1$ position of imidazole of the histidine residue and investigation of its effects upon the nature of the heme complex generated upon acidification can answer the question of NH-hydrogen bonding. However, the imidazole with methyl at the $N\delta_1$ position is known to coordinate to ferric ion, but the result is a high-spin, hexa-coordinated heme. I feel it is not the hydrogen bonding of imidazole NH alone that results in penta-coordinated heme configurations, but rather, the unique location of the histidine imidazole in heme c systems.

CORRELATION BETWEEN NEAR-INFRARED, ESR SPECTRA AND OXIDATION-
REDUCTION POTENTIALS IN LOW-SPIN HEME PROTEINS

Abel Schejter, Ida Vig
Sackler Institute of Molecular Medicine, Sackler Medical
School, Tel-Aviv University, Tel-Aviv, Israel

William A. Eaton
Laboratory of Chemical Physics, National Institute of
Arthritis, Diabetes, Digestive and Kidney Diseases,
National Institute of Health, Bethesda, Maryland 20205, USA

Introduction

The molecular orbital theory of low-spin ferric heme compounds
predicts (1,2) a series of optical absorption bands in the
near-infrared, corresponding to charge transfer transitions
from the highest-filled porphyrin orbitals to the hole in the
d_{yz} orbital of the iron (Fig. 1).

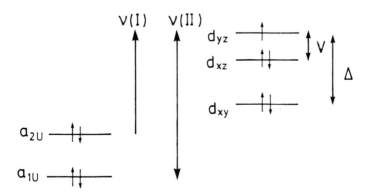

Fig. 1: The porphyrin and iron orbitals involved in the charge
transfer near-infrared transitions in low-spin ferric heme
compounds.

Optical Properties and Structure of Tetrapyrroles
© 1985 by Walter de Gruyter & Co., Berlin · New York

The presence of these bands in heme proteins was first reported
for several complexes of hemoglobin and myoglobin, and their
assignment was corroborated by their lack of natural circular
dichroism and the sign and magnitude of their magnetic circular dichroism (3,4). The frequency of the transitions corresponding to the lowest energy band can be decomposed in several
terms:

$$V(I) = \left[\varepsilon(d_{xy}) - \varepsilon(a_{2U})\right] + C_{JK} + \Delta + V/2 \qquad (1)$$

The first term of equation (1) is the energy difference between
two orbitals (Fig. 1): the d_{xy} metal orbital, and the a_{2u} porphyrin orbital. Neither d_{xy}, which is non-bonding, nor a_{2u},
which is too far away from the axial ligands, should be affected by changes in axial coordination. These qualitative assumptions are supported by the results of iterative extended Huckel
calculations (1,2). Thus, in first approximation, in a series
of complexes in which only one of the axial ligands is changed,
the energy difference $\varepsilon(d_{xy}) - \varepsilon(a_{2u})$ should remain constant.

The second term arises from interelectronic repulsions; again,
this may be considered constant in first approximation for a
series of heme protein complexes in which only one axial ligand is varied.

The crystal field components, Δ and V, can be deduced from the
g values observed in electron spin resonance spectra (5,6,7).
They are expressed in terms of λ, the spin-orbit coupling
constant of ferric ion. The tetragonal distorsion, Δ, depends
mainly on the nature of the axial ligands, while the rhombic
distorsion, V, results from asymmetry in the interactions between the axial ligands and the d_{yx} and d_{xz} orbitals. The remaining term in Equation (1) can be conveniently expressed in
the following way: $\lambda[(\Delta+V/2)/\lambda]$, leading to:

$$V(I) = \left[\varepsilon(d_{xy}) - \varepsilon(a_{2U})\right] + C_{JK} + \lambda\left[(\Delta + V/2/\lambda\right] \qquad (2)$$

and to similar expressions for the bands due to transitions
originating in the other porphyrin orbitals, e.g.:

$$V(II) = \left[\epsilon(d_{xy}) - \epsilon(a_{1U})\right] + C_{J,K} + \lambda\left[(\Delta + V/2)/\lambda\right] \quad (3)$$

The assumed constancy of the first and second terms leaves only
the last term of Equations (2) and (3) as a linear variable,
since $(\Delta + V/2)/\lambda$ can be obtained experimentally. In this way,
two entirely independent spectroscopic results can be correlated.

Equations (2) and (3) predict that $\nu(I)$ and $\nu(II)$ will be linear functions of $(\Delta + V/2)/\lambda$, with slopes λ, and intercepts
that can be estimated from theory. We present here (see also
reference [8]) a study of the near-infrared spectra of a series
of low-spin ferric heme protein complexes that corroborates in
first approximation the hypothesis represented in Equations (2)
and (3) and the assumptions underlying them. Some thermodynamic
implications of this interpretation of spectral data are also
discussed.

Experimental

Horse heart cytochrome c, type VI, was obtained from Sigma Chemical Co., St. Louis, and purified by column chromatography on
Amberlite CG-50. Dicarboxymethyl cytochrome c was prepared by
carbomethylation of the side chains of methionines 65 and 80
(9). Sperm whale skeletal muscle myoglobin (Sigma Chemical Co.,
St. Louis) was purified chromatographically on DEAE-cellulose.
The proteins were dissolved in 99.9% D_2O and lyophilized; this
procedure was repeated three times. Salts used as buffers and
imidazole were also lyophilized three times in D_2O.

Near-infrared spectra were measured between 800 and 1800 nm in
a Cary 17 spectrophotometer interfaced to a Hewlett-Packard
9825A calculator. The spectral contributions due to HOD and to
vibrational absorption of the protein were subtracted from the
spectrum of cytochrome c by reducing the sample with dithionite,
since in the ferrous state the electronic absorption in this

spectral range is very weak. In the case of myoglobin, the CO
complex of the reduced state was used as a blank for subtraction.

Extinction coefficients were determined by reference to the
α band of ferrous cytochrome c, $\varepsilon(550nm)=29.9mM^{-1}c^{-1}$ (10). The
azide and imidazole complexes of cytochrome c were formed using
the dicarboxymethylated derivative, which has a much larger
affinity for these ligands than the native protein (9). The
concentration of the proteins was between 0.7 and 2.0 mM. At
pD 7.0 the buffer was 0.1 M potassium phosphate. For the spectrum at extreme alkaline conditions, cytochrome c was dissolved
directly in 1.0 N NaOD.

Results

The spectra of cytochrome c and its cyanide and azide complexes, shown in Fig. 2, are representative of the low-spin ferric
heme protein spectra that were observed in all the compounds

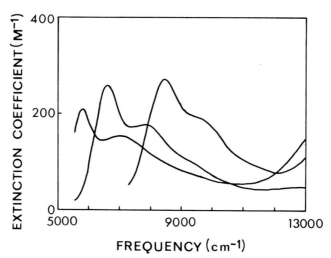

Fig. 2: Near-infrared optical absorption spectra. From left to
right: native ferric cytochrome c (pD 7.0); cytochrome c -
cyanide; alkaline cytochrome c ($\bar{1}$N NaOD).

investigated (Table I). Two bands in each spectrum are clearly observed, corresponding to $a_{2u}(\pi) \rightarrow d_{yz}$ and $a_{1u}(\pi) \rightarrow d_{yz}$ promotions, while the presence of the third band of the predicted series is suggested by the spectral shape. The bands are weak with extinctions between 200 and 300 $M^{-1}c^{-1}$. Native cytochrome c has the weakest extinction, and the band with the lowest frequency. Values of Δ and V were calculated by applying the relations derived by Kotani (6) to g values obtained from the literature (See Table I).

Table I: Near-Infrared Optical Spectra and Ligand Field Parameters derived from ESR Spectra for Ferric Low-Spin Heme Proteins and Heme Protein Complexes.

Compound	ESR Parameters		Ref.	Optical Maxima (c^{-1})	
	Δ	V			
Cytochrome c, pD7	2.56	1.48	(11)	5780	6825
Cytochrome c, 1N NaOD	6.50	3.47	(12)	8430	9520
Cytochrome c-CN	3.34	0.93	(12)	6410	7750
Cytochrome c-N$_3$	4.16	2.77	(13)	7870	9000
Cytochrome c-Imidazole	4.35	1.95	(13)	6580	7840
Myoglobin-CN	3.34	0.93	(14)	6410	7750
Myoglobin-N$_3$	4.63	2.38	(14)	7820	
Myoglobin-Imidazole	4.31	1.93	(14)	6670	7690
Hemoglobin-N$_3^a$	4.93	2.35	(15)	7820	8800

a   Optical maxima from reference (4).

In Fig. 3, the data of Table I are plotted accordingly to Equation (2). As predicted by the hypothesis represented by Equation (1), the plots were approximately linear, and almost parallel. The slopes and intercepts of these lines, estimated from a least-squares analysis of the data, are listed in Table II.

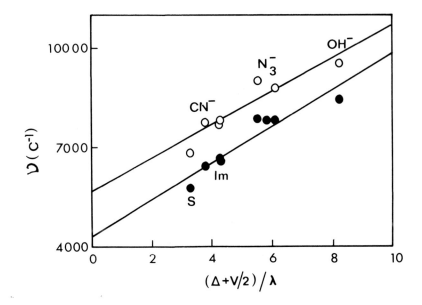

Fig. 3: The near-infrared transition frequencies of low-spin ferric heme proteins plotted as a function of $(\Delta+V/2)/\lambda$. Filled circles: first band; open circles: second band.

Table II: Slopes and intercepts of the lines obtained by plotting the frequencies of the near-infrared bands, $\nu(I)$ and $\nu(II)$, as a function of $(\Delta+V/2)/\lambda$.

Band	Slope (c^{-1})	Intercept (c^{-1})
I	550 ± 70	4320 ± 370
II	500 ± 80	5680 ± 410

Discussion

Since the beginning of their research, heme proteins have often been used as models for the experimental and theoretical development of spectroscopic techniques. While each spectroscopy provides a unique contribution to the understanding of molecular properties, none is able to answer all the possible questions raised by the investigator. The particular case of

low-spin ferric heme proteins discussed here (see also Ref.(8))
exemplifies the explanatory enhancement that results from com-
bining the results of optical and esr spectroscopies.

Cytochrome c is a "closed crevice" heme protein, in which the
iron is essentially coplanar with the porphyrin nitrogens and
the axial ligands are a histidine imidazole and a methionine
sulfur. In all the states described in this article, the methionine ligand is displaced either by an exogenous ligand, or
by a different protein residue. In the low-spin ferric myoglobin and hemoglobin complexes, the proximal imidazole of histidine F8 and the exogenous ligand are the axial ligands. Thus,
we may consider that the compounds investigated belong to a
series structurally represented by Fe-(porphyrin)-(imidazole)-
(L), where L is the changing axial ligand. Equations (2) and
(3) clearly apply to this ensemble of heme derivatives. The
linearity of the data plotted in Fig. 3 in term of Equations
(2) and (3) agrees with the theoretical expectations. The values of the slope (Table II), are well within the order of
magnitude expected from the value of λ for free ferric iron,
460 c^{-1} (16). The difference between the intercepts corresponds
to the orbital energy difference.

This was estimated as 2400 c^{-1} for low-spin ferric porphin
cyanide (1), which is also in reasonable agreement with the
observed difference of 1400 c^{-1}. The results are thus in keeping with the theoretical predictions of Equations (2) and (3),
and the assumptions underlying the derivation of these equations appear as valid in first approximation. Deviations should
not be unexpected, for a number of reasons. In the first place,
there is a source of error arising from the fact that esr experiments were performed at liquid nitrogen or lower temperatures,
while the optical spectra were measured at room temperature.
The effects of low temperatures on spin state (17,18) and the
effects on esr measurements due to sample freezing (19) suggest
that the ligand field of the sample may be different in the two

types of measurement. Furthermore, the width of the optical
bands indicates the presence of vibrational components, and a
better definition of the optical maxima should be achieved by
measuring the near-infrared spectra at low temperatures: experiments are now being undertaken for this purpose.

The energies of the charge transfer bands appear to be directly
related to the nature of the changing ligand, and in this respect methionine occupies an extreme position, resulting in the
lowest energy observed for these transitions. This seems to be
a property of the ligand unrelated to the constraints imposed
by the protein: a preliminary measurement on the acylated
horse cytochrome c heme undecapeptide (20) coordinated by N-
acetyl-methionine ethyl ester shows the $\nu(I)$ band at 1700 nm
$(5880 \ c^{-1})$.

A classical aim of heme protein research is the understanding
of the relationship between the structure of heme environment
in a particular heme protein and the specific properties of
the latter. For the cytochromes, the most important physico-
chemical property is their oxidation-reduction potential.

The magnitude of the oxidation-reduction potential results from
the addition of several independent contributions, including
those due to the net charge of the protein (21) and to the effects of the iron ligands on the electronic state of the metal
(22). Until recently, a quantitative assessment of the relative
magnitudes of electrostatic and ligand field effects in heme
proteins was not available. Lately, a study of the effect of
pH and of random chemical modification of the net charge on
cytochromes of the c type (23) provided empirical support to
the assumption that the electrostatic contribution can be estimated from the free energy change resulting from the addition
of an electron to a uniformly charged sphere with the radius R
and net charge Z of the heme protein. For soluble cytochromes,
which have on the average a radius of 1.6 nm, in aqueous solution the electrostatic contribution to the difference in oxi-

dation-reduction potentials is given by $\Delta E = 18(\Delta Z/R(nm))$ mvolt, 11 to 12 mvolt per unit difference between the net charges of two cytochromes.

The spectroscopic correlation presented above suggests a direct way to estimate the ligand field contribution. The reduction of a heme protein can be divided into two different steps: the excitation of an electron from the $a_{2u}(\pi)$ porphyrin orbital to the d_{yz} orbital followed by the addition of an electron to the $a_{2u}(\pi)$ orbital to complete the process. The second step does not depend on the nature of the axial ligands, since the energy of the $a_{2u}(\pi)$ orbital, in terms of the model on which Equation (1) is based, is not affected by them, nor by the nature of the protein environment; essentially the second step is an electrostatic contribution. Thus, the ligand field contribution to the differences in oxidation-reduction potentials between two low-spin heme proteins is the difference between the energies of their charge transfer $a_{2u} \to d_{yz}$ bands. From the data of Table I it follows that substituting a histidine imidazole for a methionine sulfur as the sixth iron ligand in a heme protein results in raising the energy of the charge transfer band by 800 c^{-1}, which is equivalent to lowering the oxidation-reduction potential by 100 mvolt. Experimental support for this calculation is found in the fact that heme model compounds, one with two imidazoles and the other with imidazole and thioether sulfur as axial ligands, differ in their potentials by 150 mvolt (24).

It follows from the preceding paragraphs that the thermodynamic effect of substituting an imidazole for a methionine in a low-spin cytochrome is approximately equivalent to the effect of changing the net charge of the protein by -10 units. Replacing the methionine by an amino group should have a similar effect, since preliminary results show that for cytochrome c complexed with n-butylamine $\nu(I) = 6800$ c^{-1}, equivalent to a decrease of 130 nm volt with respect to native cytochrome c. The limits of net charge at neutral pH for known respiratory and photosyn-

thetic c-type cytochromes vary between +8, for mammalian cytochrome c, and -8, for algal cytochrome c_6, spanning a range of
electrostatic contributions to the oxidation-reduction potential of 180 mvolt. It seems, therefore, that electrostatic and
ligand field contributions to the thermodynamic properties of
low-spin cytochromes have similar orders of magnitude.

Acknowledgements

This work was initiated while one of the authors (A.S.) was a
Visiting Scientist at the National Institutes of Health, Bethesda, Maryland, and was partly supported by a grant from the
U.S. - Israel Binational Science Foundation. We are grateful
to James Hofrichter for the data acquisition computer programs,
to Attila Szabo for very helpful discussions, and to Mitsuo
Sato for his comments.

References

1.   Zerner, M., Gouterman, M., Kobayashi, H.: Theoret. Chim.
     Acta 6, 363-400 (1966).

2.   Eaton, W.A., Hofrichter, J.: in (E. Antonini, L. Rossi-
     Bernardi and E. Chiancone, editors) Methods in Enzymology,
     Vol. 76, 175-261 (1981).

3.   Cheng, J.C., Osborne, G.A., Stephens, P.J., Eaton, W.A.:
     Nature (London) 241, 193-194 (1973).

4.   Stephens, P.J., Sutherland, J.C., Cheng, J.C., Eaton, W.A.:
     in (J.B. Birks, editor) The Excited States of Biological
     Molecules, 434-442 (1976).

5.   Griffith, J.S.: Nature (London) 180, 30-31 (1957).

6.   Kotani, M.: Suppl. Prog. Theoret. Phys. No. 17, 4-13 (1961).

7.   Taylor, C.P.S.: Biochim. Biophys. Acta 491, 137-148 (1977).

8.   Schejter, A., Eaton, W.A.: Biochemistry 23, 1081-1084 (1984).

9.   Aviram, I., Schejter, A.: J. Biol. Chem. 245, 1552-1557
     (1970).

10. Margalit, R., Schejter, A.: Eur. J. Biochem. 32, 494-499
     (1973).

11. Salmeen, I., Palmer, G.: J. Chem. Phys. <u>48</u>, 2049-2052 (1968).

12. Brautigan, D.L., Feinberg, B.A., Hoffman, B.M., Margoliash, E., Peisach, J., Blumberg, W.E.: J. Biol. Chem. <u>252</u>, 574-582 (1977).

13. Ikeda-Saito, M., Iizuka, T.: Biochim. Biophys. Acta <u>393</u>, 335-342 (1975).

14. Hori, H.: Biochim. Biophys. Acta <u>251</u>, 227-235 (1971).

15. Yonetani, T., Iizuka, T., Waterman, M.R.: J. Biol. Chem. <u>246</u>, 7683-7689 (1971).

16. Dunn, T.M.: Trans. Faraday Soc. <u>57</u>, 1441-1444 (1961).

17. Iizuka, T., Kotani, M.: Biochim. Biophys. Acta <u>194</u>, 275-286 (1969).

18. Iizuka, T., Kotani, M.: Biochim. Biophys. Acta <u>194</u>, 351-363 (1969).

19. Blumberg, W.E., in (Antonini, E., Rossi-Bernardi, L. and Chiancone, E., editors): Methods in Enzymology, Vol. <u>76</u>, Academic Press, New York, p. 312-329. (1981).

20. Margoliash, E., Frohwirth, N., Wiener, E.: Biochem. J. <u>71</u>, 559-572 (1959).

21. George P., Hanania, G.I.H., Eaton, W.A.: in (Chance, B., Estabrook, R.W. and Yonetani, T., editors): Hemes and Heme Proteins, Academic Press, New York, 267-271 (1966).

22. Williams, R.J.P.: in (Falk, J.E., Lemberg, R. and Morton, R.K., editors): Haematin Enzymes, Pergamon Press, Oxford, 41-55 (1961).

23. Schejter, A., Aviram, I., Goldkorn, T.: in (Ho, C., Eaton, W.A., Collman, J.P., Gibson, Q.H., Leigh, J.S., Margoliash, E., Moffat, J.K. and Scheidt, W.R., editors): Electron Transport and Oxygen Utilization, Elsevier, Biomedical Press, New York, 95-99 (1982).

24. Marchon, J.C., Mashiko, T., Reed, C.A.: in (Ho, C., Eaton, W.A., Collman, J.P., Gibson, Q.H., Leigh, J.S., Margoliash, E., Moffat, J.K. and Scheidt, W.R., editors): Electron Transport and Oxygen Utilization, Elsevier, Biomedical Press, New York, 67-72 (1982).

Received July 3, 1984

CIRCULAR DICHROISM CONTRIBUTIONS OF HEME-HEME AND CHLOROPHYLL-CHLOROPHYLL INTERACTIONS

Robert W. Woody

Department of Biochemistry, Colorado State University
Fort Collins, Colorado 80523, USA

Introduction

The interactions of heme groups in multiheme proteins (e.g., hemoglobin, cytochrome oxidase, cytochrome c_3) and of chlorophyll groups in complex photosynthetic systems (antenna proteins, reaction centers) have important biological consequences. For example, the cooperative binding of oxygen to hemoglobin can be described as a heme-heme interaction, although it is known not to be direct but mediated by protein conformational changes. Because circular dichroism (CD) is sensitive to longer range interactions than other forms of optical spectroscopy, the CD of such systems has frequently been discussed in terms of interactions between chromophoric centers. For example, attempts have been made to infer the proximity of the heme a and a_3 groups of cytochrome oxidase from the CD spectra (1). However, other studies (2,3) have led to the opposite conclusion. The multiple components observed in the low temperature CD and absorption spectra of a bacteriochlorophyll protein from Prosthecochloris aestuarii were used to infer the presence of at least five (4) or six (5) bacteriochlorophyll molecules per subunit. X-ray crystallography (6) has shown that there are seven bacteriochlorophyll molecules per subunit, but attempts to account for the observed CD and absorption by exciton (7) and by degenerate ground state (8) theory have not been successful.

Optical Properties and Structure of Tetrapyrroles
© 1985 by Walter de Gruyter & Co., Berlin · New York

Because of the interest in detecting and characterizing heme-
heme and chlorophyll-chlorophyll interactions, theoretical
calculations of the contributions of such interactions to the
CD of several well-characterized systems have been performed.
The systems chosen for this study were hemoglobin, cytochrome
c_3, and an artificial system in which chlorophyll derivatives
have been introduced in the heme pockets of apohemoglobin
chains (9-12). High resolution crystal structure data are
available for each of the first two systems (13-18). While
such data are not yet available for the chlorophyll-
substituted globins, NMR and fluorescence energy transfer results (11, 12) are consistent with the chlorophyll taking
up a well-defined orientation which can be inferred from the
known structure of deoxyhemoglobin (15).

Theoretical Background and Methods

The interactions of hemes or chlorophylls in systems containing two or more such chromophores can be described by exciton
theory. However, the form of exciton theory which has generally been applied to biopolymer systems (19, 20) assumes that
the chromophores are strongly coupled (21), i.e., that the
dipole-dipole coupling energy between the chromophores (V_{ij})
is much larger than the vibronic band width of the monomer
absorption band(s) (Δ). This is not the case for a system
such as hemoglobin, where the calculated coupling energies
are of the order of 10 cm^{-1}, while the band widths are ca.
2000 cm^{-1}. Even in the case of cytochrome c_3, where some
interheme distances are less than half those in the hemoglobin tetramer, the largest calculated V_{ij} is less than 100
cm^{-1}.

It is clear that these systems fall in the category of weak
coupling, according to the Peterson and Simpson (21)

criterion that $|V_{ij}| << \Delta$. Under these circumstances, the classical polarizability theory developed by DeVoe (22, 23) is useful. The classical polarizability theory treats the electronic transitions in each monomer as oscillators with complex polarizabilities. Each polarizable oscillator responds to both the applied electromagnetic field of the light and to the field created by the induced moments in other oscillators. Thus, the response of the aggregate to the electromagnetic field of the light is described by a set of coupled linear equations, relating the induced dipole moment of each oscillator to that of all the other oscillators. The optical properties can be calculated from the induced moments, which are obtained from the set of linear equations by inverting a matrix.

DeVoe (22, 23) showed that the molar extinction coefficient, ε, and the molar ellipticity, $[\theta]$, of the aggregate are related to a matrix $\underset{\sim}{A}$ as follows:

$$\varepsilon \ (\tilde{\nu}) = \frac{8 \ \pi^2 \ \tilde{\nu} \ N_o}{6909} \ \underset{i,j}{\Sigma} \ \mathrm{Im} \ A_{ij} \ \underset{\sim}{e}_i \ \cdot \ \underset{\sim}{e}_j \tag{1}$$

$$[\theta] \ (\tilde{\nu}) = \underset{i,j}{\Sigma} \ C_{ij} \ \mathrm{Im} \ A_{ij} \tag{2}$$

where $C_{ij} = 24 \ \pi^2 \ \tilde{\nu}^2 N_o \ (\underset{\sim}{e}_i \ \times \ \underset{\sim}{e}_j) \ \cdot \ \underset{\sim}{R}_{ij}$ (3)

Here $\tilde{\nu}$ is the frequency in cm^{-1} and N_o is Avogadro's number, $\underset{\sim}{e}_i$ is a unit vector in the polarization direction of the ith oscillator, and $\underset{\sim}{R}_{ij} = \underset{\sim}{R}_j - \underset{\sim}{R}_i$ is the vector between groups i and j. The matrix $\underset{\sim}{A}$ is obtained by inverting the matrix B:

$$\underset{\sim}{A} = \underset{\sim}{B}^{-1} = [\delta_{ij}/\alpha_i + G_{ij}]^{-1}.$$

Here δ_{ij} is the Kronecker δ, α_i is the complex polarizability of the ith oscillator and $G_{ij} = V_{ij}/\mu_i\mu_j$ is a measure

of the interaction energy between oscillators i and j. Two
types of information are needed to apply the classical polarizability theory to a given aggregate. First, the oscillators
which constitute the chromophores must be assigned complex
frequency-dependent polarizabilities. The imaginary part
of the polarizability is directly proportional to the molar
extinction coefficient of the chromophore, while the real
part of the polarizability is obtained by a Kramers-Kronig
transform (24) of the imaginary part.

The absorption spectrum of an appropriate model chromophore
was resolved into individual absorption bands. For calculations on oxy- and deoxy-hemoglobin, myoglobin was taken
as the monomer. The spectrum of the appropriate myoglobin
derivative (25) was decomposed into the Q, B (Soret), N, and
L band regions (26). In regions of band overlap, the resolution of the bands and the extrapolation to zero are somewhat
arbitrary, but the effects of these uncertainties on the
final results should be minimal, except near the extremes of
the bands. The heme $\pi\pi^*$ absorption bands arise from nearly
degenerate transitions, so each band corresponds to two orthogonally polarized oscillators and the intensities are assumed to be divided equally between the two oscillators. Cytochrome c was chosen as the model for cytochrome c_3. The
out-of-plane ligands in cytochrome c_3 are two histidines (27),
so, in this respect, cytochrome b_5 (28) would be a more suitable model than cytochrome c, the latter having histidine/
methionine coordination (29). However, cytochrome c_3 and
cytochrome c have in common saturated sidechains resulting
from addition of cysteine sulfhydryl groups, and this makes
cytochrome c a better model than cytochrome b_5 in terms of
band positions and intensities (30). The absorption data of
Margoliash and Frohwirt (31) for reduced horse heart cytochrome c were used. Polarizability data for the pyrochlorophyllide chromophore utilized by Boxer and coworkers (9-12)

were derived from the spectrum of pyrochlorophyll (32).
The data only extend to 350 nm, so only the Q_x, Q_y and the
nearly degenerate B_x, B_y transitions were considered.   The
other major factor entering the polarizability theory calculations is the geometry, which includes the location of the
chromophores and the polarization of the transitions.   The
protein structural data used in this study were obtained
from the Protein Data Bank (33) at the Brookhaven National
Laboratory.   The structures used were those for horse deoxyhemoglobin (13), horse aquomethemoglobin (14), cytochrome c_3
(18), and, for the pyrochlorophyllide-substituted hemoglobin,
human deoxyhemoglobin (15).   The centers of the chromophores
were taken to be at the average location of the four porphyrin nitrogens.   The N-N vectors were chosen as the polarization directions.

An effective dielectric constant is included in calculations
of the interactions between oscillators and is a source of
uncertainty.   Cech et al. (34) performed extensive classical
polarizability theory calculations on polynucleotides.   They
found that a value of ε_D = 2.0 gave the best results.   This
value can be justified heuristically as a reasonable approximation to the dielectric constant of typical organic substances at optical frequencies.   The value recommended by Cech
et al. has been used in all of the calculations described
here.   To a first approximation, the amplitude of the calculated CD spectra is inversely proportional to ε_D.

Some exciton calculations were also performed, using parameters comparable to those used in the polarizability theory
calculations.   In the exciton calculation, each transition
is treated as a Dirac δ function of frequency.   After the
rotational strengths are calculated by standard methods
(20, 35), CD spectra are generated by assigning each transition a Gaussian band shape.   In the present calculations,

the band widths of the Gaussians have been assumed to have
a uniform value for all transitions, equal to 20 nm for
oxyhemoglobin, 14.5 nm for cytochrome c$_3$, and 30.5 nm for
deoxyhemoglobin.

Results and Discussion

Hemoglobin.   The calculated contributions of the heme-heme
interactions to the CD of oxyhemoglobin (HbO$_2$) and deoxyhemoglobin (Hb) are shown in Figure 1.   Both the polarizability
and exciton theories predict that a positive couplet (36),
i.e., a positive CD maximum followed by a negative maximum
as one goes toward shorter wavelengths, results from the heme-
heme interaction.   The general shape of the curves for HbO$_2$
and Hb are very similar in the Soret region, but that for Hb
is red-shifted, corresponding to the red shift in absorption.
The heme-heme CD contribution for Hb is somewhat lower in
amplitude in the polarizability calculations and much lower
according to the exciton method.   (The latter effect is

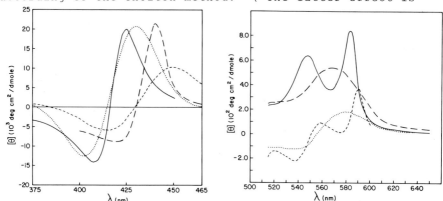

Figure 1.   Calculated contributions of heme-heme interactions
to the CD of HbO$_2$ and Hb.   (a) Soret region. Polarizability
theory for Hb (— —), HbO$_2$ (——).   Exciton theory for HbO$_2$(...),
Hb (---)   (b) Visible region.   Polarizability theory including both Q and B bands for HbO$_2$ (——) and Hb (— —).   Including only the β bands for HbO$_2$ (---) and Hb (...).

largely due to the significantly greater band width of the
assumed Gaussian. This is to some extent artifactual because
of the unusual shape of the Soret band in deoxymyoglobin.)
The smaller amplitude predicted by the polarizability method
results from two factors: the broader Soret band in Hb vs.
HbO_2, and the greater average heme-heme distance in Hb (13,
14).

Experimental CD spectra of Hb and HbO_2 are shown in Figure 2
(37).   The spectra show a strong positive band in the Soret
region for both HbO_2 and Hb, with only a very weak negative
lobe on the short wavelength side of the Soret region.   These
distinctly nonconservative CD bands are dominated by the interactions of the hemes with their immediate environment,
especially with aromatic side chains (38).   However, the predicted amplitude of the CD due to heme-heme interactions is
significant in comparison with the observed CD.   For HbO_2, the
observed maximum ellipticity at 418 nm is ca. 6×10^4 deg cm/

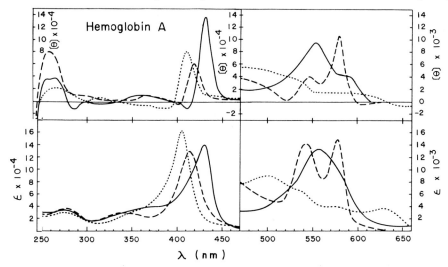

Figure 2.   CD and absorption spectra of human hemoglobin in
0.1 M phosphate buffer, pH 7.5 and 25° C.   Deoxy Hb (———),
HbO_2 (- - -), metHb (...).   Reproduced with permission from
Sugita et al., J. Biol. Chem. 246, 393-388 (1971). Copyright
1971 by the American Society of Biological Chemists, Inc.

dmole, while the present calculations predict a positive maximum at 426 nm of 2×10^4. In the case of Hb, the observed
positive maximum is 1.3×10^5, while that calculated as
arising from heme-heme interaction is ca. 2.1×10^4. The
calculated amplitude of the negative branch of the couplet is
considerably larger than that observed in the case of HbO_2,
while there is reasonable agreement in the case of Hb, where
both the calculated and experimental CD are ca. 10^4 deg
cm^2/dmole.

There are several possible explanations of the negative lobe
observed on the short-wavelength side of the Soret CD band in
HbO_2 and Hb other than heme-heme interactions. The α and β
subunits in HbO_2 and Hb could have opposite signs for the
Soret CD bands. The Soret band for each heme consists of
two components, so that sufficiently negative rotational

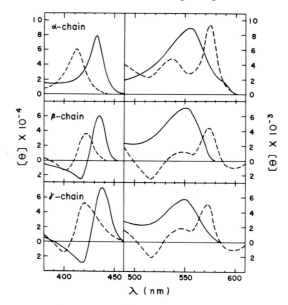

Figure 3 (a) CD spectra of isolated α, β, and γ chains of
human hemoglobin. Deoxy (——), Oxy (---). All spectra were
obtained in 0.1 M phosphate buffer, pH 7.5 at 25°C. Reproduced with permission from Sugita et al., J. Biol. Chem. 246,
383-388 (1971). Copyright 1971 by the American Society of
Biological Chemists, Inc.

strength associated with the high-energy component of either
α- or β-heme (or both) could account for the observed CD
of hemoglobin. The case for attributing the negative lobe of
the Soret CD in hemoglobin to heme-heme interactions is
strengthened by the observation that in the CD spectra of the
mammalian globins which are monomeric (myoglobin and α chains
of hemoglobin) there is no negative feature on the short-
wavelength side of the Soret band, while in those of the β_4
and γ_4 tetramers as well as the native $\alpha_2 \beta_2$ tetramer such a
feature is observed. This is illustrated in Figure 3, where
the CD spectra of isolated α, β and γ chains of hemoglobin
are shown (37). The spectrum of myoglobin resembles that of
Hb α chains in the Soret region. Thus, although the evidence
is only circumstantial, there is support for the assignment
of the weak negative lobe to heme-heme interactions. In any
case, the present calculations indicate that such interactions can be significant even when the hemes are separated by
2.5 - 4.0 nm, at least in the case of the intense Soret band.

The effects of heme-heme interaction in the visible region of
the CD spectrum of HbO_2 and Hb are shown in Figure 1. Two
sets of polarizability theory calculations are shown--one
considering both the visible and Soret transitions. Inclusion of the Soret bands clearly has a strong effect on the
calculated CD of the visible bands. By contrast, the calculated Soret CD is influenced only slightly by inclusion of
mixing with the visible bands. This is because of the order-
of-magnitude difference in the intensities of the two bands.
The more complete calculations give band shapes which more
closely resemble those of the experimental CD spectrum (Figure
2), but the calculated amplitudes are at least an order of
magnitude weaker than those observed. The relatively weak
visible bands do not permit significant heme-heme interactions over the distances involved in hemoglobin.

In addition, several minor features are missing in the calculated CD spectra. The experimental CD spectrum for Hb shows
distinct features corresponding to the O-O and O-l transitions of the Q band. The calculated spectrum fails to reproduce this, giving only a single featureless band, as expected
from the absorption spectrum. The experimental CD spectrum
of HbO_2 has a negative band on the long-wavelength side of
the Q_{OO} or α band. This feature has been assigned as a d→d
transition of the iron (39, 40). Consistent with this assignment, the calculated CD curve for HbO_2 remains positive and
decays smoothly to zero at wavelengths between 600 and 650
nm.

Cytochrome c_3. The calculated contributions of the heme-
heme interaction to the CD of cytochrome c_3 are shown in
Figure 4. A strong positive couplet is predicted in the
Soret region, with a total amplitude of ca. 1.5×10^5 deg
$cm^2 dmole^{-1}$. The results from the polarizability theory do
not depend strongly upon whether the Q and N bands

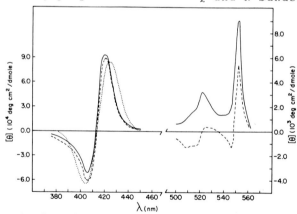

Figure 4. Calculated CD contributions of heme-heme interactions in cytochrome c_3 from <u>Desulfovibrio</u> <u>vulgaris</u>. Polarizability theory including the heme Q, B and N transitions,
——; polarizability theory including only the B transition
(Soret region) or Q transition (visible region), ---; exciton
theory including Q, B, N and L transitions, ··· .

are included. The exciton theory gives results which closely
resemble those from the polarizability theory. In the
visible region, the effect of mixing with the Soret band is
important, converting a complex spectrum showing alternating positive and negative extrema into a spectrum which is
positive from 500 to 460 nm and has a band shape like that of
the absorption spectrum. The experimental CD spectrum of
cytochrome c_3 from <u>Desulfovibrio vulgaris</u> (30) is shown in
Figure 5. In contrast to the theoretical curve, both the
Soret band and the Q_{oo} (α) band have strong negative couplets.
Experimentally, the total amplitude of the Soret CD couplet
is <u>ca</u>. 6.5 x 10^5 deg cm^2/dmole, several times that calculated
and of opposite sign. The α-carbon coordinates (17) were
checked carefully. The region from residues 90-99 was indeed
a right-handed helix, ruling out the possibility that the
wrong absolute configuration was being used in the calculations. Using the exciton theory various deviations from

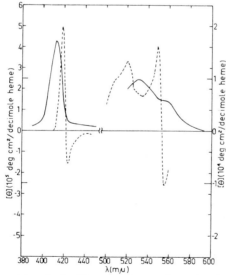

Figure 5. Experimental CD spectra of cytochrome c_3 from
<u>D. vulgaris</u> measured in 0.01 M Tris HCl, pH 8.1, 20° C.
Reproduced with permission from Drucker <u>et al.</u>, Biochemistry
<u>9</u>, 1519-1527 (1970). Copyright 1970 by the American Chemi-
cal Society.

degeneracy of the heme Soret transitions were considered,
but these had only minor effects on the calculated amplitude
of the couplet and left the sign unchanged. Deviations from
the dipole-dipole approximation were considered by using the
monopole-monopole (38, 41) approximation. This also had
little effect on the predicted CD curve, reducing the amplitude of the couplet from 1.5 to 1.3 x 10^5 deg cm^2/dmole.
Thus, even in this system where the shortest heme-heme distance is ca. 1.1 nm, the dipole-dipole approximation works
very well.

It is clear that although heme-heme interactions make a
substantial contribution to the CD of cytochrome c_3, the CD
is dominated by other factors, which cause a reversal in the
sign of the Soret and α-band couplets. As noted above, heme-
heme interaction is only one of several possible causes for
band splitting in heme proteins. The recently reported
high-resolution structure of cytochrome c_3 from D. vulgaris
(18) suggests two possible features which may be responsible
for the strong negative couplet. The porphyrin rings in cytochrome c_3 are substantially distorted from planarity, with
individual pyrrole rings deviating from the mean plane of
the porphyrin by up to 15°. Thus, the porphyrin rings are
inherently chiral chromophores. In addition, each of the
hemes has an aromatic group oriented perpendicular to the
plane of the porphyrin and nearly parallel to a liganding
histidine side chain. Detailed calculations on this system
are in progress to elucidate the origins of the strong CD
bands.

Chlorophyll-globin Complexes. The calculated contributions
of interchromophoric interactions to the CD of a fully substituted pyrochlorophyllide derivative of hemoglobin (α_2^{chl}
β_2^{chl} and of hybrid heme-chlorophyll proteins ($\alpha_2^{chl}\beta_2^{h}$ and

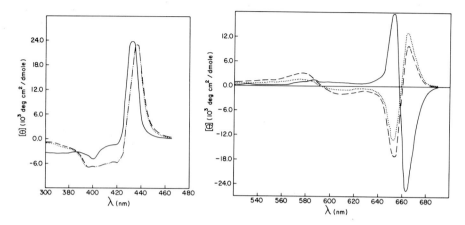

Figure 6.   Calculated contributions of interchromophoric interactions in pyrochlorophyllide-substituted hemoglobins. $\alpha_2^{chl} \beta_2^{chl}$, ———; $\alpha_2^{chl} \beta_2^{h}$, --; $\alpha_2^{h} \beta_2^{chl}$, ... (a) Soret region; (b) visible region.   Note that the calculated molar ellipticities are per chromophore so those reported for the hybrid tetramers should be multiplied by two to put them on a per chlorophyll basis.

$\alpha_2^{h} \beta_2^{chl}$) are shown in Figure 6.   In the Soret region, all three molecules are predicted to exhibit a positive peak between 430 and 440 nm, followed by a broad but weak negative region below 425 nm, with several inflections.   The predicted spectra for the two hybrid systems are nearly identical.

In the visible region of the spectrum, the $\alpha_2^{chl} \beta_2^{chl}$ is qualitatively different from the two hybrid systems.   The $\alpha_2^{chl} \beta_2^{chl}$ tetramer has a strong negative couplet in the region of the Q_y band, while the $\alpha_2^{chl} \beta_2^{h}$ and $\alpha_2^{h} \beta_2^{chl}$ have positive couplets of similar amplitude.   The reversal in sign on going from the hybrid systems to the completely substituted tetramer results from the introduction of new interactions between identical Q_y transitions in chlorophyll chromophores and the loss of Chl Q_y - heme B_x, B_y interactions.   The interactions between the α_1 and β_2 subunits (and the equivalent $\alpha_2 \beta_1$ interactions) are especially important because these

involve the closest approach between heme sites (ca. 2.5 nm).

The experimental CD data of Kuki
and Boxer (11) for these systems
in the visible region are shown
in Figure 7. In all three systems,
the Q_y 0-0 band is associated with
a strong positive band. The fully
substituted $\alpha_2^{chl}\ \beta_2^{chl}$ does not
show any evidence of the predicted
negative couplet, although the CD
on the long-wavelength side of the
Q_y band does appear to drop more
sharply than that of the hybrid
tetramers, and there is a feature
at ca. 670 nm which might result
from the interchromophoric interactions considered here. The α_2^h
β_2^{chl} hybrid has a negative band
at 625 nm which correlates with
the Q_y 0-1 band and is at too short
a wavelength to be associated with
the negative branch of the predicted couplet.

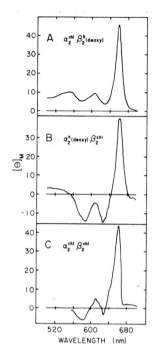

Figure 7. Experimen-
tal CD spectra of pyro-
chlorophyllide-substitu-
ted human hemoglobins in
the visible region. Reproduced with permission
from Kuki and Boxer, Bio-
chemistry, 22, 2923-2933
(1983). Copyright 1983
by the American Chemical
Society.

Thus, the visible region of the CD
spectrum of these pyrochlorophyll systems is dominated by
chlorophyll-protein interactions. This is consistent with
the strong interactions between protein groups and the chlorophyll chromophore noted by Boxer and coworkers (9, 10). These
interactions reverse the sign of the relatively strong CD of
the isolated chlorophyll when it is incorporated in globin.
Wright and Boxer (10) concluded that the analysis of natural
chlorophyll-protein complexes by CD requires consideration

of the role of the protein. It seems likely that this accounts for the problems encountered (37, 38) in interpreting
the complex CD of the bacteriochlorophyll protein from Prosthecochloris estuarii in terms of the arrangement of the
bacteriochlorophylls alone. This implies that while CD is a
powerful qualitative technique in the study of systems such
as reaction centers--to determine the number of strongly
interacting chromophores, for example--a quantitative interpretation of the CD will only be possible if interactions
with protein groups can be included.

Conclusions

The interaction of hemes and chlorophylls in proteins containing several such chromophores can lead to significant
contributions to the CD spectrum, even when the groups are
separated by distances of 3.0 nm or more, at least for
intense transitions such as the heme Soret and chlorophyll
Q_y bands. Polarizability theory, which is applicable to
weakly coupled chromophores, is useful in calculating these
interactions because it takes into account the explicit
band shape. Although these systems violate the conditions
for strong coupling, strong coupling exciton theory gives
results qualitatively consistent with the polarizability
theory, but information about the band shape is lost. It
has been shown that the contributions studied here are generally weak compared with those due to the interactions of
heme or chlorophyll with the surrounding protein groups.
This is true even in the case of cytochrome c_3 where pairs of
hemes are 1.1-1.2 nm apart. Only in the case of hemoglobin
is there some evidence of spectroscopic features which may
be attributable to heme-heme interactions.

Acknowledgements

I am grateful to Dr. Marvin W. Makinen, who provided the
absorption spectra of myoglobin derivatives used in this
study, and to Dr. S. Devarajan who assisted in the preparation of this manuscript. Permission to reproduce previously
published experimental spectra was kindly granted by the
authors and by the copyright holders, the American Society
of Biological Chemists (Figures 2,3) and the American Chemical Society (Figures 5, 7). This work was supported in part
by U.S. Public Health Service Grant GM22994.

REFERENCES

1. Urry, D.W., van Gelder, B.F.: In "Structure and Function
    of Cytochromes", (Okunuki, K., Kamen, M.D., Sekuzu, I.,
    eds.) pp. 210-214, University Park Press, Baltimore, Mary-
    land (1968).

2. Myer, Y.P., King, T.E.: Biochem. Biophys. Res. Commun.
    34, 170-175 (1969).

3. Myer, Y.P., Pande, A.J.: In "The Porphyrins" Dolphin, D.,
    ed.) Vol. IIIA, pp. 271-322. Academic Press, New York
    (1978).

4. Philipson, K.D., Sauer, K.: Biochemistry 11, 1880-1885
    (1972).

5. Olson, J.M., Ke, B., Thompson, K.H.: Biochim. Biophys.
    Acta 430, 524-537 (1976).

6. Matthews, B.W., Fenna, R.E., Bolognesi, M.C., Schmid, M.F.,
    Olson, J.M.: J. Mol. Biol. 131, 259-285 (1979).

7. Fenna, R.E., Matthews, B.W.: Brookhaven Sympos. Biol.
    28, 170-182 (1976).

8. Pearlstein, R.M., Hemenger, R.P.: Proc. Natl. Acad. Sci.,
    USA 75, 4920-4924 (1978).

9. Boxer, S.G., Wright, K.A.: J. Amer. Chem. Soc. 101, 6791-
    6794 (1979).

10. Wright, K.A., Boxer, S.G.:   Biochemistry 20, 7546-7556
     (1981).

11. Kuki, A., Boxer, S.G.:   Biochemistry 22, 2923-2933 (1983)

12. Moog, R.S., Kuki, A., Fayer, M.D., Boxer, S.G.:   Bio-
     chemistry 23, 1564-1571 (1984).

13. Bolton, W., Perutz, M.F.:   Nature 228, 551-552 (1970).

14. Ladner, R.C., Heidner, E.G., Perutz, M.F.:   J. Mol. BioL
     114, 385-414 (1977).

15. Fermi, G: J. Mol. Biol. 97, 237-256 (1975).

16. Haser, R., Pierrot, M., Frey, M., Payan, F., Astier, J.
     P., Bruschi, M., LeGall, J.:   Nature 282, 806-810 (1979).

17. Higuchi, Y., Bando, S., Kusunoki, M., Matsuura, Y.,
     Yasuoka, N., Kakudo, M., Yamanaka, T., Yagi, T., Ino-
     kuchi, H.:   J. Biochem. (Tokyo) 89, 1659-1662 (1981).

18. Higuchi, Y., Kusunoki, M., Matsuura, Y., Yasuoka, N.,
     Kakudo, M.:   J. Mol. Biol. 172, 109-139 (1984).

19. Moffitt, W.:   J. Chem. Phys. 25, 458-478 (1956).

20. Tinoco, I., Jr., Woody, R.W., Bradley, D.F.:   J. Chem.
     Phys. 38, 1317-1325 (1963).

21. Simpson, W.T., Peterson, D.L.:   J. Chem. Phys. 26,
     588-589 (1957).

22. DeVoe, H.:   J. Chem. Phys. 41, 393-400 (1964).

23. DeVoe, H.:   J. Chem. Phys. 43, 3199-3208 (1965).

24. Moscowitz, A.:   Adv. Chem. Phys. 4, 67-112 (1962).

25. Makinen, M., private communication (1978).

26. Caughey, W.S., Deal, R.M., Weiss, C., Gouterman, M.:
     J. Mol. Spectr. 16, 451-463 (1965).

27. Dobson, C.M., Hoyle, N.J., Geraldes, C.F., Wright, P.E.,
     Williams, R.J.P., Bruschi, M., LeGall, J.:   Nature 249
     425-429 (1974).

28. Matthews, F.S., Argos, P., Levine, M., Cold Spring
     Harbor Symp. Quant. Biol. 36, 387-396 (1972).

29. Dickerson, R.E., Takano, T., Eisenberg, D., Kallai, O.B.,
     Samson, L., Cooper, A., Margoliash, E.: J. Biol. Chem.
     246, 1511-1535 (1971).

30. Drucker, H., Campbell, L.L., Woody, R.W.: Biochemistry
     9, 1519-1527 (1970).

31. Margoliash, E., Frohwirt, N.: Biochem. J. 71, 570-572
     (1959).

32. Pennington, F.C., Strain, H.H., Svec, W.A., Katz, J.J.,
     J. Amer. Chem. Sco. 86, 1418-1426 (1964).

33. Bernstein, F.C., Koetzle, T.F., Williams, G.J.B., Meyer,
     E.F., Brice, M.D., Rodgers, J.R., Kennard, O., Shimanou-
     chi, T., Tasumi, M.: J. Mol. Biol. 112, 535-542 (1977).

34. Cech, C.L., Hug, W., Tinoco, I., Jr.: Biopolymers 15,
     131-152 (1976).

35. Bayley, P.M., Nielsen, E.B., Schellman, J.A.: J. Amer.
     Chem. Soc. 93, 29-37 (1971).

36. Schellman, J.A.: Accts. Chem. Res. 1, 144-151 (1968).

37. Sugita, Y., Nagai, M., Yoneyama, Y.: J. Biol. Chem.
     246, 383-388 (1971).

38. Hsu, M.-C., Woody, R.W.: J. Amer. Chem. Soc. 93,
     3515-3525 (1971).

39. Eaton, W.A., Hanson, L.K., Stephens, P.J., Sutherland, J.
     C., Dunn, J.B.R.: J. Amer. Chem. Soc. 100, 4991-5003
     (1978).

40. Case, D.A., Huynh, B.H., Karplus, M.: J. Amer. Chem.
     Soc. 101, 4433-4453 (1979).

41. Tinoco, I., Jr.: Adv. Chem. Phys. 4, 113-160 (1962).

Received July 10, 1984

Discussion

Blauer: Could different ligands bound to the heme iron in cytochromes
account for asymmetry producing optical activity and how would you estimate such contributions?

Woody: The effects of different ligands could be estimated from the detailed structure, using the type of techniques which have been used in previous calculations of heme protein CD, at least for the bis-histidine liganding found in cytochrome c_3. Your question suggests a possibly important contribution which I had not considered in this system. Hsu and I (Ref. 38) found that the proximal histidine of myoglobin and hemoglobin gives large contributions (ca. ±0.4 DBM) to the rotational strengths of the individual Soret components, but these are opposite in sign and nearly equal in magnitude. In cytochrome c_3 there are 8 such histidine-heme interactions, so although there will probably be some cancellation, this could be an important factor in determining the strength of the CD couplet.

Blauer: Would the energies involved in deviations from planarity of the tetrapyrrole system mentioned by you not be too high and which factors could bring about such distortions?

Woody: The porphyrin ring system is much more flexible than was previously thought. X-ray diffraction studies of simple porphyrins almost always show some significant deviations from planarity. In every heme protein whose structure has been refined wihtout prior constraints on the heme, some deviations from planarity have been inferred, although these cannot be characterized very accurately. Thus, deviations of a heme from planarity apparently do not cost much free energy. The heme binding site in a protein is certainly not symmetrical so, in the absence of a large energy barrier, some distortions from planarity are to be expected.

Song: Could you attribute the disagreement in the Cotton effect sign between the calculated and observed Soret CD bands specifically to an environment induced distortion of the porphyrin plane or to a chiral disposition of aromatic residues around the heme?

Woody: Until detailed calculations are completed, utilizing the X-ray structural data, it will not be possible to definitively distinguish these possibilities.

Scheer: I have two questions. First, you mentioned in your introduction the problem of distinguishing exciton interaction and accidental occurrence of chromophores with opposite CD-sign. This is just the situation often observed in photosynthetic systems. Could you suggest means to distinguish these two possibilities spectroscopically?

Woody: I know no general solution to this problem but supplementary linear dichroism measurements would be useful. In the exciton effect in a dimer, the two exciton branches are orthogonally polarized. The same is true for infinite helices. In oligomers of low symmetry, there are no rigorous selection rules but generally the two strongest absorption features can be expected to be polarized at or near right angles. Observation of orthogonally polarized regions in absorption would support an exciton assignment, although two nonidentical chromophores which are coupled might coincidentally have their transition moments oriented orthogonally. Observation of an angle substantially different from 90^0 would make an exciton assignment implausible.

Scheer: The second question concerns your last remarks. Did you try to estimate the effect of distortions of the porphyrin ring on exciton coupling, or, the other way around, is a strong deviation from the exciton type spectrum an indication for induced dissymmetry in the chromophore?

Woody: I would not expect deviations from planarity to significantly alter the exciton interaction as such. However, such deviations can lead to large contributions to the CD which can alter the simple pattern predicted by exciton theory. Unfortunately, inherent chirality of the chromophore is not the only possible source of such distortions. Nondegenerate coupled oscillator interactions can be of equal or even greater importance. In the case of myoglobin and hemoglobin, we know that the latter type of interactions is comparable to the observed magnitude, although some contributions from inherent chirality of the heme cannot be excluded.

Gouterman: a) Are there large circular dichroism effects observed in the monomeric hemoproteins due to the local interactions?
b) Does the phenomenon of local CD add amplitude-wise or intensity-wise to the interheme CD?

Woody: a) Monomeric heme proteins such as myoglobin and invertebrate hemoglobins have Soret band rotational strengths with magnitudes of the order of several tenths to half a Debye-Bohr magneton. Thus local interactions are quite significant in these systems.
b) Such interactions would be additive to the contributions considered here.

Gouterman: Could there be contributions due to (d,d) transitions which are not apparent in absorption? We calculate (d,d) transitions throughout the region.

Woody: d-d Transitions and also charge transfer transitions almost certainly contribute to the CD. The CD of globins is dominated in the visible and near ultraviolet by heme π,π^* transitions, but there are minor features which are probably due to other transitions. In the case of cytochrome c, there is an abundance of CD bands, many of which are probably charge transfer bands and some of which (in the near infrared) have been assigned by Eaton and Charney to d-d transitions.

Falk: As the distance between the interacting chromophores is rather large, how do you take account of some sort of dielectric constant, using the dipole-dipole approximation?

Woody: The question of whether and how a dielectric constant should be included in this type of calculation is still open. Although I failed to mention the point, an effective dielectric constant of 2 was used throughout these calculations. This value was used because Tinoco and coworkers had found it to give the best results in their calculations on nucleic acids. To the extent that the medium between the chromophores can be treated as a continuum with uniform dielectric constant, a value of ca.2 appears to be a reasonable upper limit for the types of interaction being considered. The coupled oscillator interactions involve electronic motions at optical frequencies, so $\varepsilon_{eff} \sim n^2$. The considerably larger static dielectric constant would not be appropriate.

Bode: Having the X-ray coordinates, it should be possible to approximately calculate the complete CD spectrum. Have you done such calculations
for these proteins, and are accordingly the contributions of the protein
environment large?

Woody: Such calculations have been done for hemoglobin and for myoglobin
(Ref. 38) with results that agree semiquantitatively with experiment. We
intend to perform similar calculations for cytochrome c_3.

OPTICAL PROPERTIES OF CONJUGATED PROTEINS WHICH TRANSPORT
HAEM, PORPHYRINS AND BILIRUBIN

Brian Ketterer.
Courtauld Institute of Biochemistry, Middlesex Hospital
Medical School, London W1P 7PN, Great Britain.

1.   Introduction

Haem (iron protoporphyrin IX) is almost exclusively an intracellular
component where it is a prosthetic group, strongly bound to protein.
The major source of haem is haemoglobin which occurs in the erythrocyte
at the remarkably high concentration of 5 mM and in the average man
accounts for about 35 g of haem.   Another important source is myoglobin
which occurs in muscle and yet another are haem enzymes such as
mitochondrial and microsomal cytochromes, catalase etc.   This paper is
concerned not with proteins associated with the biological function of
haem, but with proteins concerned with the transport of haem and compounds
metabolically related to it.   The turnover of erythrocytes is therefore
of interest since they have the potential to release such large quantities
of haem.   No such release occurs under physiological circumstances, but
after a life span of some 120 days erythrocytes are removed from the
circulation by such organs as the spleen, liver and bone marrow where
globin is degraded and haem is converted to the linear metal-free
tetrapyrrole bilirubin.   The bilirubin so formed is conjugated with
glucuronic acid in the liver and excreted via the bile. About 75% of
the daily excretion of bilirubin results from the destruction of
senescent erythrocytes: the rest is derived from other sources of
haem most notably the rapid turning over of hepatic cytochrome P_{450}
(1, 2).   A small proportion of haem may enter the circulation. It may

be derived from haemoglobin released as a result of intravascular
haemolysis or perhaps the release of myoglobin from bruised muscle (3).
Both bilirubin and haem which enter the circulation become protein bound.
Such binding ensures that these lipophilic compounds are solubilized
and also protects the organism from their toxicity. In certain disease
states, the biosynthesis of haem is at fault and porphyrin precursors
of haem accumulate: porphyrins are photosensitizing agents and likewise
there is good cause for their strong binding to protein and rapid
elimination (4, 5).

All these tetrapyrroles are eliminated via the liver (1-5). In the
present paper the optical properties of haem-, porphyrin- and bilirubin-
binding proteins in the extracellular space and the liver parenchyma
are described and the role of these proteins in the binding, transport
and excretion of these tetrapyrroles is discussed.

2.   Optical Properties of Haem, Protoporphyrin and Bilirubin
      Binding Proteins of Plasma

2.1   Human Serum Albumin

Human serum albumin has long been studied as a haem-binding protein.
The spectrum of ferrohaem-albumin demonstrates a red shift and sharpening
of the Soret band associated with the conversion of the haem dimer which
occurs in aqueous solution to the haem monomer and also an absorption
maximum at 620 nm indicative of high-spin iron. A high spin state
implies that co-ordinate bonds between the iron atom and protein do not
occur (6-8).

Although haem has no asymmetric centre the circular dichroism spectrum of
haem-albumin shows optical activity in the region of the Soret band
absorption indicating dissymmetry induced by the binding site of the

albumin. This circular dichroism spectrum changes little on the addition
of cyanide ions confirming the conclusion drawn from the absorption
spectrum that co-ordinate bonds between the iron atom and protein are not
involved (6).

Fluorescence spectra of human serum albumin as a function of haem concentration show quenching by haem which affects first tryptophan and then tyrosine emissions. On the basis of fluorescence quenching data, a K_{assoc} of 5×10^7 M^{-1} was determined for the binding of haem-human serum albumin at a high affinity site and additional sites of lower affinity were also detected. Human serum albumin also binds fatty acids with high affinity, and spectrophotometric titrations show that fatty acids can not compete with haem at its high affinity site but can displace it from its low affinity sites. Human serum albumin also binds bilirubin and similarly haem and bilirubin do not compete at their high affinity sites. This is clearly shown by the minimal effect of haem on the circular dichroism spectra of the bilirubin-bound human serum albumin. There is evidence however, for competition between haem and bilirubin at their weaker binding sites (6).

Serum albumins in general may have no significance for the transport of haem. Only primate albumins have been shown to have high affinity and even these albumins may not have sufficient binding activity to compete effectively with another plasma haem-binding protein namely hemopexin (7, 9). Serum albumins do, however, have considerable importance in the transport of bilirubin (2). The binding of bilirubin by human serum albumin affects the absorption spectrum of bilirubin. There is a slight red shift and partial resolution of two absorption bands with maxima at approximately 410 and 460 nm. The ratio of the intensities of the two bands differs with the binding protein, the absorption at approximately 460 nm being greater in bilirubin-human serum albumin and accounting for the red shift (10, 11).

Bilirubin bound to human serum albumin gives a bilobate circular dichroism spectrum in the region of its optical absorption, with negative ellipticity at the shorter wavelength and positive ellipticity at the longer wavelength (10, 12). The rotational strengths are greater than would be expected from a symmetrical arrangement perturbed by dissymmetry at the binding site, such as has been suggested for haem, and are more likely due to dissymmetry of the chromophore as a whole (10). The sign of the ellipticity differs with the source of the serum albumin, both bovine and rat serum albumin being the opposite to human serum albumin

(11, 12).   It also varies with the pH, the sense being reversed in the region of the N–F transition (12).

The structure of serum albumin is well understood.   It is a prolate ellipsoid of overall dimensions 38 x 150 Å (17) and has a mol. wt. of 66,248.   It is remarkable in having 17 disulphide bridges which define its structure and which give rise to long and short double loops numbered 1 to 9 in which 1, 3, 4, 6, 7 and 9 are long and 2, 5 and 8 are short. These loops form three domains 1 to 3, 4 to 6 and 7 to 9 (14).   Circular dichroism spectra show that the secondary structure of bovine serum albumin consists of 68% α–helix and 18% β–structure.   Limited tryptic or peptic hydrolysis gives fragments which largely retain their secondary structure and original ligand-binding properties.   Thus the strong bilirubin site is retained by fragments containing residues 1–385, 1–306 and 186–306 but not 239–306 suggesting that residues 186–238 constitute the essential region for bilirubin binding (15).   This is the region of loop 4.   Involvement of this part of the molecule in high affinity bilirubin binding is confirmed by experiments with the agent carbodiimide which under carefully controlled conditions couples one of the carboxyl groups of bilirubin with the ε–amino group of Lysine 240 (16) which is on loop 4 (as is also the tryptophan residue the fluorescence of which is quenched on bilirubin binding).   The principal fatty acid binding site is far removed.   In bovine serum albumin it is in the region of loops 7 and 8 while the weaker sites are associated with loops 6 and 7 and also loop 4 (15).

In addition to an important physiological role in the transport of bilirubin, serum albumin also binds protoporphyrin IX which appears in the plasma in cases of congenital protoporphyria and severe acute cases of lead poisoning.   The binding of protoporphyrin IX to serum albumin results in a red shift in excitation and emission maxima and in the case of human serum albumin, a much increased fluorescence yield.   Fluorometric titration gives data from which a K_{assoc} of 3 x 10^9 M^{-1} has been calculated (17).

2.2   Hemopexin

Haem-hemopexin gives an absorption spectrum which is quite different
from that of haem-human serum albumin and more related to that of
haemoglobin and haem enzymes where the iron atom is co-ordinated with
the protein(s).

The spectrum is characteristic of haem in which the iron is in low spin
in both oxidized and reduced states.   This has been confirmed by
examination of electron spin resonance, Mössbauer and magnetic circular
dichroism spectra (7, 8).   From magnetic circular dichroism spectra in
the reduced and oxidized states it was concluded that the oxidized
spectrum closely resembles dihistidyl model compounds and cytochrome b_5
which is known to be hexa-co-ordinated through two histidyl residues
(7). Further evidence for the involvement of histidine comes from
photo-oxidation experiments using apo-hemopexin and either deuteroporphyrin or Rose Bengal in which experiments the the ability to bind
haem and give absorption spectrum characteristic of haem-hemopexin is
lost in parallel with loss of histidine.   Similar results were obtained
after the chemical modification of histidine residues by diethylpyro-
carbonate (7). The two co-ordinate links with histidine are associated
with very strong binding. Competition with methaemoglobin indicates a
very high association constant for haem of approximately 10^{13} M^{-1} (18).

A study of difference absorption spectra of haem-hemopexin and
apohemopexin shows that interaction of apohemopexin with haem produces
distinct changes in the ultraviolet with sharp maxima near 285 and
292 nm which are interpreted as being due to an alteration by haem
binding of the environment of tryptophan and possibly also of tyrosine
(19). Similar conclusions have been drawn from circular dichroism
spectra which show a positive ellipticity at 231 nm ascribable to the
involvement of tryptophan or tyrosine residues in the interaction with
haem.   Treatment with N-bromosuccinimide which brings about a loss in
tryptophan residues causes a concomitant loss in $[\theta]_{231}$ and haem
binding (19). Although haem-hemopexin resembles cytochrome b_5 with
respect to the hexa-co-ordination of its haem iron with two histidyl
residues it differs in its circular dichroism spectrum and the ease

with which its iron is reduced. Thus haem-hemopexin has a low standard electrode potential ($E° = 0.064$) and is found to be 99% in the oxidized form (20).

Hemopexin binds porphyrins in addition to haem, but in the absence of iron the association constants are 4-5 orders of magnitude lower (17). Bilirubin is also bound to hemopexin with an affinity similar to that of the porphyrins (K_{assoc} 10^6 M^{-1}). The absorption spectrum of hemopexin-bound bilirubin also gave two bands at approximately 410 and 460 nm but the ratio of their intensities of absorption were opposite to that observed with bilirubin-human serum albumin, i.e there was a slight blue shift overall (11).

3.   Optical Properties of Haem-, Porphyrin- and Bilirubin-bound
      Proteins of the Rat Liver Cytosol

3.1 Ligandin

Ligandin is a name given to a protein fraction, which was isolated from the rat liver soluble supernatant and which possessed high affinity binding for haem, bilirubin and a range of other substances with lipophilic moieties (21, 22, 23). It is now known that the term ligandin referred to a mixture of two glutathione (GSH) transferases (24). These enzymes are a large family of isoenzymes which are dimers of subunits differing in their primary structures and enzymic specificities.
According to the current nomenclature, as each subunit is characterized, it is given a number.   So far eight GSH transferases have been named as follows: GSH transferases 1-1, 1-2, 2-2, 3-3, 3-4, 4-4, 5-5 and 6-6

(25). The GSH transferases all catalyse the reaction of the hydrophilic nucleophile GSH with compounds with an electrophilic centre and a lipophilic moiety. Certain subunits e.g. 1,2 and 5 in addition to GSH transferase activity possess GSH peroxidase activity towards lipophilic peroxides (26, 27). Subunit 1 not only possesses both these enzymic activities but also a unique lipophile binding site either identical to, or overlapping with, the electrophilic substrate binding site. This site has high affinity for haem, bilirubin etc. and is responsible for the binding properties associated with ligandin preparations. Thus GSH transferases 1-1 and 1-2 are ligandins the former possessing two binding sites per mole and the latter one (24, 28).

Ligandin gave spectra for bound-haem resembling those obtained with human serum albumin, thus ligandin gave an absorption spectrum for haem in the Soret region characteristic of the ligand in its monomeric state and was able to form co-ordination complexes with cyanide but not azide or fluoride, indicating that protein-binding did not involve co-ordination complexes with the haem iron atom. Subunit 1 also contains a tryptophanyl residue and, as might be expected, haem quenches its fluorescence. Data from fluorescence quenching gave a value for haem binding of 2×10^7 M^{-1}. Extrinsic Cotton effects were also obtained in the Soret region, indicating dissymmetry generated by haem-protein binding (23).

As with human serum albumin the binding of bilirubin to ligandin results in a red shift. The spectrum shows partial resolution of two absorption bands with maxima in the region of 400 and 460 nm, the intensity of absorption of the former being less than that of the latter. The binding of bilirubin by ligandin also induces unusually large Cotton effects with positive ellipicity at approximately 410 nm and negative ellipicity at 460 nm (22). The bilirubin binding site of ligandin seems to be labile unlike that of serum albumin. Its stability seems to depend to a large extent on the presence of GSH, presumably at the GSH binding site. This is very apparent when separations by gel filtration of soluble supernatant from the Gunn rat in the presence and absence of GSH are compared. Bilirubin binding to the ligandin fraction is much reduced in the absence of GSH (29). The Gunn rat has a congenital deficiency in bilirubin conjugation and consequently has abnormally high levels of bilirubin in its hepatic cytosol. Evidence for an effect of GSH on bilirubin binding is also obtained from circular dichroism spectra. Addition of GSH to bilirubin-bound ligandin produces spectral shifts, new peaks occurring at 490, 449 and 400 nm (30).

The K_{assoc} for ligandin and bilirubin in vivo is not known with
certainty.   Mixtures of Gunn rat soluble supernatant with Wistar rat
serum indicate that ligandin has a higher affinity for bilirubin
than does serum albumin, and, bearing in mind that the concentration of
albumin in plasma is somewhat greater than that of ligandin in cytosol,
the in vivo K_{assoc} for ligandin and bilirubin is probably greater
than 10^8 M^{-1}.   Values reported for ligandin purified in the absence
of GSH ranging from 7×10^6 M^{-1} to 5×10^7 M^{-1} are presumed to be low
and to be regarded as artifacts of purification (29, 31).   It is
appropriate to point out that when the affinity is high, accurate
values for K_{assoc} are difficult to determine due to the difficulty
in obtaining accurate values for the very low concentrations of aqueous
free ligand which result.   This is particularly so in the case of
ligands such as haem, porphyrins and bilirubin which tend to aggregate
in free solution.

Ligandin also binds porphyrins and in doing so gives spectra characteristic
of a monomeric species.   That is the broad spectral band characteristic of
aggregated porphyrins in aqueous solutions changes on binding to a sharp
intense Soret band at longer wavelengths.   Protoporphyrin IX, haematoporphyrin, coproporphyrins I and III and uroporphyrins 1 and III have
all been studied. K_{assoc} for haematoporphyrin, coproporphyrins and
uroporphyrins are 6×10^5, 3×10^5 and 3×10^4 M^{-1} respectively.
In the case of uroporphyrins the tendency to aggregate in aqueous
solution is weak and the binding affinity for ligandin so low that
binding has little effect on the spectrum (32).

Progress has been made in structural studies of GSH transferase subunit
1, namely the subunit responsible for ligandin activity.   DNA clones
complementary to mRNA for this subunit have been constructed in several
laboratories and the primary structure deduced from them (33-35).
Titration with sulphydryl reagents suggests that, in the formation of the
dimer, otherwise identical subunits take up non-identical conformations.
Thus there are two sets of cysteinyl residues in identical sequences in
each dimer, but in the dimer one not two is titrated by iodoacetamide,
two are titrated by N-ethylmaleimide and three by p-chloromercuribenzoate.
The fourth has yet to be titrated by a sulphydryl reagent and is either

blocked or inaccessible to such reagents. Only when the third cysteine is blocked is there any effect on either binding affinity for bilirubin or enzymic activity (36). Reference to the primary structure of subunit 1 shows one cysteinyl residue (Cysteine 18) to be at a point where a hydrophilic sequence joins a hydrophobic sequence and the other (Cysteine 112) in a somewhat more hydrophobic area. It is possible that one Cysteine 18 is more readily titrated in the dimer by iodoacetamide than the other; that both of these cysteines are titrated by N-ethylmaleimide and that only one of the two possible Cysteine 112 residues is available to more hydrophobic reagents. This residue is sufficiently close to a binding site for its substitution to cause some loss of activity.

3.2   Fatty Acid Binding Protein

Purified hepatic fatty acid binding protein (37) [Carcinogen-Binding Protein A (38), Z-protein (39)] also binds a number of ligands with lipophilic moieties including tetrapyrroles, but its most important physiological ligand is believed to be long chain acyl CoA (38, 40). It can be isolated in three forms with different isoelectric points (38, 41). Form I (pI 7.6) is fatty acid free (41) and when used in studies of haem and bilirubin binding is shown to bind both these ligands to give spectral changes very similar to those already reported for

ligandin (22, 23). The most obvious difference between ligandin and fatty acid binding protein being in the circular dichroism spectra of the bilirubin-bound proteins which are opposite in sign (22). Fatty acid binding protein has been reported to have a K_{assoc} for palmitoyl CoA, as determined by equilibrium dialysis, of 6.7×10^6 M^{-1} and for haem and bilirubin, as determined spectroscopically, of 4×10^6 M^{-1} and 10^7 M^{-1} respectively (38, 22, 23). However recent evidence suggests that, although all three isoelectric forms occur intracellularly, Form I is a minor component and forms II (pI 6.0) and III (pI 5.0) are major components and occupied by fatty acid derivatives. (41). Available evidence suggests that all three forms have the same primary structure (38, 41). Amino acid sequences have been obtained by the techniques of protein chemistry (41) and deduced from the nucleotide sequence of cDNA clones (42, 43). The secondary structure of Form I as determined by circular dichroism shows the presence of very little α-structure (44). It is possible that conformational changes occur on binding: they would account for the large differences in isoelectric point between the fatty acid free form (I) and the fatty acid bound forms (II and III). Changes in isoelectric point believed to be associated with conformational changes, occur with the binding of fatty acid by serum albumin (45).

4. The Physiological Significance of these Binding Proteins

Any plasma protein which binds a ligand *in vivo* transports it by virtue of the pumping action of the heart. Whether or not a particular protein which binds a ligand *in vitro* also binds it *in vivo* depends on its binding activity in relation to that of other candidate proteins, binding activity being the product of binding capacity, (i.e. the concentration of binding sites), and binding affinity (31). The transport of bilirubin in plasma is clear cut. There is only one real candidate protein, namely serum albumin. Its concentration in plasma is approximately 0.62 mM and its binding affinity is approx. 10^8 M^{-1}. Hemopexin also binds bilirubin, but its K_{assoc} is lower than that of albumin and its concentration is only 15 µM. It therefore has so much less binding capacity than albumin that its contribution to plasma bilirubin transport is negligible. All serum albumins so far studied have affinities for bilirubin of a similar order. The situation with respect to haem resolves differently. The binding affinity of hemopexin is 5-6 orders of magnitude greater than that of albumin which more than compensates for the fact that its concentration is 40 times less. Plasma haem is therefore transported as haem-hemopexin.

Intracellular transport is somewhat more complex. In a system such as the cytoplasm of the hepatocyte, in which it is assumed there is no mechanical movement, net transport should be by diffusion down a concentration gradient. In the hepatocyte there is a large lipid phase in the form of the phospholipid bilayer of the endoplasmic reticulum and other membrane structures. Proteins such as ligandin and fatty acid binding proteins, which together provide a concentration of lipophile binding sites in the cytosol of approximately 1 mM, greatly facilitate the diffusion of lipophiles in the aqueous phase and, in so doing, break down the barrier for the passage of lipophiles from one part of the membrane system to another, which is created by the aqueous phase. The result is greatly to increase the speed of transport through the cell as a whole (46). The construction of mathematical models of the hepatocyte has suggested that the binding affinity of a lipophilic compound (e.g. bilirubin or haem) for binding proteins such as ligandin and fatty acid binding protein is a much more important determinant of rate of transport through the cell than partition into the membrane (31). Therefore, the present evidence favours the view that ligandin and fatty acid binding protein have an important intracellular transport function.

Gradients exist for bilirubin transport between the plasma membrane and the glucuronidating enzymes of the endoplasmic reticulum; for haem transport between its site of entry into the cell bound to hemopexin and its conversion to biliverdin by haem oxygenase in the endoplasmic reticulum; and haem transport from hepatic mitochondria where it is synthesized and its utilization by apocytochrome P_{450} in the endoplasmic reticulum. The question remains as to which protein transports which ligand. As indicated above both ligandin and fatty acid binding protein occur at high concentration. GSH transferase subunit 1 is approximately 0.6 mM and fatty acid binding protein, (Forms I, II and III combined) is apparently 0.4 mM. Purified ligandin has a higher binding affinity for haem in vitro, and apparently also in vivo since in experiments in which the binding of haem synthesized intrahepatically from labelled 5-amino-laevulinic acid is studied, labelled crystalline haem can be recovered from ligandin, but not from fatty acid binding protein. Ligandin also induces the release of haem or its analogue cobalt deuteroporphyrin from isolated mitochondria (47). Thus ligandin is the hepatic intracellular haem transporting protein in vivo: it has the dual role of releasing haem from the mitochondria and also transporting it to its site of utilization. The situation is more complex with respect to bilirubin which has a higher affinity for purified fatty acid binding protein than purified ligandin.   However, it has been mentioned above that there is evidence that unfractionated ligandin in the presence of GSH has much more affinity than purified ligandin.   An indication of the relative importance of these two proteins _in vivo_ can be obtained from the separation of soluble supernatant from the Gunn rat in the presence of GSH.   Twenty three per cent of the bilirubin was fatty acid binding protein-bound (29).   Thus both ligandin and fatty acid binding protein appear to be involved in the transport of bilirubin _in vivo_.   Ligandin has also been shown to bind bilirubin glucuronides (48) and may also transport conjugated bilirubin to the bile canaliculus for excretion.

5.   Conclusions

With one exception all transport described involves proteins with K_{assoc} in the region of 10^7-10^8 M^{-1} and therefore appreciable dissociation rate constants.   The necessary high levels of bound ligand and negligible concentrations of free ligand are achieved by the presence of high concentrations of binding proteins.   Transport of this sort occurs extracellularly for bilirubin and intracellularly for both bilirubin and haem.   The one exception concerns the transport of extracellular haem. In this case, namely that of haem-hemopexin, the haem is bound very strongly and the dissociation rate constant is negligible.   It would seem that extracellular haem is a hazard from which the body requires particularly effective protection.

The major port of exit for haem and its metabolites is the liver (2, 7, 17).   An area where knowledge is limited is that involving the transfer of the ligand from the plasma to the hepatic cytosol.   For example the hepatic uptake of bilirubin is **more efficient** than the experimentally determined dissociation rate constant of bilirubin from serum albumin and the sinusoidal transport time would suggest.   To accommodate this finding previous hypotheses that free bilirubin diffuses through the lipid phase of the membrane or is taken across by a specific membrane carrier have now been joined by a model whereby bilirubin-albumin makes direct contact with the hepatocellular plasma membrane such that dissociation rate constants in free solution are no longer limiting (49). There is less doubt about the mechanism of the entry of free haem into the liver. Double labelling experiments indicate that both the haem and the hemopexin enter the liver together perhaps by endocytosis involving a specific hemopexin receptor. There is evidence that some hemopexin turnover occurs and some is returned to the plasma (3). The transport of haem to the liver is therefore a costly process in energy terms.

Finally there is the question of the mechanism of the release of haem from mitochondria. We have shown that serum albumin binds haem in a similar fashion to ligandin but this does not enable albumin to release haem which after biosynthesis is accumulated in mitochondria against a concentration gradient (47). That ligandin does release haem suggests that in addition to the haem binding site, there may be on the surface of ligandin, but not albumin, a site which is essential for the interaction of ligandin with mitochondrial membranes and which enables the release of haem.

It should be noted that as happens so often in biochemistry and bio-physics species differences are recorded, but information does not exist which enables a fundamental sequence to be followed throughout in one species. For example most information concerning serum albumin comes from human and bovine sources, that concerning hemopexins is mostly from the rabbit and the human and the only detailed information on the intracellular binding proteins is from the rat. Sufficient is known about rat serum albumin and hemopexin to suggest that the sequences of extracellular and intracellular transport indicated here apply to the rat. Whether or not a similar mechanism of intracellular transport could occur in other species remains to be determined. The existence of ligandin (50) and something like fatty acid binding protein (51) in human liver suggests that it may do.

Acknowledgment

The author is a Life Fellow of the Cancer Research Campaign which he thanks for its generous support.

REFERENCES

1. Harris, J.W., Kellermeyer, R.W.: The Red Cell. Production Metabolism, Destruction: Normal and Abnormal. Harvard University Press, Cambridge, MA. (1970).

2. Gollan, J.L., Schmid, R.: in "Progress in Liver Diseases, vol VII", (Popper, H., Schaffner, F. eds.) pp 261-283, Grune & Stratton, New York (1982).

3. Muller-Eberhard, U.: In "Transport by Proteins", (Blauer, G., Sund, H. eds.) pp 295-310, Walter de Gruyter, Berlin (1978).

4. Doss, M.: in "Progress in Liver Diseases, vol VII", (Popper, H., Schaffner, F. eds.) pp 573-597, Grune & Stratton, New York (1982).

5. Spikes, J.D.: Ann. N.Y. Acad. Sci.: 244, 496-508 (1974).

6. Beaven, G.H., Chen, S.-H., D'Albis, A., Gratzer, W.B.: Eur. J. Biochem. 41, 539-546 (1974).

7. Muller-Eberhard, U., Morgan, W.T.: Ann. N.Y. Acad. Sci. 244, 624-650 (1974).

8. Bearden, A.J., Morgan, W.T., Muller-Eberhard, U.: Biochem. Biophys. Res. Commun. 61, 265-272 (1974).

9. Bunn, H.F., Jandl, J.H.: J. Biol. Chem. 243, 465-475 (1968).

10. Blauer, G., King, T.E.: J. Biol. Chem. 245, 372-381 (1970).

11. Morgan, W.T., Muller-Eberhard, U., Lamola, A.: Biochim. Biophys. Acta 532, 57-64 (1978).

12. Beaven, G.H., D'Albis, A., Gratzer, W.B.: Eur. J. Biochem. 33, 500-510 (1970).

13. Squire, B.G., Moser, P., O'Konski, C.T.: Biochemistry 7, 4261-4272 (1968).

14. Brown, J.R.: Fed. Proc. 34, 591 (1975).

15. Reed, R.G., Feldhof, R.C., Clute, O.L., Peters, T. Jr: Biochemistry 14, 4578-4583 (1975).

16. Jacobsen, C.: Biochem. J. 171, 453-459 (1978).

17. Lamola, A.A., Asher, I., Muller-Eberhard, U., Poh-Fitzpatrick, M.: Biochem. J. 196, 693-689 (1981).

18. Hrkal, Z., Kalousek, I., Vodrazka, Z.: Eur. J. Biochem. 43, 73-78 (1974).

19. Morgan, W.T., Sutor, R.P., Muller-Eberhard, U.: Biochim. Biophys. Acta 434, 311-323 (1976).

20. Hrkal, Z., Suttner, J., Vodrazka, Z.: Studia Biophys. 63, 55-58 (1977).

21. Litwack, G., Ketterer, B., Arias, I.M.: Nature 234, 466-467.

22. Tipping, E., Ketterer, B., Christodoulides, L., Enderby, G.: Biochem. J. 157, 211-216 (1976).

23. Tipping, E., Ketterer, B., Christodoulides, L., Enderby, G.: Biochem. J. 157, 461-467 (1976).

24. Beale, D., Ketterer, B., Carne, T., Meyer, D., Taylor, J.B.: Biochem. Soc. Trans. 10, 359-360 (1982).

25. Jakoby, W.B., Ketterer, B., Mannervik, B.: Biochem. Pharmacol. 33, 2539-2540 (1984).

26. Prohaska, J.R., Ganther, H.E.: J. Neurochem. 27, 1379-1387 (1976).

27. Meyer, D.J., Christodoulides, L.G., Tan, K.H., Ketterer, B.: FEBS Lett. 173, 327-330.

28. Bhargava, M.M., Ohmi, N., Listowsky, I., Arias, I.M.: J. Biol. Chem. 255, 718-723 (1980).

29. Meuwissen, J.A.T.P., Zeegers, M., Srai, K.S., Ketterer, B.: Biochem. Soc. Trans. 5, 1404-1407 (1977).

30. Kamisaka, K., Listowsky, I., Gatmaitan, Z., Arias, I.M.: Biochemistry, 14, 2175-2180 (1975).

31. Meuwissen, J.A.T.P., Ketterer, B., Heirwegh, K.P.M.: In "Chemistry and Physiology of Bile Pigments" (Berk, P.O., Berlin, N.I. eds.) pp 323-337. U.S. Department of Health, Education and Welfare, Washington, D.C. (1977).

32. Tipping, E., Ketterer, B., Koskelo, P.: Biochem. J. 169, 509-516 (1978).

33. Kalinyak, J.E., Taylor, J.M.: J. Biol. Chem. 257, 523-530 (1982).

34. Tu, C.-P.D., Weiss, M.J., Karakawa, W., Reddy, C.C.: Nucleic Acids Res. 10, 5407-5419 (1982).

35. Taylor, J.B., Craig, R.K., Beale, D., Ketterer, B.: Biochem. J. 219, 223-231 (1984).

36. Carne, T., Tipping, E., Ketterer, B.: Biochem. J. 177, 433-439 (1979).

37. Ockner, R.K., Manning, J.A., Kane, J.P.: J. Biochem. 257, 7872-7878 (1982).

38. Ketterer, B., Tipping, E., Hackney, J.F., Beale, D.: Biochem. J. 155, 511-521 (1976).

39. Trulzsch, D., Arias, I.M.: Arch. Biochem. Biophys. 209, 433-440 (1981).

40. Mishkin, S., Stein, L., Fleischner, G., Gatmaitan, Z., Arias, I.M.: Am. J. Physiol. 228, 1634-1640 (1975).

41. Takahashi, K., Odani, S., Ono, T.: Eur. J. Biochem. 136, 589-601 (1983).

42. Gordon, J.I., Alpers, D.H., Ockner, R.K., Strauss, A.W.: J. Biol. Chem. 258, 3356-3363 (1983).

43. McTigue, J., Taylor, J.B., Craig, R.K., Ketterer, B.: unpublished information.

44. Ketterer, B., Carne, T., Tipping, E.: in "Transport by Proteins" (Blauer, G., Sund, H., eds.) pp 79-94. Walter de Gruyter, Berlin (1978).

45. Basu, S.P., Rao, N.S., Hartsuck, J.A.: Biochim. Biophys. Acta 533, 66-73 (1978).

46. Tipping, E., Ketterer, B.: Biochem. J. 195, 441-452 (1981).

47. Husby, P., Srai, S.K.S., Ketterer, B., Romslo, I.: Biochem. Biophys. Res. Commun. 100, 651-659 (1981).

48. Wolkoff, A.W., Ketley, J.N., Waggoner, J.G., Berk, P.D., Jakoby, W.B.: J. Clin. Invest. 61, 142-149 (1978).

49. Weisiger, R.A., Gollan, J.L., Ockner, R.K.: in "Progress in Liver Diseases vol. VII" (Popper, H., Schaffner, F. eds.) pp 71-85, Grune and Stratton, New York (1982).

50. Kamisaka, K., Habig, W.H., Ketley, J.N., Arias, I.M., Jakoby, W.B.: Eur. J. Biochem. 60, 153-161.

51. Kornguth, M.L., Monson, R.K., Kunin , C.M.: Arch. Biochem. Biophys. 176, 339-343 (1976).

Received August 12, 1984

Discussion

Beychok: You said that heme does not bind to the bilirubin site of serum albumin. Does bilirubin bind to the heme site?

Ketterer: There is evidence that bilirubin binding to a weaker secondary site might compete with haem.

Beychok: Are the stoichiometries of bilirubin binding to ligandin and FA binding protein determined? Are they 1:1?

Ketterer: The stoichiometry of high affinity binding of bilirubin to FA binding protein is 1:1. Values of 1.7 to 1 and 2 to 1 have been obtained for the stoichiometry of binding of bilirubin to purified homodimer gluta-thione transferase 1-1, i.e. the ligandin which contains two high affinity binding subunits per mole. Exact stoichiometry may depend on the degree to which native structure survives purification.

Beychok: Is heme present at levels of $10^{-5}M$?

Ketterer: The extracellular and intracellular concentration of uncom-plexed haem are not known to me. Since hemopexin concentrations vary from 5 to 15 µM it is reasonable to assume that plasma levels are $10^{-6}M$ or less.

Blauer: In the absorption spectrum of bilirubin bound to albumin, the observed spectral shift is very likely due to exciton interactions between the two dipyrromethenone halves, each of the two split bands having different intensities. The CD spectrum of bilirubin-human serum albumin is of opposite sign to the bilirubin-ligandin complex at excess protein and pH 7. This has been explained by an opposite chirality in bilirubin binding being produced by free rotation around the bilirubin methylene bond during fixation of the molecule to the binding site [see G. Blauer, Isr.J.Chem. 23, 205 (1983) and Ref. 25 cited therein].

Ketterer: This could well be so. It is a fair assumption that binding immobilizes bilirubin in such a manner that the two exciton bands make different contributions to absorption and have opposing ellipticities. This is a very reasonable extension of earlier conclusions you drew as a basis of your CD studies of bilirubin bound albumin.

Buchler: You mentioned that in albumin the binding of bilirubin caused a bathochromic shift of the optical spectrum. This could inter alia be caused by complexation to a metal ion. Which metal ions can be found in albumin?

Ketterer: There are single binding sites for Cu^{2+}, Ni^{2+} and Mn^{2+}. Other metals, e.g. Co, Zn, Cd, are bound weakly.

Buchler: When hemin is transported by hemopexin, there is a low spin ferric ESR signal, as you showed. This implies at least binding to one nitrogenous base, if not two. Are there histidines in the protein chain? Does the ferrous system bind carbon monoxide?

Ketterer: There is considerable supporting data for the view that binding is to imidazole (there are in fact 16 histidines per mole of hemopexin). For example, magnetic circular dichroism indicates that a hexacoordinate bis-imidazole complex is involved. Yes, ferrohaem-hemopexin binds CO. The rate is low compared with CO-binding to pentacoordinate haemoproteins, e.g. haemoglobin, cytochrome P_{450} etc. Cytochrome b_5 which is also hexacoordinated through imidazoles, unlike hemopexin, does not appear to bind CO.

Woody: You mentioned that as far as is known, the hemopexin-heme complex does not undergo redox reactions. I wonder if intracellular reduction might not provide a mechanism for releasing the heme?

Ketterer: No. Haem hemopexin has a low standard electrode potential (hemopexin has $E^0=0.064$ V compared with 0.175 V for haemoglobin and 0.26 V for cytochrome c) and should not reduce under physiological conditions known to exist in the hepatocyte. Furthermore, even when it is reduced with dithionite its reaction with CO is slow suggesting that the coordinate linkages are still strong. Denaturation, perhaps even proteolysis of the hemopexin may be necessary to release the haem.

Fuhrhop: How easy is it to get hemopexin?

Ketterer: Hemopexin is readily purified from serum by the successive use of polyethyleneglycol precipitation followed by DEAE-Sepharose CL-6B chromatography. This gives a fraction containing mostly hemopexin and trans-

ferrin. The former can be separated from the latter by affinity chromatography using wheat germ lectin as the affinity ligand (Vretblad & Hjorth,
Biochem.J. 167, 759-764 (1977).

Lamola: You have told us about the pools of tetrapyrroles in blood and
inside liver cells. Can you review what is known about how the various porphyrins and bile pigments go from blood to liver cell cytosol?

Ketterer: Bilirubin enters the cell by an energy independent process.
Bilirubin may pass through the phospholipid bilayer by transmembrane diffusion or, as suggested by some workers, there may be a specific membrane
carrier. Haem, on the other hand, appears to require a process which is
quite costly with regard to energy. It is carried into the cell with hemopexin, presumably via a receptor. It is not known how the haem is released.
Haem-hemopexin uptake involves some hemopexin degradation; but it is not
known whether all the haem-hemopexin endocytosed is involved. Porphyrins
are also taken up. If there is not a specific bilirubin carrier in the
plasma membrane there is apparently an organic anion carrier. The more
rigid a porphyrin, amphipathic substances such as the porphyrins may use
this method of entry in the same manner as bromosulphophthalein is believed
to do so.

SECTION III
BILE PIGMENTS

*Chairmen: R.P.F. Gregory, H. Sund
and R.W. Woody*

LIGHT ABSORPTION OF BILATRIENES-ABC
AND 2,3-DIHYDROBILATRIENES-ABC

Heinz Falk

Institut für Analytische, Organische und Physikalische Chemie
der Johannes-Kepler-Universität, A-4040 Linz, Austria

Introduction

In naturally occurring systems of low abundance, light absorption and associated phenomena, like chiroptical properties or
light emission, sometimes are the only means to investigate
structural details of the chromophore and its vicinity. This
is still the case with phytochrome, the ubiquitous photoreceptor of photomorphogenesis in the plant kingdom. It is now
firmly established that the chromophoric unit of this system
is a 2,3-dihydrobilatriene-abc covalently bound to a protein
(1). Therefore it seemed to be appropriate to investigate and
summarise the influences of internal and external structural
variables on the light absorption properties of model compounds of the bilatriene-abc and 2,3-dihydrobilatriene-abc
series systematically. A systematic view of this kind could
be a guidance to acquire and discuss correlations of structural features of the phytochrome chromophore in its thermal
and photodynamic forms and their light absorption properties.

Results and Discussion

The most prominent influences on the absorption spectra of
verdinoid bile pigments may be exerted by the length of the
conjugated π-electron system, the electronic properties of
substituents, tautomerism, configuration, certain conforma-

tions, protonation, deprotonation, chelation with metal ions,
homo - and hetero - association, external point charges, and
hydrophobic or polar interactions with the chromophore's "environment". It should be noted that most of these influencing
factors belong to the hierarchy of structural aspects which
is correlated to the energies necessary to change them, i.e.
constitution - tautomerism - configuration - conformation -
non bonded interactions. This hierarchical order with respect
to energy should as well be followed in the course of structural investigations, as e.g. one should not draw conclusions
about conformation before one is sure about the constitution
of a compound (2). In the following the correlations of the
factors given above to the light absorption of verdinoid bile
pigments representing model compounds for the phytochrome
chromophore are discussed:

1. The length of the conjugated π-electron system is determining the absorption spectra of given linear polypyrroles profoundly. This can be derived immediatly comparing the "main"
absorption maxima of several typical linear polypyrrolic systems given in table 1. Elongation of the system by a pyrrolic ring yields a bathochromic shift of the long wavelength absorption band of nearly 100 nm. Additional double
bonds or hydrogenation of double bonds located outside the
primary conjugation path have only minor effects on the wavelength of the absorption bands: Compare the shifts on going
from aetiobiliverdin-IV-γ to biliverdindimethylester or a
2,3-dihydrobilatriene-abc in table 1.

It may be noted that the general features of the absorption
spectra may easily be understood on the basis of a qualitative
"electron in a box" model (4): More or less "one-dimensional"
systems are characterised by one main absorption band - on
becoming "two dimensional" a two band system is observed.

Table 1: Approximate absorption maxima (nm) of typical linear
         polypyrroles in chloroform solutions

polypyrrole	absorption	maxima	(nm)
bilenes-a, bilenes-b, pyrromethenes, pyrromethe- nones, biladienes-ac		420	
biladienes-ab	320	550	
bilatrienes-abc (e.g. aetio- biliverdin-IV-γ)	365	630	
biliverdindimethylester	380	660	
2,3-dihydrobiliverdines-abc	350	590	
"pentapyrrin" [1](3)	320	450	730

[1]

With regard to the changes brought about for the absorption
spectra of 2,3-dihydrobilatrienes-abc as model systems for the
phytochrome chromophore, addition reactions in positions 4,5
and 9,10 could be documented (5,6). These additions result in
a change of the chromophoric system from the verdinoid to the
violinoid and the rubinoid type respectively. Although such
additions are not very favoured under bimolecular reaction
conditions, they may become candidates for thermal reaction
sequences under intramolecular conditions, which may be present within a protein - chromophore - system (6).

2. The influence of <u>electronic properties of a substituent</u> attached to a verdinoid pigment is less documented. The main problem here is to separate electronic effects from an extension of conjugation in case of unsaturated substituents on one hand, and from steric interactions which may exert an influence on conformational equilibria on the other hand.

As an example (7), 10-cyano-aetiobiliverdin-IV-γ shows a long wavelength band at 726 nm. This is a bathochromic shift compared to the unsubstituted system of about eighty nanometers. As cross conjugation of this kind does not yield a shift in this order of magnitude and steric interactions should lead via enhanced torsion at the 9,10-single bond to a hypsochromic shift, we may assign most of the observed shift to the electronic properties of the cyano group. However such strong influences will be of minor interest for molecules of biological importance and we are rather safe to ignore this influence in discussions related to the phytochrome chromophore.

3. There are two kinds of <u>tautomerism</u> one has to take into account for verdinoid bile pigments:

The first one may be termed <u>methene tautomerism</u> - the proton within this partial structure of bilatrienes-abc may be situated on N-22 or N-23. The position of this proton can either be fixed as it is the case with 2,3-dihydrobilatrienes-abc like [2], or a rapid dynamic process leads to a mixture of the two tautomeric species (8). The latter is observed for bilatrienes-abc. This type of tautomerism is of minor influence on the absorption spectra: Shifts in the order of ten nanometers may be associated with the transfer of the proton as judged from PPP-calculations (9,10). Experimentally one is not able to isolate such a shift from more prominent shifts stemming from conformational changes induced by shifting the tautomeric equilibrium.

The second kind is the <u>lactam - lactim tautomerism</u> of the terminal rings. From a series of studies using coupled protonation equilibria and spectroscopic techniques it is now beyond
doubt that the lactam form predominates the lactim form for all
bile pigments investigated so far (11, 12) by several orders of
magnitude. By fixation of the lactim form as its methyl ether
it has been possible to gain experimental information on the
spectroscopic shifts associated with this kind of proton transfer even for the 2,3-dihydrobilatrienes-abc: The absorption
bands of the 2,3-dihydrobilatriene-abc [2] at 346/586 nm are
shifted to 350/610 nm on forming the lactim ether at ring D [3]
and to 346/671 nm at ring A [4] (12).

As in case of aetiobiliverdin-IV-γ (13) a minor bathochromic
shift of about twenty nanometers is observed as well, this
seems to be typical of the lactim form of the unsaturated
ring. However it should be noted that the hydrogen bonding
scheme becomes more efficient in this type of compounds and
therefore a conformational influence will be contained in this

shift.   On forming the lactim ether of the saturated lactam
ring (compound [4]), a dramatic bathochromic shift of about
eighty nanometers is observed.   Here the elongation of the
conjugated π–system together with a possible flattening of
the helical molecule due to a more efficient hydrogen bonding
system are mainly responsible for the spectroscopic shift.

4.   The exocyclic double bonds of verdinoid bile pigments give
rise to diastereomerism.   The influence of configuration at
these bonds per se is neglegible.   However different steric requirements in the vicinity of the double bond, which is responsible for the diastereomers, will induce in these isomers different dihedral angles at the exocyclic single bond adjacent to
the double bond.   This difference in conformation will then be
followed by a hypsochromic shift in the order of forty nanometers on going from a diastereomer of configuration (4Z,9Z,15Z)
to one of (4Z,9Z,15E) in case of a 2,3-dihydrobilatriene-abc
(14). This shift is caused by stronger twisting of the single
bond adjacent to the isomerised double bond, i.e. 14 - 15.

5.   A very prominent influence on the absorption spectra of verdinoid bile pigments will be exerted by the conformations at
the exocyclic single bonds.   Comparison of the spectra of
aetiobiliverdin-IV-γ and its 5-nitro derivative points to a
shift corresponding to the one observed on going from bilatrienes-abc to biladienes-ab (table 1).   As has been shown (15),
the 5-nitro group in such derivatives leads to a torsion of the
5-6 - single bond of about ninety degrees, thereby decoupling
the terminal lactam ring from the conjugation of the π-system.

Another aspect of spectral influence can be derived comparing
the spectra of aetiobiliverdin-IV-γ, its 21,24-methano bridged derivative (16) and a phorcabilinester (17).   The fixed
stretched conformation of the latter yields an inversion of
relative intensities of the two main absorption bands compared
to the fixed cyclic derivative.

Moreover all of the factors discussed so far obviously give
rise to an indirect influence on the absorption spectra. They
tend to induce conformational changes or, at least, they are
accompanied by such changes that critically influence the absorption spectra.

6. Protonation of the basic pyrrolenine nitrogen of verdinoid
bile pigments results in a bathochromic shift of about thirty
nanometers which is accompanied by an intensity enhancement of
the long wavelength absorption band. This shift is comparable
for bilatrienes-abc and 2,3-dihydrobilatrienes-abc (12). Although the prominent part of this shift is due to the electronic modification of the chromophore, as can be judged from
the protonation of a N_{21}-N_{24}-methylene bridged aetiobiliverdin-IV-γ (16), shifts stemming from a slightly changed conformation cannot be excluded.

7. Of the three possible deprotonation sites at acidic centers
in verdinoid bile pigments only the one of the saturated lactam
moiety of 2,3-dihydrobilatrienes-abc is of interest: It is
only this NH which may be abstracted under physiological conditions, i.e. its pK_a value is measured to be below thirteen
(18). The absorption spectroscopic consequence of deprotonation of the saturated lactam fragment of 2,3-dihydrobilatri-
enes-abc is a very pronounced bathochromic shift of the long
wavelength absorption band. It is found to be nearly 180 nanometers. At the same time it is intensified and the short
wavelength absorption band is split into two bands and shifted bathochromically by more than fifty nanometers (18). So
deprotonation of 2,3-dihydrobilatrienes by far is the most
prominent influence on the absorption spectra of verdinoid
systems observed so far. Although most of the shift might
come from the change in the electronic structure of the chromophore, a certain amount of it may be due to a somewhat flattened helix initiated by a better hydrogen bonding situation
within the anion. Again a conformational change may be involved to a certain extent.

8. On chelation of the well suited ligands which are represented by verdinoid bile pigments (19), a spectral pattern resembling that of deprotonation (18) is observed. The bathochromic
shift is much less pronounced and is somewhat dependent on the
cation. Using zinc ion and a 2,3-dihydrobilatriene-abc it
amounts to approximately eighty nanometers. As the helix of
the free chromophore will be flattened by chelation, a conformational shift equivalent will be contained in the observed
shift value.

9. Association may be a very strong influence upon the absorption spectra of bile pigments:

The self - association (mostly dimerisation) of rubinoid systems has become a well studied field (20, 21). With verdinoid pigments from the bilatriene-abc and 2,3-dihydrobilatri-
ene-abc series no evidence of such processes could be observed
so far (22) using differential vapour pressure osmometry and
Lambert - Beer studies in a concentration range of nearly five
orders of magnitude and between 80 and 300 K.

However in case of hetero - association with electron deficient
molecules an influence on the absorption spectra can be documented. E.g. addition of tetracyanoethylene to solutions of aetiobiliverdin-IV-γ or of 2,3-dihydrobilatrienes-abc in dichlorome-
thane results in a bathochromically shifted two - band absorption in the red region and a rather small shift of the short
wavelength absorption band. Obviously a charge transfer band
is added to the absorbing system which possibly is changed
somewhat in its conformational situation by the interaction of
the associating species.

10. The influence of external non conjugated point charges on
the absorption spectra of verdinoid bile pigments has been suggested on the basis of semiempirical MO calculations (23) and is

paralleled by the discussions of this aspect in the fields of
the visual pigments and of chlorophyll - protein complexes.
We were able to document a bathochromic shift of about thirty
nanometers in case of protonation of an amino group covalently
attached to a bilatriene-abc (in position 8 at a distance of
eight bonds (24)).

However only minor spectroscopic differences were observed for
2,3-dihydrobilatrienes-abc bearing anions or cations analogous-
ly attached in position 3 or 12 (25). To our impression the
influence of an external point charge on the absorption spectra via interaction of this charge with excited states or the
ground state is of minor relevance. It is suspected that even
the shift observed in one case (24) is due mainly to conformational changes induced by interactions between the charge and
partial permanent dipole moments of the chromophore.

11. Another more or less indirect influence on the absorption
spectra of verdinoid bile pigments may result from hydrophobic
or polar interactions between the chromophore and its "environment":

To test for the influence of hydrophobic interactions a cholesteryl moiety is attached to position 3 of a 2,3-dihydrobila-
triene-abc via an acetic ester linkage to yield exclusively
one of the two possible diastereomers ([5],R = cholesteryl).
Comparison with the corresponding methylester derivative ([5],
R = OCH_3) reveiled no significant spectral changes. By application of an advanced NMR method (NOESY), a "compact" conformation of this compound was deduced: The chromophore and cholesteryl residues are placed face to face, stabilised by the hydrophobic interaction of the two moieties. However the interaction energy is too low to discriminate markedly between the
two enantiomeric helix conformations of the chromophore, as is
deduced from circular dichroism measurements (26), or even

to yield shifts of absorption bands due to conformational changes.

[5]

To create a model for <u>polar interactions</u>, amino acid residues
were attached to a 2,3-dihydrobilatriene-abc by means of a linkage between the amino group of an amino acid and an acetic acid
ligand in position 3 of the chromophore ([5],R = amino acid residue). It may be noted that in such derivatives the number of
atoms between the chiral center of the amino acid and position
3 is the same as in phytochromopeptides. Both diastereomers
were obtained in approximately equal amounts and could be separated by chromatography. Small but distinct spectroscopic
shifts are observed between the diastereomers and the corresponding underivatised chromophore, respectively. However the chiroptical properties of the two main absorption bands reach values observed for native systems. E.g. $\Delta\varepsilon$ values for the long-
and the short - wavelength bands of about fifty and eighty, respectively, are obtained for [5],R = NHCH(COO-tert.-butyl)(CH$_2$)$_4$
NHCO-tert.-butyl). For the two diastereomers the circular dichroism curves resemble mirror images.

By applying a simplified Wood - Fickett - Kirkwood treatment to
the temperature dependent chiroptical data it can be concluded,
that in these derivatives up to 90% enantiomeric preponderance
of one helix enantiomer can be achieved (27). An approximately
linear dependence of the chiroptical property on the inverse of
the solvents dielectricity constants before showing a saturation
phenomenon points to an interaction between partial dipole moments of the amino acid moiety and of the chromophore as the

main reason. Steric interactions obviously play only a modulati-
ve role as can be derived from comparison of various amino acid
derivatives (e.g. alanine ethylester, tryptophane ethylester,
etc.). It follows that polar interactions between the permanent
dipole moments of chromophore and the peptide backbone of a bi-
liproteide may result in a stabilisation of a certain conformer
by several kilojoules per mole. This is at least sufficient to
discriminate between enantiomeric arrangements of the chromophore, as observed for natural systems or associates between bile
pigments and proteins (28), but could as well be the main source
of fixation of a certain chromophore conformation in a native
biliprotein. The dominant influence of polar interactions on
the absorption spectra of 2,3-dihydrobilatrienes-abc is due to
conformational changes induced by these interactions.

To summarise, a whole series of factors exist which may profoundly influence the absorption spectra of verdinoid bile pigments. These factors themselves will influence the absorption
spectra of bilatrienes-abc and 2,3-dihydrobilatrienes-abc to a
certain degree - but at the same time they will also change
the conformational situation of the rather flexible bile pigments (29). In turn this will be followed by spectral shifts
accordingly.

As the absorption spectrum (wavelengths, intensities, polarisa-
tions) and its associated phenomena like chiroptical properties
and emission are only single observables, extreme care has to be
exercised in correlating them with structural details which are
numerous and have comparable influences on these observables.
Therefore absorption spectroscopic informations should be used
only as single puzzle stones which have to be complemented by
pieces of informations derived from other methods. However
careful examination of model systems may result in a better understanding of absorption spectroscopic measurements of natural
systems and help in their sensible and meaningful interpretation.

References

1. For an overview see the collection of reviews in: "Photo-morphogenesis", Encyclop.Plant Physiol. 16 (1983).

2. Falk H. in: "Bilirubin, Bile Pigments and Jaundice" (D. Ostrow, Ed.), M.Dekker, N.Y., in press.

3. Falk H., Flödl H.: to be published elsewhere in detail.

4. Heilbronner E., Bock H.: Das HMO-Modell und seine Anwendung, Verlag Chemie, Weinheim, 1970.

5. Falk H., Müller N., Schlederer T.: Monatsh.Chem. 111, 159-175 (1980).

6. Falk H., Zrunek U.: Monatsh.Chem. 115, 101-111 (1984).

7. Falk H., Schlederer T.: Monatsh.Chem. 109, 1013-1015 (1978).

8. Falk H., Grubmayr K., Magauer K., Müller N., Zrunek U.: Isr. J.Chem., 23, 187-194 (1983).

9. Falk H., Gergely S., Hofer O.: Monatsh.Chem. 105, 853-862 (1974). For an extensive PPP investigation of bilatrienes-abc see: Falk H., Höllbacher G.: Monatsh.Chem. 109, 1429-1449 (1978).

10. Falk H., Hubauer E., Müller N.,: to be published elsewhere in detail.

11. Falk H., Gergely S., Grubmayr K., Hofer O.: Ann.Chem. 1977, 565-581; Falk H., Schlederer T.: Ann.Chem. 1979, 1560-1570.

12. Falk H., Zrunek U., Monatsh.Chem. 114. 983-998 (1983).

13. Falk H., Grubmayr K., Thirring K.: Z.Naturforsch. 33b, 924-931 (1978). For isolation and characterisation of the aetio-biliverdin-IV-γ-monolactim see: Falk H., Schlederer T.: Ann.Chem. 1979, 1560-1570; so far, all experiments to iso-late a lactim form of 2,3-dihydrobilatrienes-abc failed (12).

14. Falk H., Grubmayr K., Kapl G., Müller N., Zrunek U., Mo-natsh.Chem. 114, 753-771 (1983);
For the first documentation of (Z)/(E) diastereomers of bilatrienes-abc see: Falk H., Grubmayr K.: Angew.Chem. 89, 487-488 (1977).

15. Bonfiglio J.V., Bonnett R., Buckley D.G., Hamzetash D., Trotter J.: J.Chem.Soc., Perkin I, 1982, 1291-1292 (1982).

16. Falk H., Thirring K.: Tetrahedron 37, 761-766 (1981).

17. Bois-Choussy M., Barbier M., Heterocycles 9, 677-680 (1978). For a recent example of the spectral properties of a stret-ched linear tetrapyrrole compare: Leumann C., Eschenmoser A.: J.Chem.Soc.Chem.Commun. 1984, 583-585. The intensity cri-terion for stretched vs. compact conformations of conjuga-ted linear tetrapyrrols has been proposed from a HMO inve-stigation: Burke J.M., Pratt D.C., Moscowitz A.: Biochemi-stry 11, 4025-4031 (1972).

18. Falk H., Zrunek U.: Monatsh.Chem. 114, 1107-1123 (1983);
    Falk H., Wolschann P., Zrunek U.: Monatsh.Chem. 115, 243-
    249 (1984).

19. Eichinger D., Falk H.: Monatsh.Chem. 113, 355-364 (1982).

20. Falk H., Müller N.: Monatsh.Chem. 113, 111-112 (1982).

21. Trull F.R., Ma J., Landen G.L., Lightner D.A.: Isr.J.Chem.
    23, 211-218 (1983).

22. Falk H., Schlederer T., Wolschann P.: Monatsh.Chem. 112,
    199-207 (1981); Falk H., Gsaller H.W.: to be published.

23. Suzuki H., Sugimoto T., Ishikawa K.: J.Phys.Soc.Jap. 38,
    1110-1153 (1975); Sugimoto T., Ishikawa K., Suzuki H.: J.
    Phys.Soc.Jap. 40, 258-266 (1976).

24. Falk H., Müller N., Purschitzky A.: Monatsh.Chem. 115, 121-
    124 (1984).

25. Falk H., Zrunek U.: Monatsh.Chem., in press.

26. Edinger J., Falk H., Müller N.: Monatsh.Chem., in press.

27. Edinger J., Falk H., Jungwirth W., Müller N., Zrunek U.:
    Monatsh.Chem., in press; Falk H., Medinger W.: to be pub-
    lished elsewhere in detail.

28. Blauer G.: Isr.J.Chem. 23, 201-209 (1983).

29. For a discussion of flexibility and conformational states
    of bilatrienes-abc using force field calculations see:
    Falk H., Müller N.: Tetrahedron 39, 1875-1885 (1983) and
    Monatsh.Chem. 112, 1791-800 (1981).

Received May 28, 1984

Discussion

Buchler: What are the arguments that allow to regard the 4,5 and 15,16
double bonds in the formulae [2] of your paper as clean double bonds, and
the 5,6 and 14,15 single bonds as clear single bonds? These are part of a
conjugated system and therefore should have lower bond order or higher bond
order, respectively.

Falk: First, there is simple evidence from the possibility to prepare,
isolate and characterize (Z)/(E) diastereomers with respect to the double
bonds in question. Their thermal activation parameters indicate that their
double band character is reduced compared to isolated double bonds, i.e.
E_a being around 60 kJ mole^{-1} and log A ~8,4 in case of a compound comparable to [2]. [An account of this work was published: Falk H., Kapl, G.,
Müller, N., Monatsh.Chem. 114, 773 (1983)]. Secondly, an X-ray structural
analysis of [2] (Kratky, C., Falk, H., Zrunek, U., Monatsh. Chem., in press)

clearly shows the bond alternation of a canonically conjugated system.

Fuhrhop: Would you please compare the cis-trans diastereomers in respect
to spectroscopic effects and stability with polyenes, e.g. retinal.

Falk: The thermal stability of the thermodynamically more unstable (E)-
diastereomers (we should use this nomenclature instead of the older cis/
trans, as the latter are now used to assign relative stereochemistry of
substituents in ring systems) might be comparable to the unstable (E)-
carotenoids. There is, however, a pronounced enhancement of reversion under catalytic conditions.

Concerning the spectroscopic shifts one can say that a change in configuration of an exocyclic double bond in bilatrienes itself has only marginal
spectroscopic consequences unless this isomerisation causes conformational
changes of the molecule. In this case, shifts of the order of 40 nm are
observed for the long wavelength band. This has to be compared with the
development of a secondary band (the "cis-band") in the carotenoid series.

Lamola: I would not like to see the development of the impression that
the story for polyenes is fundamentally different from that for the bile
pigments vis a vis configurational isomerism and spectral properties. It
is clear that for the real systems of both sorts, configuration and conformation are coupled with the degree of interdependence dependent upon
details of the particular case. In both families, conformation (twists
about single bonds) appears to have the major effect upon the spectra.
This gives "mother nature" great opportunities to use associated protein
structure to subtly tune absorption spectra.

The lower stabilities of the higher energy configurational isomers of the
bile pigments compared to the polyenes should not be taken to indicate any
fundamental difference in double bond strengths. Rather, there are many
more low activation barrier mechanisms for thermal isomerization available
for the bile pigments compared to the polyenes, mechanisms that involve
acid/base catalysis and the richness of protonation and deprotonation sites
in the bile pigments compared to polyene hydrocarbons.

Scheer: You mentioned the possibility of close lying ^1S and ^2S levels
after rotation around C-10, which would give rise to a split long-wavelength band. Could you comment on other causes for such split bands (vibronic structure, tautomerism), because the zinc-complex, for example, has
a split band, but is certainly fixed in a cyclic conformation.

Falk: In the case of the zinc complex I would think it is a vibronic
progression. Our studies on the dihydro system have clearly shown that the
hydrogen is frozen at N-23, and that tautomerism can be ruled out.

Scheer: In the case of the zinc-complex the vibronic structure is rendered visible by restriction of conformers. The same mechanism would of
course also apply if certain conformers were to be fixed by the protein,
e.g., in the biliproteins.

Falk: Yes, this is right, and it shows again the difficulty of getting
unambiguous answers from the limited data available.

Holzwarth: I have a comment on the intensity of the second electronic
transition in the visible range of phytochrome. If one compares the mirror
image of the phytochrome fluorescence spectrum with the absorption, it becomes clear that the majority of the short-wave length shoulder arises from
a vibronic $S_0 \rightarrow S_1$ transition.

Song: I have a comment regarding the choice of 10-11-s-bond conforma-
tion(s) as a spectral probe for the phytochrome Q_x shoulder at ~610 nm.
The experimental oscillator strength of the shoulder is such (~25-30 %
of the visible band in Pr) that 14-15-s-bond conformations may well account for the spectrum.

Do you have any examples showing photochemical transformation of dihydro-
derivatives via isomerization and/or tautomerization, analogous to the
phytochrome phototransformation, with a bathochromic shift of ~1,900 cm^{-1}?

Falk: No, the only shifts observed in this order of magnitude are observed for deprotonation (cf. Chapter 7 of my report).

Eilfeld: Do you think that the influence of a point charge upon the
absorption spectrum of a tetrapyrrole may be different in a stretched
conformation than in a cyclic conformation?

Falk: I would like to stress that although we have obtained a significant shift in one experimental case (cf. Chapter 10 of my account), that
I doubt that all of this shift really stems from an interaction of the
point charge with the chromophore. Instead, I think the interaction of
this charge with the permanent dipole moment of the chromophore may result in conformational changes and therefore in spectroscopic shifts.
From our studies as well as of those of others I have the impression that
these point charges function much as a *deus ex machina,* as their influence can easily be calculated using semiempirical calculations. There is
still no convincing experimental model available to settle this question.
Therefore, I would like not to speculate too much about this point. With
regard to your question I would think that given this influence, it
should be comparable for all conformations.

Gouterman: Can you explain the difference between the words "configuration" and "conformation"?

Falk: Within the IUPAC recommendations on stereochemical nomenclature
of organic compounds the term "configuration" denotes in our context the
relative arrangement of ligands at the exocyclic double bonds of the bile
pigments. It is designated as Z(usammen) or E(ntgegen) depending on the
orientation of the two, according to the "priority rule", largest ligands
on either side of the respective double bond. "Conformation" specifies a
certain torsional arrangement at a methine single bond characterized by
a certain dihedral angle. It may be denoted by using the Klyne-Prelog
nomenclature.

MECHANISM OF PHOTOTHERAPY OF NEONATAL JAUNDICE.   REGIOSPECIFIC
PHOTOISOMERIZATION OF BILIRUBINS

Antony F. McDonagh

Department of Medicine and the Liver Center, University of California, San
Francisco, California 94143, U.S.A.

David A. Lightner

Department of Chemistry and The Cellular and Molecular Biology Program,
University of Nevada, Reno, Nevada 89557, U.S.A.

Introduction

The optical properties of bilirubinoid pigments have long been important
in the diagnosis of jaundice and in clinical measurements of bile pigments
in body fluids. But it was the introduction of phototherapy for treatment
of neonatal jaundice some 15-20 years ago that stimulated particular
interest in the photochemistry of these pigments. Bilirubin is a lipophilic metabolite that is poorly excreted and can become toxic if it
accumulates in the body. Normally, it does not accumulate to any marked
degree because it is converted enzymically to glucuronide metabolites
that are less lipophilic and are excreted readily via the liver into the
duodenum in bile. However, the ability of newborn infants to metabolize
bilirubin is very low compared to adults. This, coupled with a relatively
high bilirubin production rate, causes the pigment to accumulate and the
infant to develop transient jaundice. A high blood level of bilirubin is
associated with increased risk of brain damage and clinical intervention
is sometimes necessary to prevent it from becoming excessive. The two
most common methods are exchange blood transfusion and phototherapy with
visible light. What exchange transfusion does physically, phototherapy
achieves photochemically. It stimulates the expulsion of bilirubin from
the body and leads to a decrease in the concentration of the pigment in
the blood. The complete mechanism of phototherapy is still unknown. But
it is now fairly well established that three distinct competing photochemical reactions of bilirubin make a contribution (1). These are

Optical Properties and Structure of Tetrapyrroles
© 1985 by Walter de Gruyter & Co., Berlin · New York

oxidation, configurational isomerization, and structural isomerization.
In this paper, we review some of our evidence for the occurrence of the
photoisomerization reactions in vivo. We also present evidence for
intramolecular energy transfer in the bilirubin molecule and describe two
novel and rather striking examples of regiospecific photochemistry.

Photoisomerization of bilirubin

Bilirubin (1) is a bichromophoric molecule which exhibits a broad
absorption band in the visible spectrum ($\lambda_{max} \sim$ 450 nm) that is not
markedly sensitive to solvent. The bridging exocyclic double bond
in each of the two pyrromethenone chromophores has a Z configuration
(Fig. 1) (2). This is the most stable configuration for this type of
pyrromethenone chromophore and it is further stabilized in bilirubin by
intramolecular H-bonding between propionic acid side-chains and pyrrolic
NH and CO groups. The most quantum efficient photochemical reaction of

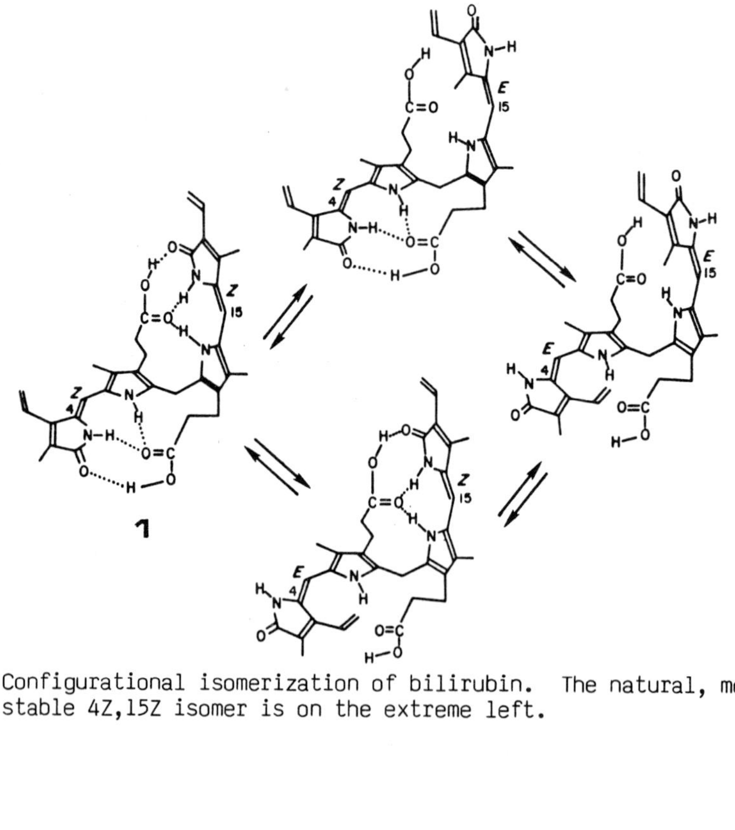

Fig. 1. Configurational isomerization of bilirubin. The natural, most
stable 4Z,15Z isomer is on the extreme left.

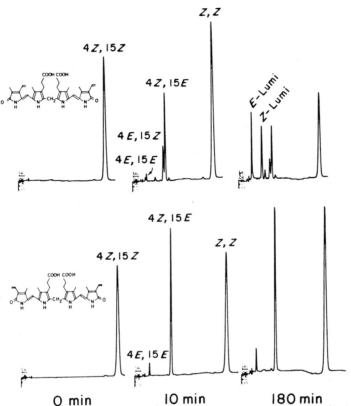

Fig. 2   Photolysis of bilirubin IXα (top) and bilirubin IIIα (bottom) in
          CHCl₃/Et₃N (1:1). Solutions (0.25 mM) were photolysed with
          Westinghouse Special Blue fluorescent tubes under argon and
          evaporated aliquots were chromatographed by isocratic reversed-
          phase HPLC (C-18, 1 ml/min) with a mobile phase consisting of
          0.1 M di-n-octylamine acetate in MeOH (92%) and H₂0 (8%).

bilirubin appears to be configurational isomerization about the methene

bridges to give thermally unstable E isomers (Fig. 1) (3, 4). The reaction

is photochemically reversible and the isomers have overlapping absorption

spectra (4, 5). Therefore, a photoequilibrium state is eventually reached

under anaerobic conditions, as readily demonstrated by difference

spectroscopy. We have developed a series of isocratic reversed-phase

solvent systems, based on methanolic solutions of long-chain dialkylamine

acetates, which allow rapid separation of all four possible configura-

tional isomers of bilirubin with baseline resolution and in which the E

isomers are sufficiently stable to allow accurate analysis (Fig. 2) (6).

Because of their more polar nature, E isomers of bilirubin can be
separated from the parent Z,Z isomer by solvent extraction with MeOH,
and small quantities of the individual isomers can be isolated by HPLC.
However, their instability has so far precluded their complete purification and characterization. Evidence for their identity is based on
chemical and physical studies of partially purified preparations,
including proton NMR (6), and conversion of a mixture of the isomers
to the corresponding biliverdin dimethyl ester isomers (7) which are of
established constitution. One characteristic property of E isomers of
bilirubin, which is useful in identifying them, is their instantaneous and
quantitative reversal to the Z,Z isomer on treatment in $CHCl_3$ with a trace
of trifluoroacetic acid (6, 8). The composition of the mixture of
configurational isomers obtained at photoequilibration is markedly
dependent on the wavelength emission of the light source (9), but the
relative proportion of the two diastereomeric E,Z isomers is only weakly
wavelength dependent, with a modest regiospecificity for the 4Z,15E isomer
in organic solvents. Thus, in $CHCl_3/Et_3N$, the 4Z,15E:4E,15Z ratios at
photoequilibrium are 2.3, 2.4, 2.0, and 1.7 for irradiation at 403, 453,
478, and 503 nm, respectively.

Anaerobic photolysis of bilirubins and dipyrromethenones that do not have
an endo-vinyl substituent (e.g. mesobilirubins III, IX, and XIIIα,
bilirubin IIIα, endo-ethyl-bilirubin IXα, xanthobilirubic acid, etc.)
rapidly generates a true photostationary mixture of configurational
isomers (Fig. 2), the composition of which is maintained on extended
photolysis until there is eventual slow (oxidative?) destruction of the
chromophore (8). However, with dipyrromethenones and bilirubins that
do bear an endo-vinyl substituent (e.g. bilirubins IXα and XIIIα, and
exo-ethyl-bilirubin IXα) configurational isomerization is accompanied by a
somewhat slower irreversible intramolecular cyclization of the endo-vinyl
group (Fig. 3). Thus, with these compounds, a true photostationary state
is not achieved and on extended photolysis additional peaks appear in the
HPLC (Fig. 2) and there is loss of isosbestic points in the difference
spectrum. Assignment of the cyclic structure 2 was based on proton NMR
and, more recently, C-13 NMR (10-13). An alternative structure involving

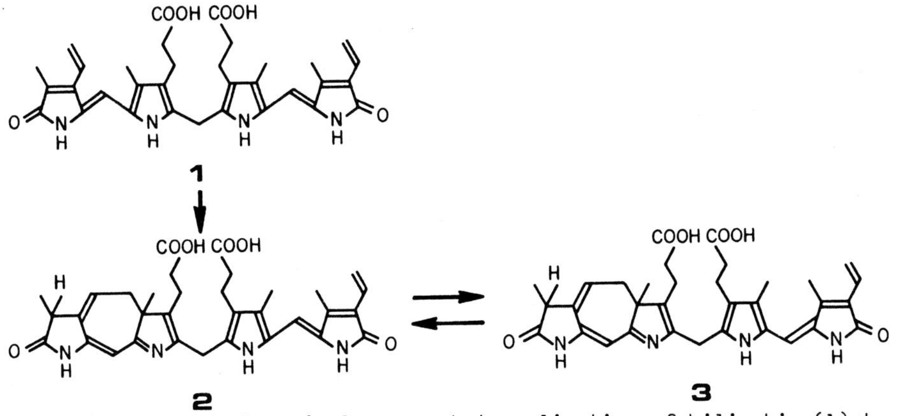

Fig. 3   Intramolecular vinyl group photocyclization of bilirubin (1) to
          Z-lumirubin (2) and configurational photoisomerization of the
          product.

intramolecular cyclization of a propionic acid side chain to produce a
lactone ring (11, 14) has been ruled out on chemical grounds. Thus,
bilirubin dimethyl ester exhibits a similar reaction, and hydrolysis of
the product yields material identical to that formed directly from
unesterified bilirubin. In addition, the cyclic product from bilirubin
reacts with diazomethane to form a dimethyl ester that is isomeric with
bilirubin dimethyl ester by FD mass spectrometry (10).

Cyclization of the endo-vinyl group introduces two chiral centers.
Therefore, there are four possible diastereoisomers of 2. These do not
separate on the reversed phase HPLC systems used in this paper, but they
do separate cleanly into two pairs of enantiomers on silica TLC. These
two pairs readily interconvert in base via epimerization at C-2 (10).
For the remainder of this paper, no distinction will be made between the
four diastereoisomers of Z-2.

Anaerobic photolysis of 2 and similar compounds results in rapid Z → E
isomerization of the non-cyclized dipyrrylmethenone half of the molecule
(Fig. 3). If this half lacks an endo-vinyl group, as in 2 and 3, a true
photoequilibrium is reached, as shown by HPLC analyses and the appearance
of a characteristic difference spectrum with tight isosbestic points (10).

Treatment of the mixture with trifluoroacetic acid catalyses complete reformation of the Z isomer. Because Z → E isomerization of the dipyrromethenone chromophore is faster than the structural endo-vinyl isomerization, the appearance of Z-configuration structural isomers during photolysis of rubins and dipyrromethenones is invariably accompanied by the appearance of the corresponding E isomers.

Studies in vivo

As expected, both types of bilirubin photoisomers, structural and configurational, are more polar than the natural Z,Z isomer. Unlike the natural isomer they are excreted readily in bile when they are injected intravenously into Gunn rats, which are a strain of rats that are congenitally unable to convert bilirubin to its glucuronides and are consequently jaundiced. Z-lumirubin (2), the structural isomer of bilirubin, is excreted slightly faster than the Z,E/E,Z configurational isomers. Interestingly, the 4E,15Z isomer is excreted somewhat faster than the 4Z,15E isomer. Therefore, the hepatobiliary system of the rat is sensitive to very subtle structural differences, acting in effect like a low-efficiency reversed-phase HPLC column.

When Gunn rats were irradiated with blue fluorescent phototherapy lights, there was a prompt increase in the concentration of yellow pigment excreted in bile (15, 16). With continuous irradiation, the concentration of pigment increased over a period of about 90 min and then became constant. This in vivo photostationary state may not be due to achievement of a photochemical stationary state since increasing the intensity of the light increased the pigment concentration in the bile to a new constant level. When the light was switched off, pigment excretion returned gradually to the normal pre-light levels. HPLC of bile during the photostationary period, with precautions to obviate adventitious secondary reactions of the photoisomers after their excretion, showed the presence of all four configurational isomers of bilirubin and both Z- and E-lumirubin structural isomers (Fig. 4) (1, 10, 17). The major products

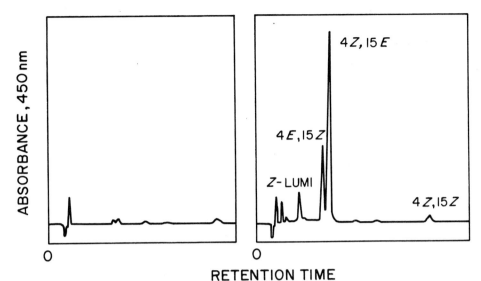

Fig. 4   HPLC of bile from a jaundiced rat before (left) and at the
         photostationary state during phototherapy with blue fluorescent
         light (right).

were the two 4Z,15E/4E,15Z isomers and their relative ratio was similar to
that observed in photolyses of bilirubin in organic solvents in vitro.

Structural and configurational isomerization of bilirubin was also
demonstrated to occur in human infants during phototherapy by HPLC of
serum and of bile obtained by aspiration through a naso-gastric tube (1,
10, 18). Surprisingly, however, the 4Z,15E:4E,15Z ratio in blood serum
was extremely high (~ 32:1). There were two possible explanations for
this observation. Either the less prevalent 4E,15Z isomer is excreted
much faster than the predominant 4Z,15E isomer, or the photochemistry in
the baby is much more regioselective than in the rat or in organic
solvents. Evidence for the latter was obtained by further in vitro
photolyses of bilirubin and model compounds in water in the presence of
albumin.

Regiospecificity

Most of the bilirubin in infants with neonatal jaundice is associated
reversibly and noncovalently with albumin. When solutions of bilirubin

in serum or in buffered (pH 7.4) serum albumin (pigment:albumin ratio <1)
were irradiated with blue phototherapy lights both configurational and
structural isomerization were observed. Both the rates of the two reactions and the regiospecificity of the configurational isomerization were
found to be species-dependent (1, 10). In particular, the configurational
isomerization in the presence of human albumin was highly regioselective
for the 4Z,15E isomer. Thus, photolysis of bilirubin in human or monkey
serum gave a 4Z,15E:4E,15Z ratio of about 100:1 at photoequilibrium,
whereas photolysis in rat serum or rabbit serum gave ratios of 1.3:1 and
0.5:1, respectively. As observed with organic solvents the composition of
the photoequilibrium mixture in the presence of human serum albumin was
highly wavelength-dependent, but the 4Z,14E:4E,15Z ratio was not.

Although bilirubin is not symmetrically substituted, the two chromophores
of which it is composed have almost mirror-image symmetry. This led us to
surmise that the pigment might be able to dock into its binding site on
albumin in either of two nearly isoenergetic orientations. If steric
factors at the binding site were to inhibit photoisomerization in one
specific half of the bound pigment, then the variable regiospecificity
seen with different albumins could be explained by differences in the
affinity of each type of albumin for each of the two orientations of
bilirubin. To explore this possibility, we are studying the photochemistry of several analogues of bilirubin with slightly modified
side-chains. Our most striking results so far have been obtained with
compound 4 which is an isomer of dihydrobilirubin.

The photochemistry of 4 in CHCl$_3$/Et$_3$N and in aqueous human albumin is
qualitatively similar to that of bilirubin. However, in the organic
solvent (λ_{max} 428 nm), the configurational isomerization is highly
specific for one half of the molecule with an E,Z:Z,E isomer ratio ranging
from 9:1 for photolysis at 405 nm to 11:1 for photolysis at 455 nm.

Photolysis with a blue phototherapy light produced an isomer ratio of 9.8:1 at close to photoequilibrium. (Presently we are unable to identify which half of the molecule preferentially isomerizes.) In marked contrast, in the presence of human albumin (λ_{max} 444 nm) the isomerization is highly regioselective for the other half of the molecule (isomer ratio ~ 1:43 for photolysis with the blue phototherapy light).

The regioselectivity observed with 4 in organic solvents is most plausibly explained by intramolecular energy transfer between the two coupled chromophores of the molecule in the excited state. Unsymmetrically substituted bilirubins consist of two loosely coupled and non-identical pyrromethenones that interact through geometry-dependent exciton coupling. The excited state of these molecules can be represented by a double-minimum potential energy well (Fig. 5) in which each of the wells represents an energy minimum corresponding to excitation of either the left- or the right-hand half of the molecule. Excitation into either well should also populate the other well by intramolecular energy transfer. Therefore, excitation of either chromophore can lead to photochemistry in the other. If it is reasonably assumed that the rates of decay from either minimum

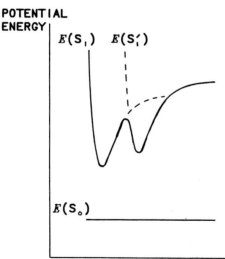

Fig. 5   Double minimum potential-well diagram for the first excited state of bilirubins. Derived from a pictorial representation of two coupled diatomic molecules. For simplicity, the ground state energy levels, $E(S_0)$, are represented by a horizontal line.

are similar, then the ratio of diastereomeric E,Z isomers obtained at
photoequilibrium will depend on the relative populations in the two halves
of the double minimum potential well, which in turn will depend on the
relative depths of the two minima. Calculations based on the present data
indicate that the energy difference between the two minima is of the order
of 1 kcal/mole. Since compound 4 and bilirubin are structurally very
similar, small structural changes and small energy differences clearly can
have a dramatic effect on the regioselectivity of the photochemistry.

The high regioselectivity of the Z → E isomerization for bilirubins
bound to human serum albumin is not so readily explained. It would appear
that both steric and electronic interactions are important. Steric
interactions to select only one of the two ''degenerate'' bilirubin
orientations; electronic interactions to alter the relative heights of the
two potential energy minima in the excited state so that the order is
reversed compared to the situation in organic solvents.

In conclusion, these findings demonstrate the marked effect of protein
binding on the stereochemical course of bilirubin photochemistry. They
suggest that, even if the rates of Z → E isomerization are the same for
each pyrromethenone unit of a bilirubin molecule, the isomeric E,Z product
ratios are independent of the absorption cross-section of each half of the
molecule and may vary considerably according to the nature and location of
the β-substituents and the nature and extent of solute-solute interactions.
Lastly, these studies provide an explanation for the observation that the
absorption and excitation spectra of bilirubin are not identical (19).

Acknowledgment

We acknowledge support of this work by grants from the National Institutes
of Health [AM-26307, AM-26703, AM-11275 (AFM); HD-17779, HD-09026 (DAL)].

References

1. Lightner, D.A., McDonagh, A.F.: Acc. Chem. Res. (In press).

2. Sheldrick, W.S.: Isr. J. Chem. 23, 155-166 (1983).

3. Lamola, A.A., Flores, J.: J. Am. Chem. Soc. 104, 2530-2534 (1982).

4. McDonagh, A.F., Palma, L.A., Lightner, D.A.: Science 208, 145-151
    (1980).

5. Lightner, D.A., Wooldridge, T.A., McDonagh, A.F.: Proc. Natl. Acad.
    Sci. USA 76, 29-32 (1979).

6. McDonagh, A.F., Palma, L.A., Trull, F.R., Lightner, D.A.: J. Am.
    Chem. Soc. 104, 6865-6867 (1982).

7. Falk, H., Muller, N., Ratzenhofer, M., Winsauer, K.: Monat. Chem.
    113, 1421-1432 (1982).

8. McDonagh, A.F., Lightner, D.A., Wooldridge, T.A.: J.C.S. Chem. Comm.
    110-112 (1979).

9. Ennever, J.F., McDonagh, A.F., Speck, W.T.: J. Pediatr. 103, 295-299
    (1983).

10. McDonagh, A.F., Palma, L.A., Lightner, D.A.: J. Am. Chem. Soc. 104,
    6867-6869 (1982).

11. Stoll, M.S., Vicker, N., Gray, C.H., Bonnett, R.: Biochem. J. 201,
    179-188 (1982).

12. Bonnett, R., Buckley, D.G., Hamzetash, D., Hawkes, G.E., Ioannou, S.,
    Stoll, M.S.: Biochem. J. 219, 1053-1056 (1984).

13. Yokoyama, T., Ogino, T., Onishi, S., Isobe, K., Itoh, S.,
    Yamakawa, T.: Biochem. J. 220, 377-383 (1984).

14. Onishi, S., Itoh, S., Isobe, K., Sugiyama, S.: Photomed. Photobiol.
    3, 59-60 (1981).

15. Ostrow, J.D.: J. Clin. Invest. 50, 707-718 (1971).

16. McDonagh, A.F., Palma, L.A.: J. Clin. Invest. 66, 1182-1185 (1980).

17. McDonagh, A.F.: in New Trends in Phototherapy (Rubaltelli, F.F. and
    Jori, G., eds), Plenum Press, New York (In press).

18. Onishi, S., Isobe, K., Itoh, S., Kawade, N., Sugiyama, S.:
    Biochem. J. 190, 533-536 (1980).

19. Bonnett, R., Dalton, J., Hamilton, D.E.: J.C.S. Chem. Comm. 639-640
    (1975).

Received August 3, 1984

Discussion

Falk: There are further examples that the regiospecificity of isomerization of rubins with non-identical pyrromethene units is due to the different energy of their excited states, e.g. on addition of nucleophiles to
position 10 of 2,3-dihydrobilatrienes-abc the resulting rubinoid system is
isomerized in position 15 exclusively - this has to be compared to the easy
isomerization of 3,4-dihydropyrromethenones. We have found as well that in
case of dichromophoric systems with one exocyclic single bond having a
stronger torsion, only the double bound adjacent to this site will be isomerized.

Is the occurrence of bilirubin in the excretion a sign that the dose used
in phototherapy is too high and are the wavelengths used too far in the UV?

McDonagh: I think not. However, since we do not yet know the relative
efficiencies of the E-isomer and lumirubin pathways in human infants in
terms of their contribution to the overall lowering of serum bilirubin,
it is not possible to answer that question at present.

With respect to your comment, the examples you cite are certainly interesting and relevant. However, in those cases there are rather large differences between the structures and excitation energies of the two chromophores in the molecule. What is remarkable about the dihydrobilirubin XIIIα
system is that marked regiospecificity is seen in a system in which the two
chromophores are very similar.

Lamola: (Re: Dr. Falk's question and Dr. McDonagh's answer)
Were the configurational isomer of bilirubin the overwhelming contributor
to pigment excretion during phototherapy it would not be useful to increase
the therapeutic light intensity to a level greater than that required to
achieve an isomer ratio close to the photoequilibrium ratio. Any more light
just cycles the isomers without increasing the excretable isomer. Were irreversible photoreactions such as lumirubin production and bilirubin photooxidation important contributors to pigment excretion, then it would be
worthwhile to increase light intensity up to the point that the rates of
those reactions in the region of light penetration are saturated, intensities expected to be much higher than those necessary to achieve photoequilibration of configurational isomers.

Scheer: You find no lumirubin in the serum, is this due to the fast excretion?

McDonagh: In human infants only relatively small amounts of lumirubin are
detectable in the serum during phototherapy. Two factors appear to be responsible for this: a) relatively fast excretion of lumirubin and b) relatively slow photochemical formation of lumirubin.

Scheer: Could there also be a chemical reason besides the physiological
one, e.g., are there possibilities that lumirubin is formed not only photochemically but also from one of the photoisomers in the liver?

McDonagh: This is conceivable, but there is no experimental evidence that
it occurs.

Gouterman: Are black babies more subject to hyperbilirubin diseases? Do
they respond to phototherapy?

McDonagh: Black babies are not more subject to hyperbilirubinemia than
other babies, and their response to phototherapy is not measurably different from the response of lighter complexioned babies.

Holzwarth: Is the reaction mechanism leading to lumirubin known? Does it
require one or two photons to form lumirubin?

McDonagh: It is almost certainly a concerted one-photon reaction.

Vogler: How efficient are the photoisomerizations? What are the quantum
yields?

Question referred to Lamola for reply: Quantum yields of photoreactions
of bilirubin bound to human serum albumin in physiologic buffer at room
temperature at 430 nm are as follows: Z,Z to Z,E isomerization, 0.2;
photodestruction (photooxidation), 0.001; lumirubin formation, <0.01.

Fuhrhop: Do you think that irradiation of shaved rats with visible light
may increase the amount of heme which is converted to bile pigments? This
would be in accordance with your finding of a raise in the steady state of
bilirubin concentration in light as compared to the dark.

McDonagh: There is no evidence that visible light increases bile pigment
production from heme and our experiments indicate that it does not occur.
We do see an increase in the steady state concentration of bilirubin in
bile during phototherapy, but this increase is due to E and lumi isomers
and not to the Z,Z isomer. The light does not cause an increase in the serum concentration of bilirubin which would be expected if light increased
bilirubin formation.

Blauer: It is well known that bilirubin IX is only sparingly soluble in
water at physiological pH. With reference to the increased rate of excretion of photobilirubin, have any experiments been made to determine quantitatively the solubility of PBR in water?

McDonagh: There are no quantitative data on the aqueous solubility of
bilirubin photoisomers, but it is clear that they are more soluble than
bilirubin at pH 7.4-8.5. Solubility measurements on the E isomers in water
would be difficult because they undergo thermal E → Z reversion in water.

Sund: With respect to the isomers and to the binding of bilirubin to
albumin, I would like to ask how many binding sites for bilirubin can be
determined for one albumin molecule and if there is more than one binding
site, how have these binding sites been characterized.

McDonagh: Human serum albumin has a single high-affinity binding site
for bilirubin. Additional molecules of bilirubin can bind with lower,
though nevertheless high, affinity but it is not known whether these bind
at specific sites. The structure of the high-affinity site is not known,
but there is some evidence in the literature indicating the approximate

location of the site on the albumin molecule.

Rapoport: Can bilirubin XIII be converted to the bis-lumi isomer?

McDonagh: The bis-lumi compound has not yet been synthesized. When bilirubin XIIIα is photolyzed anaerobically in the presence of equimolar human
serum albumin it is eventually converted quantitatively to a mixture of the
E and Z lumirubin XIIIα isomers without any apparent appearance of a bis-
lumi isomer. However, it should be possible to make the bis isomer by
further photolysis of Z-lumirubin.

Rapoport: Can the lumi isomer be oxidized to the corresponding cyclo-
heptatriene?

McDonagh: Lumirubin does undergo ready autoxidation to a purple product,
but the structure of this product is unknown.

EFFECTS OF ENVIRONMENT ON PHOTOPHYSICAL PROCESSES OF BILIRUBIN

Angelo A. Lamola

AT&T Bell Labs
Murray Hill, NJ, USA 07974

Introduction

Under the sorts of conditions in which it is found _in vivo_, $4Z,15Z$-bilirubin-IXα (BR) exhibits both very low quantum yields (<0.01) of fluorescence and very low yields (<0.01) of triplet state formation when excited with blue or near ultraviolet light. The efficiency of photodestruction of this pigment is also low (<0.01) and so the major pathway(s) for relaxation of photoexcited BR involves radiationless conversion from an excited singlet state(s) to the ground state. It is now well established that this radiationless relaxation channel involves twisting of the excited molecules about one of the so-called _meso_ carbon-carbon double bonds. This twisting can be very rapid (~10 ps). In the twisted geometry the excited state and ground state potential surfaces are brought close together and conversion from the upper to the lower surface becomes ultrafast (< 10 ps). Subsequent vibrational relaxation of the ground state molecule returns it to the original geometry or to the configurational ($\underline{Z} \rightarrow \underline{E}$) isomer. The overall photoisomerization can proceed with relatively high quantum efficiency (≥ 0.2). BR undergoes other photochemical reactions but all of those occur with very much lower yields.

In this report I review work in my laboratory and in others concerning the relaxation modes of optically excited BR and the effects of the microscopic environment, especially protein binding, upon them.

Optical Properties and Structure of Tetrapyrroles
© 1985 by Walter de Gruyter & Co., Berlin · New York

Particularly interesting and clinically important are the effects of binding of BR to human serum albumin (HSA) (1,2). This most abundant of the plasma proteins possesses a specific high affinity ($\sim 10^7 M^{-1}$) binding site for BR and can be considered to have two functions with respect to the pigment. HSA transports BR to the liver for processing and excretion. In addition HSA serves to sequester BR in the HSA-space, keeping the cytotoxic pigment out of cells. This sequestering function is especially important for protecting neonates with unconjugated hyperbilirubinemia from bilirubin encephalopathy. HSA can also bind the configurational isomer of BR formed during phototherapy. This photoisomer is bound at the BR binding site and is thereby stabilized as well as sequestered and transported to the liver. Finally, binding to the primary site on HSA has important consequences for relaxation of photoexcited BR. Radiationless relaxation is slowed so that fluorescence is increased greatly over that for unbound BR or BR bound at other sites in the blood, providing a basis for fluorimetric assays of binding status (2). Configurational photoisomerization of BR is also rendered regiospecific through binding to HSA (3).

Bilirubin: A Bichromophore

The primary structure of BR is given below for reference. Although usually classified as tetrapyrroles, chemists and spectroscopists have recognized that bilirubins are best described as pairs of interacting dipyrromethenones (4,5). However, little more can be said definitively about the optical spectra of bilirubins because: (a) the spectra themselves (as

BILIRUBIN (IX α)

well as those of dipyrromethenones) are broad with little resolved structure, (b) change of solvent or chemical modification (e.g., esterification
of propionic acid side chains) alters conformation, hydrogen bonding, state
of aggregation and specific solvent interactions, precluding unique interpretation of spectral response, and (c) the only well defined three-
dimensional structure for a bilirubin is that of crystalline BR and no
optical spectra have been recorded for it.

For the purpose of this review the question is asked, what are the consequences of the bichromophoric nature of BR for electronic excitation and
relaxation in the pigment? Assume that the intense longwave (450 nm)
absorption band represents overlapping (and/or mixed) essentially single
electronic transitions in each of the dipyrromethenone halves. The diagram
of Fig. 1 would then roughly describe the situation. From the reported
absorption spectra of Z,Z-bilirubin-IIIα and XIIIα in various solvents, it
is suggested that the long-wavelength transition in the dipyrromethenone
half with the exo vinyl group (through conjugation) lies 300 cm^{-1} lower
than that of the half with the endo vinyl group (cross conjugation) (6).
Of course, interactions with the microenvironment, say in a protein, could
easily reverse this state ordering. An upper limit for the spectral
breadth (at half height) is taken to be 2500 cm^{-1}, the breadth of the
long-wavelength absorption bands of a variety of dipyrromethenones in different solvents. The contribution of conformational and solvation hetero-

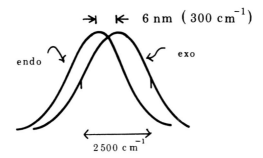

Figure 1. Schematic of the visible absorption bands of the
two dipyrromethenone halves of BR in a homogeneous solvent.
Endo and exo refer to the position of the vinyl group.

geneity (inhomogeneous broadening) is not known. Now, from the oscillator
strength obtained by integrating the visible absorption band, the dipolar
interaction between the transition moments of the two halves, taking 0.7 nm
for the distance between chromophore centers, is approximately 500 cm^{-1} x $f(\Theta)$, where $f(\Theta)$ represents the orientational (angular) dependence of
the dipolar interaction (7). Thus, it appears that for a "random"
conformation of BR for which $f(\Theta){\sim}1$, the electronic interaction energy is
comparable to the difference in the purely electronic transition energies
of the two halves but smaller than the vibronic widths of the transitions
[weak coupling limit (8)]. This means that excitation is localized in one
or the other half of the molecule but can hop back and forth between them.
The rate of excitation transfer is comparable to that of vibrational relaxation steps so that the excitation hops back and forth until it is "trapped" or destroyed in one or the other half. If the geometry remains
unchanged, the excitation, if not degraded otherwise, would localize (at
low temperatures) on the half with the lower transition energy. However,
if some excited state geometry change in either half leads to a state with
a lower energy, the excitation would "dig in" there. Therefore, it can be
seen that conformational flexibility and heterogeneity can have important
effects not only upon the absorption and fluorescence spectra of BR but
potentially also on the pathway(s) for excitation relaxation.

One way to assess conformational heterogeneity is to examine the luminescence spectrum of BR recorded using different exciting light wavelengths.
If fluorescence is faster than conformational interconversion the luminescence spectrum may change with excitation wavelength. We have observed (A.
A. Lamola and J. Flores, unpublished observations) red shifts in the
fluorescence spectrum of BR bound to its primary site on human serum
albumin (HSA) as the excitation wavelength was increased (Table 1). This
may be taken as evidence for modest conformational heterogeneity of BR/HSA.
However, because the fluorescence lifetime of BR/HSA is extremely short
under the conditions of the experiment (see below), the spectral variation
may reflect, in part, fluorescence from molecules not yet fully vibra-
tionally relaxed. Another interpretation is that the fluorescence occurs
on a time scale that is comparable to the rate of excitation transfer

TABLE 1

Wavelength Dependence of BR/HSA Fluorescence
(neutral buffer at 22°C)

Excitation wavelength(nm)	Emission maximum(nm)
420	520
440	522
460	525
470	526

between the two dipyrromethenone chromophores which have somewhat different fluorescence spectra. In any event such observations are expected when fluorescence is weak, that is fluorescence lifetimes are very short, and one or more other fast processes occur on the same time scale.

Ultrafast Radiationless Deactivation

That BR in common neutral solvents of low viscosity near room temperature exhibits an extremely low fluorescence yield ($\leq$0.001) which increases dramatically upon binding to albumins or upon decreasing the temperature was recognized a decade ago (9). It was pointed out by Matheson and workers (10) that the rigidity of the microenvironment is probably an important factor in controlling fluorescence yield.

In my laboratory the work has emphasized the HSA complex of BR not only because it is relevant to clinical issues but because it offers the benefit of providing a relatively homogeneous monodisperse form of BR. We found (11) the fluorescence yield of BR/HSA, although much higher than that of BR in water, to be quite small at room temperature (Table 2), but to increase enormously as the temperature is lowered reaching a plateau (in 50/50 ethylene glycol/water) near 120K (Fig. 2) (12). The value at 77K is 0.92.

That environmental rigidity is, indeed, a major factor is shown by the quantum yield of 0.71 obtained for BR in a polymethylmethacrylate (PMMA) film at room temperature compared to no fluorescence observed for BR in ethyl acetate taken as an appropriate fluid reference (Fig. 3). The much lower limiting (77K) fluorescence yields, ~0.1, reported for BR in rigid organic solvents (10) may well be due to operation of relaxation channels

associated with pigment aggregation during sample cooling.   One advantage
of working with BR/HSA is the assurance that the BR remains monodispersed
as the temperature is lowered.

TABLE 2

BR Fluorescence Yields (22°C)

"Solvent"	ϕ_f
H_2O	< 0.00002
$CHCl_3$	< 0.0002
HSA	0.003 (< 0.01 isc)
HSA (50% EG)	0.006
HSA (50% EG; 77°K)	0.92 (no phos.)
PMMA	0.71 (no phos.)

Although hard sought, no <u>bona</u> <u>fide</u> phosphorescence has been observed from
BR or from several dipyrromethenone even in rigid glasses at 77K (5). We
looked hard for phosphorescence from BR/HSA at 77K and found none (11).
The maximum yield of triplet states estimated for BR in benzene at room

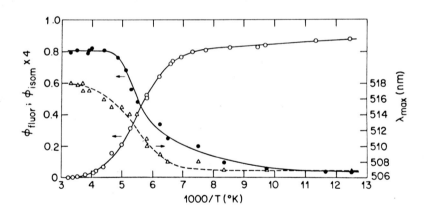

Figure 2. Properties of BR/HSA (in 1/1 v/v ethylene
glycol/water) as a function of the temperature: fluorescence
quantum yields, o; quantum yields of configurational isomerization, ●; wavelength of the fluorescence maximum excited
at 431 nm, △ .

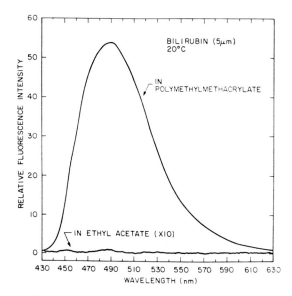

Figure 3. Luminescence signals recorded from BR in polymethylmethacrylate and in ethyl acetate near room temperature.

temperature is of the order of 0.01 (13) and the estimate is still smaller for BR/HSA and for BR in other solvents. Higher yields of triplet states (about 0.1) in several halogen-containing solvents, especially in the presence of oxygen, were estimated by Matheson and coworkers (14) based upon the photooxidaton rate of BR assuming a singlet oxygen pathway mediated by the BR triplet. However, oxygen-mediated free radical reactions of BR may be operating in these cases (15) and Matheson's interpretation may not be warranted. It appears that it is safe to conclude that the yield of triplet states from photoexcited BR/HSA is ≤ 0.08 at 77K and <0.01 at room temperature.

The fluorescence lifetime of BR/HSA at 77K was found to be 2.8 ns (11) giving a radiative lifetime of 3 ns assuming a temperature independent rate of fluorescence. The room temperature fluorescence yields of 0.003 translates into a lifetime of 10 ps. This means that the radiationless decay channel(s) for the BR/HSA singlet state operates with a rate of 10^{11} s^{-1} at room temperature. The same conclusion obtains for BR and for a variety of dipyrromethenones in in common organic solvents (5). It has

been shown by the work of Falk and coworkers (5,16), McDonagh and Lightner
and coworkers (3,17), and by us (11,18) that this fast relaxation mode
involves twisting about one or the other exocyclic carbon-carbon double
bonds which can lead to reversible configurational ($\underline{Z \to E}$) isomerization.
Work in my laboratory has shown that for BR/HSA the twisting can account
for virtually all of the radiationless decay efficiency even though the
quantum yields of isomerized pigment is about 0.2 (18). This conclusion
was based upon the results of a number of experiments which were facilitated by the fact the photoisomerization of BR/HSA gives essentially only the
$\underline{Z},\underline{E}$ isomer with little or no $\underline{E},\underline{Z}$ isomer (3), and that the protein greatly
stabilizes the $\underline{Z},\underline{E}$ isomer with respect to thermal reversion to BR (18).
Determined in these experiments at various temperatures were: isomeriza-
tion quantum yields in the forward and reverse directions (18), isomer
ratios at photoequilibrium (18), fluorescence decay rates (19), and rates
of recovery of the ground state after photoexcitation (11). All the observations were consistent with the operation of an activated (temperature
dependent) twisting about the 14,15 double bond in the excited singlet
state. The barrier is probably due primarily to viscous drag in the protein environment (see below). This twisting competes with fluorescence
which is itself faster than other deactivation modes. Consistent with this
picture are the low yields determined for other known photochemical
pathways (18) (Table 3).

TABLE 3

Photochemical Yields[*]
BR/HSA   22°C

Process	φ
$\underline{Z,Z} \to \underline{Z,E}$	0.22
→ lumiproduct	< 0.005
→ adduct	< 0.002
→ destruction (oxidation)	0.0006

*from reference 12

The inefficiency in the $\underline{Z} \rightarrow \underline{E}$ isomerization is due to partitioning of the twisted state. In terms of the schematic shown in Fig.4, $k_I >>> k_f$ and $k_1/k_2 \cong 4$. The $\underline{Z},\underline{E}$ isomer (PBR) bound to HSA exhibits a fluorescence yields at least 20 times smaller than that of BR/HSA. PRB/HSA is photoconverted to BR/HSA with a quantum yield expected on the basis of this model to be 0.8 and measured to be >0.6, a difficult measurement to carry out (18).

Dynamical studies (11) of the absorption spectrum of BR/HSA in the picosecond time regime following excitation with a 0.4 ps pulse of light revealed that $(k_1 + k_2) >> k_I$ and that the process of twisting was the rate limiting step in the relaxation process (ground state recovery) whose inverse rate was found to be 19 ±2 ps ($22^{\circ}C$). Consistent with this, direct streak camera determination of the fluorescence decay rate of BR/HSA gave 18 ± 3 ps. No longer lived fluorescence was observed during these studies.

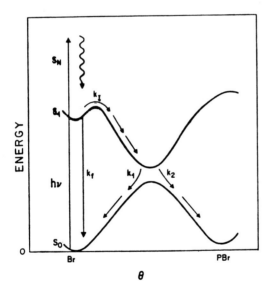

Figure 4. Schematic energy level diagram for BR/HSA. Θ designates the angle of twist about the 15,16-double bond. The various arrows indicate pathways by which photoexcited BR relaxes to BR or to PBR, its $\underline{Z},\underline{E}$ isomer. A potential barrier, corresponding to a measured Arrhenius activation energy of 5.5 kcal/mole, is depicted between the S_1 state of BR and a twisted conformation. This value reflects temperature-dependent processes in its protein microenvironment as well as the true intrapigment potential.

[The suggestion of Tran and Beddard (20) that the configurational isomers of BR exhibit fluorescence lifetimes of the order of 1 ns is not correct. In their experiments they were, perhaps, observing fluorescence from lumirubins which exhibit relatively high luminescence yields and which well could have been produced under their experimental conditions.] Thus, twisting in the excited singlet state of PRB/HSA is even faster than that for BR/HSA, of the order of 10^{12} s^{-1}. It is interesting that despite its greater structural complexity, the relaxation of photoexcited BR/HSA can be described in a way very similar to that of stilbene (21).

Studies by Falk and coworkers (5,16) and by Lippitsch and coworkers (22) have led those workers to invoke a mechanism for the Z→E photoisomerization of dipyrromethenone and some derivatives that is nearly identical to that described above for BR.

While our data would accommodate a few percent of the BR/HSA singlet states decaying radiationlessly without twisting, there is no compelling reason to include this possibility in the model of Fig. 4. Other candidates for ultrafast radiationless deactivation modes are tautomerizations that involve proton transfers such as the ones shown in Fig. 5. Such proton tautomerization provide important pathways for relaxation of photoexcited double-stranded DNA (23) and for relaxation of some photoexcited dipyrromethenes and related compounds (16). In $CHCl_3$, wherein it is likely that internally hydrogen bonded conformations of BR exist (24), configurational photoisomerization of BR to give a mixture of the E,Z and Z,E isomers proceeds with a quantum yield of the the order of 0.3 (A. A. Lamola and J. Flores, unpublished results). Thus, despite the low fluorescence yield of BR in $CHCl_3$, there is no reason to invoke additional decay paths that involve the internal hydrogen bonds such as tautomerization.

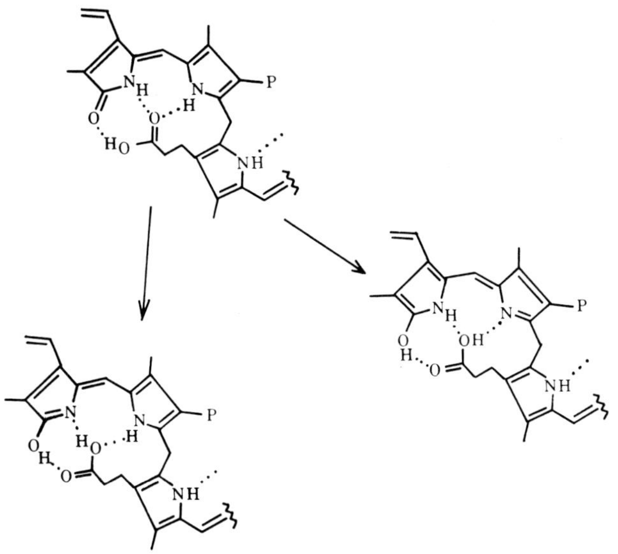

Figure 5. Part of the structure of the internally hydrogen bonded conformation of BR with two tautomers that can be formulated by transfer of protons.

Protein Binding

BR is so sparingly soluble in water that in blood it is virtually all bound to proteins, cell membranes and fat micelles (1,2). When its specific binding site on HSA is far from saturated nearly all the BR is bound to HSA. When the primary HSA site nears saturation, BR will bind to a secondary site(s) on HSA, to many other plasma proteins and to red cells. Most of these sites have affinities of the order of 10^5 M^{-1} (2).However, when bound to such lower affinity sites the fluorescence yield of BR is on the average 20 times lower than it is when bound to the primary HSA site. These observations provided the basis for a whole blood fluorimetric method for assessing the bilirubin binding capacity of HSA in blood specimens from neonates as an aide in the management of neonatal hyperbilirubinemia (2).

The binding of BR to HSA leads to an increase in the fluorescence yield over that of the unbound pigment presumably because the binding inhibits singlet state twisting. It is interesting to compare BR fluorescence yields and binding affinities for various proteins. Relative values for some albumins are given in Table 4. It was pointed out above that for proteins which bind BR with a very much lower affinity than does HSA the fluorescence yield is negligible. It is tempting to suggest that a higher binding affinity reflects a greater "tightness" with which the BR is bound which is, in turn, reflected in a higher fluorescence yield.

While it appears clear that binding to HSA slows the rate of excited state twisting, the rate is still surprisingly fast at room temperature, sufficient to result in near unit efficiency for formation of the twisted state. It is striking to compare the twisting rate for BR/HSA (5×10^{11} s^{-1}) with that for BR in $CHCl_3$ which is only sightly faster (10^{12} s^{-1}) at room temperature (11).

The Arrhenius activation energy for twisting of BR bound to HSA was found to be 5.5 kcal/mole with most of the energy assigned to barriers associated with the pigment and the protein (11). Fluorescence yields from BR/HSA which had been exchanged in D_2O for one day were higher than those from unexchanged samples and the Arrhenius barrier for twisting was increased substantially, 11 kcal/mole (A. A. Lamola and J. Flores, unpublished results).

TABLE 4

Serum Albumin Comparisons
(neutral buffer near room temperature)

Albumin	Relative ϕ_f	Relative K_a	$\underline{Z,E/E,Z}$ [b]
Human	[=1]	[=1]	>>10
Rhesus	1	∿1	-
Rat	0.4	0.05	1.3
Bovine	0.35	0.075^a	∿1
Rabbit	-	0.055^a	0.5

a. data of reference 25
b. see reference 3

Binding of BR to HSA has an important controlling effect on the regiospecificity of the configurational photoisomerization (Table 4). McDonagh,
Lightner and coworkers found that BR/HSA yields the Z,E isomer virtually
exclusively (3). In contrast, BR in $CHCl_3$ and BR bound to other albumins
(of lower affinity than HSA) yields the Z,E and E,Z isomers in comparable
amounts. That both isomers can be obtained from BR despite the expectation
that the exo-vinyl side has the lower excitation energy (at least in homogeneous solution) is in accord with the discussion of excitation hopping
given above. The interaction between the dipyrromethenone halves is sufficient to keep the excitation shuttling back and forth until twisting starts
at one or the other of the meso carbon-carbon double bonds and the excitation "digs in" on that half. On the basis of this "dynamical" model the
regiospecificity observed for BR/HSA indicates that in the specific HSA
binding site the dipyrromethenone half with the endo vinyl group is held
"tighter" than that with the exo vinyl group.

Through competition experiments monitored by CD spectroscopy (26), we were
able to estimate that HSA binds Z,E bilirubin with an affinity only about a
factor of two lower than that for the Z,Z isomer. HSA-bound Z,E bilirubin
reverts to BR/HSA upon photoexcitation with apparently no formation of the
E,E isomer. The very fast twisting rates in photoexcited HSA-bound bilirubin, the regiospecificity of photoisomerization, and the similar binding
affinities for BR and Z,E-bilirubin to HSA provide excellent support for a
model of the BR binding site on HSA discussed by Brodersen (1). In this
model only one dipyrromethenone half occupies a cleft in the protein while
the other half is exposed to solvent. The observations reviewed above
would suggest that, based upon this model, the half with the endo vinyl
group is the one that is bound in the protein cleft leaving the other half
relatively free to undergo photoisomerization. Extension of this reasoning
would suggest that binding of BR by rat and bovine albumins is quite different.

In an attempt to test the model given above, the twisting rate of excited
BR/HSA was determined as a function of the buffer viscosity (12). The
effect of buffer viscosity on the twisting rate of BR/HSA was found to be

small, accounting for only about 15% of the barrier, and did not support
the idea that the isomerizable dipyrromethenone half is well exposed to the
solvent.   Neither was the model supported by measurements of fluorescence
and isomerization of the <u>endo</u> vinyl analog of neoxanthobilirubinic acid
methyl ester ("halfbilirubin" methyl ester) (27). When bound to HSA this
dipyrromethenone undergoes reversible photoisomerization as efficiently and
as rapidly as does BR.

Polyenes can undergo a "bicycle pedal" motion to effect configurational
isomerization (Figure 6).   This motion does not sweep out a large volume
and so, for example, visual pigment photoisomerization can occur rapidly (a
few picoseconds) even at 77K (28). The configurational isomerization of BR
is like that of an ethylene, for example, stilbene, in which at least one
attached group must swing around and thus, sweep out a certain volume
during the process (21). If BR and the "halfbilirubin" are relatively
close packed by protein residues when they are bound to HSA, then substantial protein group motion must occur very rapidly to accommodate pigment
configurational isomerization.

It is obvious that much still remains to be done in order to understand the
photophysics of bilirubin bound to proteins. By proceeding with this work
one may also learn much more about the dynamics of the internal motions of
the proteins.

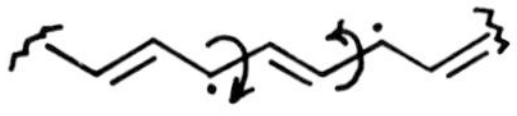

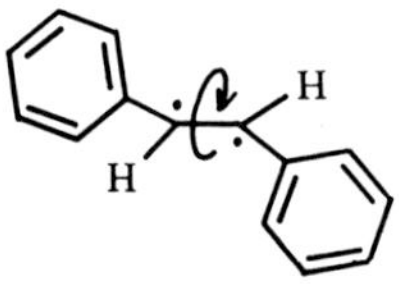

Figure 6. Representations of the lowest lying excited states of a polyene and of stilbene produced by photoexcitation.

[Note: The values of the fluorescence quantum yields listed in Table 2 are about twice as large as those reported previously by us (11,12). We had determined the yield for BR/HSA at 22 $^\circ$C relative to a reference standard of acridine orange in ethanol. We had neglected to correct for the fact that while the fluorescence from the reference was, in the absence of instrumental polarization effects, totally depolarized, the fluorescence from bilirubin had the maximum polarization because of its extremely short lifetime. The instrument-dependent polarization correction was found to be quite large, about a factor of two.]

References

1. For an overview see Brodersen, R.: CRC Critical Revs. Clin. Lab. Sci. 11, 305-399 (1979)

2. See Wells, R., Hammond, K., Lamola, A.A. and Blumberg, W.E.: Clin. Chem. 28, 432-439 (1982) and references therein

3. See references in McDonagh, A.F. and Lightner, D.A.: These Proceedings

4. Gossauer, A.: "Chemistry of Pyrroles" in Organic Chemistry in Monographs, Vol. 152, Springer-Verlag, New York, 1974

5. Falk, H. and Neufingerl, F.: Monatsch. Chem. 110, 1127-1146, 1243-1255 (1979)

6. See data in McDonagh, A.F.: "Bile Pigments: Bilatrienes and 5,15-Biladienes," in The Porphyrins, Vol. 6, Academic Press, New York, 1979

7. For a discussion of such calculations, see Murrell, J.N.: The Theory of Electronic Spectra of Organic Molecules, Methuen, London, 1963

8. Forster, Th.: "Delocalized Excitation and Excitation Transfer" in Modern Quantum Chemistry, Vol. 3, Academic Press, New York, 1965

9. Chen, R.F.: Arch. Biochem. Biophys. 160, 106-112 (1974)

10. Matheson, I.B.C., Faini, G.J. and Lee, J.: Photochem. Photobiol. 21, 135-137 (1975)

11. Greene, B., Lamola, A.A. and Shank, C.V.: Proc. Nat'l. Acad. Sci. (USA) 78, 2008-2012 (1981)

12. Lamola, A.A. and Flores, J.: J. Am. Chem. Soc. 104, 2530-2534 (1982)

13. Sloper, R.W. and Truscott, T. G.: Photochem. Photobiol. 31, 445-450 (1980)

14. Matheson, I.B.C., Curry, N.U. and Lee, J.: Photochem. Photobiol. 31, 115-120 (1980)

15. McDonagh, A.F. and Assisi, F.: Biochem. J. 129, 797-800 (1972)

16. Falk, H., Grubmayr, K. and Neufingerl, F.: Monatsch. Chem. 110, 1127-1146 (1979)

17. McDonagh, A.F., Palma, L.A. and Lightner, D.A.: Science 208, 145-151 (1980)

18. Lamola, A.A., Flores, J. and Doleiden, F.H.: Photochem. Photobiol. 35, 649-654 (1982)

19. Lamola, A.A., Junnarkar, M., Alfano, R.R. and Blumberg, W.E.: Photochem. Photobiol. 39, 88S (1984)

20. Tran, C.D. and Beddard, G.S.: Biochim. Biophys. Acta 678, 497-504 (1981)

21. Saltiel, J. and D'Agostino, J.T.: J. Am. Chem. Soc. 94, 6445-6451 (1972)

22. Lippitsch, M.E., Leitner, A., Reigler, M. and Aussenegg, F.R.: Springer. Ser. Chem. Phys. 14, 327-330 (1980)

23. Gueron, M., Eisinger, J. and Lamola, A.A.: "Excited States of Nucleic Acids" in Basic Principles in Nucleic Acid Chemistry, Vol. 1, Academic Press, New York, 1974

24. Kaplan, D. and Navon, G.: Israel J. Chem. 23, 177-186 (1983)

25. Blauer, G., Lavie, E. and Selfin, J.: Biochim. Biophys. Acta 492, 64-69 (1977)

26. Lamola, A.A., Flores, J. and Blumberg, W.E.: Eur. J. Biochem. 132, 165-169 (1983)

27. Lamola, A.A., Braslavsky, S.E., Schaffner, K. and Lightner, D.A.: Photochem. Photobiol. 37, 263-270 (1983)

28. For a discussion see Doukas, A.G., Junnarkar, M.R., Alfano, R.R. et al.: Proc. Nat'l. Acad. Sci. (USA) 81, 4790-4794 (1984)

Received August 12, 1984

Discussion

Schneider: In stilbene, the lowering of the excited state energy upon twisting is due to interaction with higher excited states (avoided level crossing). Is there theoretical evidence that a similar situation holds in BR?

Lamola: I do not know of such a theoretical treatment for the dipyrromethenone system. Prof. Falk's group has made molecular orbital calculations for dipyrromethones that clearly indicate favorable energetics for twisting in the lowest excited π,π^* state.

Gouterman: What happens to the shape of absorption and emission spectra of BR/HSA as a function of temperature?

Lamola: In 50/50 v/v ethylene glycol/water the absorption spectrum of BR/HSA shifts slightly to the red and increases in intensity as the temperature is lowered, but the effects are accounted for on the basis of the increase in refractive index of the solvent and the shrinkage of the sample. The emission increases dramatically in intensity as the temperature is lowered, the emission maximum shifts to the blue and the band narrows, both effects larger than expected for solvent refractive index effect.

Scheer: You would lower the reaction volume of the isomerization if you pre-strain the bilirubin in its bound form. Is there anything known about the energetics and the conformation of the bound bilirubin?

Lamola: The absorption spectrum of bilirubin bound to its primary site on HSA is not unusual compared to the spectra of bilirubin in various homogeneous solvents or in detergent micelles. Therefore I do not believe that binding of bilirubin to albumin is accompanied by significant strain at relevant points in the pigment molecule.

Song: What would be the effect of D_2O, substituting the H-bonded protons with deuteriums, on the twisting rate and θ_F of BR-HSA?

Lamola: We have determined the fluorescence yield of BR/HSA dissolved in a large excess of 1/1 v/v D_2O/ethylene glycol-d_6 after waiting one day for exchange. The fluorescence yield at 22°C of the deuterated sample was found to be about three times greater than that of a protonated sample. We also found that the Arrhenius barrier for temperature associated quenching of fluorescence of the deuterated systems is twice of the protonated system. That is to say, deuterium substitution (at sites which we have not determined) inhibits the rate of twisting. I cannot say whether this is due to changes in the pigment or the protein or both. Spectra were not altered by the deuteration.

Song: Is it possible that BR does not phosphoresce even at 77°K is because the localized triplet (dipyrrole $^3\pi,\pi^*$) is at higher energy than that of the stabilized exciton fluorescent state?

Lamola: Work by Truscott and Land in England on bilirubin and by Falk and his coworkers on dipyrromethenones, in which the rate of transfer of triplet excitation from various donors to the pigments was assessed, indicates that there is at least one triplet state of bilirubin that is much lower than the fluorescent state.

Blauer: The reversal in sign of the CD bands in albumin-bound photobilirubin as compared to bilirubin obviously indicates opposite chirality. Would this be due to the $Z \rightarrow E$ isomerization or would rotation around the center methylene bridge be necessary for inversion of the bands due to transition moment interactions in a different conformation of the whole molecule? In the latter case, there should be no free rotation around the methylene bond in the bound state and how would this conform with the proposed mode of binding where one dipyrromethenone should be more outside the protein?

Lamola: My prejudice is that there is little difference between HSA/PBR and HSA/BR outside of the configuration at the 15,16-double bond. I have not attempted to interpret the CD spectra in terms of structures. More "food for thought" is provided by the observation that the CD bands of BSA/PBR and BSA/BR have the same signs.

Myer: I would like to make a comment. It appears that the preferential isomerization is being interpreted considering the geometry of binding and the conformation of bilirubin in the free state. The conformation of a small molecule such as bilirubin can be significantly different when bound than when free. Although the geometry of binding may be important, it is the resulting conformation of the bound molecule which may be relatable to selectivity. Determination of the conformation in the bound state could lead to a meaningful conformation-linked interpretation.

Lamola: You misinterpret me. I am suggesting that in addition to other factors, relative flexibility of the protein environment about the two halves of the bilirubin can lead to regiospecificity. Of course, the other factors such as the conformation of the bound bilirubin may play an important rôle.

Ketterer: Do you know a site of photochemical adduct formation of bilirubin with serum albumin? This might be determined fairly readily. Extensive proteolysis should give a bilirubin-bound peptide which should be very easily separated from the rest of the polar peptides. The bilirubin-bound peptide could be purified by HPLC and its amino acid analysis could then be referred back to the known sequence of serum albumin.

Lamola: Work on photoaddition of bilirubin to albumin is due chiefly to Jori and Rubatelli and their coworkers in Padova. As far as I know they did not determine the site(s) of covalent binding of the pigment to the protein. I have determined that the quantum yield of such a product(s) must be lower than 0.002. Because it is such a low probability reaction and because little or nothing is known about mechanism I am not terribly encouraged about following up on this work.

Holzwarth: Further comment on the question of Ketterer. With regard to the formation of covalently bound products upon BR/HSA irradiation, I

should mention that we have unpublished data from a fairly extensive study
on this problem. We gave up, however, doing this because there occurs a
large number of addition products which are all formed with low quantum
yield. This chemistry is very complicated and not well-defined.

Fuhrhop: Do the photo-isomerizations of BR on albumin change the ratio
of protein-bound to water-dissolved pigment? Is there a different distribution before and after irradiation and how large is the effect?

Lamola: The affinity of Z,E-BR for albumin appears to be only a factor
of 2 or 3 lower than that for Z,Z-BR, which is $\geq 10^7 M^{-1}$, so that the amount
of free (unbound) pigment should increase with the fraction of isomerized
BR. Unless the bilirubin binding status is already quite dangerous, phototherapy does not significantly increase (≤ 10 %) the amount of unbound pigment in the circulating plasma of infants under phototherapy.

McDonagh: I think that the low fluorescence and lack of $Z \rightarrow E$ isomerization of bilirubin in water is most likely due to aggregation of bilirubin
in water. I might also add that any explanation of the regiospecificity of
the $Z \rightarrow E$ isomerization has to explain why the reaction is regiospecific,
even at bilirubin/albumin ratios of greater than unity.

In your discussion of rapid transfer back and forth between vibrational levels of diastereomeric bilirubin excited states you suggested that there
might be marked edge effects on the photochemistry. Would you elaborate on
this and explain what you mean by edge effects? Secondly, would you explain
why binding of bilirubin to albumin markedly inhibits the $E \rightarrow Z$ thermal
reversion, yet does not inhibit (even accelerates) the photochemical $Z \rightarrow E$
reaction?

Lamola: You address four points. Concerning bilirubin in water, where I
agree it is certainly aggregated, one must still find a fast relaxation
mechanism, in this case one that does not involve fluorescence. Proton
tautomerization, when it does occur, can be very fast and I suggest it as
a possibility. Reversible electron transfer is another possibility and some
argument for this may be found in the observed photoenhanced free-radical
mediated scrambling among IX, III and XIII isomers. Of course, that E-isomers are not observed may be due simply to the fact that in water they are
very unstable towards reisomerization to the Z isomers.

I can explain why regioselectivity is apparently maintained even at BR to
HSA ratios greater than unity. Let us say that in sites other than the primary binding site both Z,E and E,Z isomers are found. If these photoisomers
do not bind well to secondary sites and the E,Z isomer does not bind well
to the primary site, then the E,Z-isomer will not be stabilized with respect to thermal reisomerization and might not be observed. An experiment
to test this suggestion would be to test the relative stabilities of the
E,Z and Z,E isomers in water in the presence of HSA.

At room temperature, RT equals 200 cm^{-1}. If the difference in the purely
electronic excitation transition energies of the two chromophoric halves
of BR, say when it is bound to HSA, is significantly greater than 200 cm^{-1}
then irradiation with wavelengths at the very red edge of the absorption
band would excite only one of the chromophores and, at room temperature or
lower, the excitation would not have as much opportunity to hop to the

other half of the pigment molecule as would excitation originally of higher
energy (shorter wavelength). Depending on many details one might observe
markedly different fluorescence yield, isomerization yield and/or photoequilibrium isomer ratio for edge excitation than would be predicted on
the assumption of excitation of both chromophoric halves.

Fast thermal reversion of E isomers to Z isomers in solvents, especially
polar and hydroxylic solvents, involves acid and/or base catalysis. When
then E-isomer is protected from solvent inside the protein binding site
such catalyzed paths are precluded. Since the non-catalyzed reversion is
apparently very slow the overall reversion rate is proportional to the unbound fraction. This is nicely demonstrated by the observation we made
that the thermal reversion rates for E-isomers of dipyrromethenones corre-
late with their affinities for HSA. Lower affinity led to fast reversion.

Rapoport: Can the question of bilirubin-albumin binding be examined by
competetive binding with rigid or semi-rigid analogues?

Lamola: This would be a valid and useful approach given the apparent
great difficulty in the X-ray crystallographic approach for albumin. However, appropriate analogues do not exist and must, therefore, be designed
and synthesized, the effort required for the latter being considerable.

THE MOLECULAR MODEL OF PHYTOCHROME DEDUCED FROM OPTICAL PROBES

Pill-Soon Song

Department of Chemistry, Texas Tech University, Lubbock,
Texas 79409, USA

Introduction

Phytochrome is a tetrapyrrolic chromoprotein which plays a
vital role as the red-light photoreceptor, controlling several
diverse morphogenetic and developmental responses in plants,
such as leaf movements of _Albizzia_, flowering in shorter and
long-day plants, seed germination, stem elongation, and the
biosynthesis of chlorophyll, chloroplasts, carotenoids, and
anthocyanins [for a detailed review, see (1,2)].

Higher plants respond to red light with extreme sensitivity.
This high sensitivity is achieved by virtue of the absorption
of 660 nm light by one form of phytochrome (Pr), which is converted to the physiologically active form (Pfr) from the excited singlet state of the inactive Pr form (1,3). The photoreversible transformation of phytochrome and its relationship
to red light responses in plants are schematically shown in
Fig. 1.

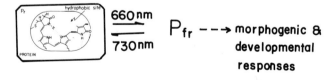

Fig. 1. A general scheme for the phototransformation of phytochrome and its functional role in higher plants.
The structure (5) and conformation (3) of phytochrome
are shown for the physiologically inactive Pr form.

In this review, the mechanism of the phytochrome phototransformation is described in terms of a molecular model based on
spectroscopic and other data. Due to page space limitation,
no attempts have been made to cover all the relevant literature.

The Chromophore and Its Topography

The chromophore of the oat Pr phytochrome is covalently linked
to an apoprotein of 124,000 mol wt via a thioether bond (4, 5).
Several proposals have been made regarding the chemical structure of Pfr (5, 6). It has been argued on the basis of the
ratios of the oscillator strengths of $Q_{x,y}$ (visible) to $B_{x,y}$
(Soret, near UV) bands of Pr and Pfr that the chromophore
structure of Pfr is not merely a conformational/configurational
isomer of the Pr chromophore. However, configurational isomerism is likely to be involved in the Pr $\longrightarrow$ Pfr phototransformation (9), even though the conformations of both Pr and
Pfr chromophores are essentially identical (3, 8, 10, 11).

The reported molecular weight of the phytochrome protein has
chronologically increased from 60,000 (pre-1973) and 118,000-
120,000 (post-1973) to the most recent value of 124,000 (oat;
12). The latest value probably represents "undegraded" phytochrome, with all previous values arising from proteolytic digestion of phytochrome during isolation and purification procedures [for review, see (1)]. Figure 2 illustrates the
terminology used for various forms of phytochrome in this
lecture.

According to the proposed model for the phototransformation of
phytochrome (1, 3, 13), the Pfr chromophore of large phytochrome is more exposed than the Pr chromophore (13-15). Recently, tetranitromethane (TNM) has been used for probing the
differential chromophore exposure of intact 124 kdalton

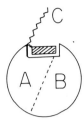

Fig. 2.   Proteolytic domains for the Pr form of 124 kDa phy-
          tochrome.   A: ca. 60,000 daltons, with chromophore,
          "small phytochrome"; B: ca. 60,000 daltons, without
          chromophore; C: 6,000 daltons, without chromophores,
          "6 kDa peptide"; A + B: ca. 118,000 daltons, with
          chromophore, "large phytochrome"; this preparation
          may contain a 114,000 dalton protein as a contami-
          nant (24); A + B + C: ca. 124,000 daltons, with
          chromophore, "intact phytochrome".

phytochrome as a more specific reagent toward tetrapyrroles
than the previously used permanganate and borohydride (11, 13-
15).   It has been reported that TNM causes a preferential
bleaching of the Pfr form of phytochrome due to a modification
of the tyrosyl residues (16).   We found that TNM is much more
specific for the tetrapyrrolic chromophores than for tyrosyl
residues.   Thus, the pseudo-first order rate constant (3.7 x
$10^{-4}s^{-1}$) for the oxidation of tyrosine (10 μM) by TNM (1.4 mM)
is 33 times lower than that for the two structurally very
different tetrapyrrolic chromophores, bilirubin and biliverdin
(15).   Furthermore, a 140 molar excess of tyrosine over the
substrate concentration showed no detectable inhibitory effect
on the rate of oxidation of the latter.

Our assumption that the differences in the rates of the TNM
oxidation of phytochrome is due to the differential accessibility of the Pr and Pfr chromophores is validated on the following grounds.   (i) The chromophore oxidation rate depends on
the size of protein and that the relative rates of oxidation
of the Pr and Pfr forms do not follow the same proportionate
dependence on the size of protein.   (ii) The model chromophores

of two distinctly different structures, i.e. bilirubin and
biliverdin are oxidized at the same rate. (iii) Kinetically,
TNM is fully specific to phytochrome chromophores over tyrosyl
residues, since the pseudo-first order rate constants reported
(15) were obtained at one-tenth of the concentration used for
tyrosine, as noted above. Furthermore, even if a critical tyrosyl residue located at the chromophore site of phytochrome
were fully accessible to TNM, its rate of oxidation would be
about 50-fold slower than that of free tyrosine. This is because of the fact that the collisional cross-section for the
tyrosyl residue in phytochrome (radius ~80 Å) (17) is reduced
by a factor of 4×10^{-3}. (iv) The chemical reactivity of the
Pr and Pfr chromophores with different chemical structures proposed for the latter in the literature remains essentially
identical, as assessed by the frontier electron and perturbation methods (18). (v) The chromopeptides of Pr and Pfr are
oxidized by TNM at about the same rate (Song, P.S., Thümmler,
F., and Rüdiger, W., to be published).

From the TNM oxidation results, it may be suggested that the
Pfr chromophore is more exposed than the Pr chromophore. However, the degree of exposure clearly depends on the size of
protein, indicating in particular that the 6000-dalton peptide
chain shields and interacts closely with the Pfr chromophore.
To what extent the 6000-dalton peptide segment enhances or re-
tards the reorientation of the chromophore in going from the
Pr to the Pfr form in vivo (19) and in large phytochrome (20)
is discussed below.

From the data shown in Table 1, it can be suggested that the
Pfr chromophore of different mol wt forms of oat phytochrome
is chemically more accessible to TNM than the Pr chromophore.
Furthermore, proton NMR peak(s) at about 6.15 ppm of the small
Pr phytochrome disappears upon its phototransformation to Pfr
in D_2O (21). The lost proton peak(s) in the Pfr form cannot
be regenerated upon photoreversion to the Pr form, suggesting

Table 1.   Relative rates of the oxidation of oat phytochrome
           and model chromophores by TNM at $2°C$.

Phytochrome	Relative rate		Rate Ratio Pfr/Pr
	Pr	Pfr	
Intact	1.00	8.41	8.3
Large	0.97	39.14	40.5
Small	5.71	88.97	15.6
Bilirubin	210.34		1.0*
Biliverdin		217.24	

*Rate (biliverdin/bilirubin) for the
model chromophores.

that a rapid hydrogen-deuterium exchange occurred during the
residency of phytochrome in its Pfr form. One of these proton
resonances has been tentatively assigned to the chromophore
(22). With large Pr phytochrome, the fluorescence quantum
yield increases in D_2O only after the protein was cycled in
D_2O:

$$Pr \xrightarrow[D_2O]{660 \text{ nm}} Pfr \xrightarrow{\text{H-D exchange}} Pr \text{ (deuterated chromophore)}$$

enhanced fluorescence
and lifetime (8, 23)

The above observations are consistent with the model that the
chromophore is significantly exposed in the Pfr form, whereas
the Pr chromophore is deeply buried in a hydrophobic pocket
(1, 8). The D_2O solvent isotope effect noted above can be explained by assuming that the Pfr-NH- protons exchanged with
deuterons during phytochrome's residency in the Pfr form.

The degree of chromophore exposure in the Pfr form is modulated by a 6 kdalton molecular mass peptide, as mentioned earlier.
Data shown in Table 1 suggest that the 124 kdalton Pfr chromophore is somewhat shielded by the peptide, compared to the Pfr

chromophore of large phytochrome.

Chromophore-Apoprotein Interactions

The absorption maxima of different molecular weight phytochromes are virtually identical for the Pr form, but the maximum for the intact Pfr is at slightly longer wavelength (730-
732 nm) than that for degraded phytochromes (720-725 nm) (24,
25). In absolute energy unit, the bathochromic shift of the
Pfr absorption maximum corresponds to less than 200 cm^{-1} or
0.6 kcal/mol, suggesting that difference in interaction forces
between the chromophore and the apoprotein among different mol
wt forms of phytochrome is small. The energy difference between the Pr and the Pfr absorption maxima of phytochromes is
approximately 1400 cm^{-1} or 4 kcal/mol. These relatively small
energy differences reflect one or more of the following interaction forces, a) hydrogen bonding, b) hydrophobic force, and
c) electrostatic forces. In addition, a small conformational
change of the chromophore as a result of interactions with
the apoprotein can also account for the above-mentioned spectral shifts. The fact that ionic, hydrophobic fluorescence
probe ANS bleaches the Pfr form of phytochrome more effectively
(26, 27) than neutral detergents (28), by forcing the chromophore to lose the anchoring forces to the apoprotein and to
resume a cyclic conformation as in free chromophores (3),
suggest the participation of hydrogen bonding and/or electrostatic forces between the chromophore and apoprotein. Although
ANS bleaches preferentially the Pfr form of large phytochrome,
as compared to the Pr form (26), it also bleaches the Pfr form
of intact phytochrome about 30% more than the Pr form (27).
At ANS concentrations greater than 2 mM, the Pfr spectrum of
intact phytochrome is completely bleached upon photocycling,
which also causes a blue shift of the absorption maximum to
620 nm, indicating that the chromophore is now forced out of
the binding crevice.

In the intact Pfr phytochrome, the chromophore appears to interact with a 6 kdalton-peptide which contains one Trp residue
(29). This may well account for the bathochromic shift and
negative circular dichroism enhancement by ca. 5% at 217 nm
(27), in going from a degraded Pfr to the intact Pfr chromophore absorption, as there is indirect evidence that the Trp
residue interacts with the Pfr chromophore in the intact phytochrome (29). In large phytochrome, lack of the interacting
Trp residue in the Pfr form has been deduced from the difference
in energy transfer efficiency from the triplet state of Trp to
the chromophores of Pr and Pfr (20). Thus, energy transfer is
minimal in the latter because of chromophore reorientation
with respect to the acceptor Trp residue at/near the chromophore crevice. On the other hand, energy transfer seems to
occur equally efficiently for both Pr and Pfr forms of intact
phytochrome (25, 27, 29), as a Trp residue in the 6-kdalton
peptide interacts with the Pfr chromophore (29).

Phytochrome preparations often contain a polyphenolic impurity,
with a phosphorescence emission maximum at 465 nm (25).
Energy transfer from this contaminant "X" to the chromophore
occurs exclusively in the Pr form, suggesting that the Pr ⟶
Pfr phototransformation is accompanied by some degree of reorientation relative to the chromophore binding proper. Similar results have been observed with exogenously added FMN.
Although FMN does not measurably bind phytochrome (30), kinetic
measurements of the FMN-sensitized phototransformation of large
phytochrome suggested that the Pfr chromophore orientation is
different from the Pr chromophore relative to the bound flavin.
Thus, the photosensitization with blue light almost exclusively
absorbed by the added FMN occurs only in the forward phototransformation, but not in the photoreversion probably due to
the reorientation of the chromophore relative to the FMN plane.

It has been mentioned earlier that ANS and other hydrophobic
substances with polar/ionic groups (e.g., saponins; 31) causes

chromophore bleaching of phytochrome, while neutral detergents
bind the Pfr protein preferentially without bleaching the visible absorbance band (28). This observation suggests that the
chromophore vicinity includes one or more electrostatic interaction sites with the apoprotein.

Although the nature of non-covalent interactions between the
chromophore and the apoprotein largely remains to be elucidated,
observations described above are consistant with the suggestion
that the chromophore planarity and its semicircular/semi-extend-
ed conformation of π-electron system are maintained via hydrophobic as well as electrostatic interactions (including hydrogen
bonding) between the chromophore and the apoprotein.

Proteins

Various studies of the protein conformation of phytochrome have
been reviewed elsewhere (1, 22, 30, 32, 33); only a limited
scope of the work carried out in this laboratory will be reviewed here due to page space restriction.

It has been pointed out that in a reversible conformational
change, proteins cannot gain or lose significant amounts of
volume. Thus, it is expected that photoreversible phytochrome
does not exhibit a large conformational change, especially
given the small energy (43 kcal/mol) of a red photon (17).

High-field NMR (21; unpublished data), quasielastic light
scattering (17), and fluorescence anisotropy and rotational
laxation time (11, 17) measurements showed that the gross conformations of degraded Pr and Pfr forms of oat phytochrome are
virtually identical. This conclusion is also valid for intact
phytochrome, on the basis of limited data available such as
high-field NMR (unpublished data obtained on Nicolet QE-300
and IBM 270 NMR spectrometers), hydrogen-tritium exchange (34)

and quasielastic light scattering data (17, 29). Stokes' radii
for intact Pr and Pfr are 83 and 80 Å, respectively, which are
not different from those for large Pr and Pfr phytochromes (80
and 81 Å respectively; 17, 29). The 6-kdalton peptide in the
former appears to play some role in determining the Stokes'
radius and hydrophobicity of the phytochrome protein, particularly at higher concentrations and ionic strengths (17, 29).
Since this peptide contains at least one Trp residue which
interacts with the hydrophobic chromophore in the Pfr form of
intact phytochrome, it is essentially free from interaction
with the Pr chromophore, as suggested earlier. It is not
surprising that intact phytochromes tend to be more hydrophobic than degraded proteins. Aggregation of the intact
proteins due to their surface hydrophobicity partially accounts
for the lack of strong spectral bleaching of the Pfr form of
intact phytochrome, as compared to the Pfr form of large phytochrome mentioned earlier (i.e. ANS bleaching).

Although the gross conformations of the intact Pr and Pfr
phytochromes are similar, there are recognizable changes in
the local conformation and/or topography of the chromophore
at/near the chromophore binding pocket. The three different
molecular weight forms (small, large and intact) of phytochrome expose an additional surface relative to the respective
Pr forms and the resulting hydrogen-tritium exchange originates
form the chromophore binding domain, i.e. domain "A" in Fig. 2
(34, 35). A major difference in hydrogen-tritium exchange
kinetics between the large and 124 kdalton phytochromes is
that the half-time for the latter is 3-4 times longer than the
former (34). This difference may be attributed to oligomerization of the intact protein and/or additional shielding of
exchangeable protons by the 6-kdalton peptide especially in
the Pfr form.

The Trp fluorescence of large phytochrome is more efficiently
quenchable with anionic quencher I⁻ ion in its Pfr form than

in the Pr form (20). The reverse is true for intact phytochrome. This is apparently due to the fact that the exposed
6-kdalton peptide contains a quenchable Trp residue. This Trp
residue is located near an electrostatically negative environment, since its fluorescence is more efficiently quenched by
cationic quencher Cs^+ ion (36). The hydrophobic quencher
acrylamide fully quenches the Trp fluorescence of both large
and intact phytochromes (20, 29, 36).

Largely based on the chemical modification of accessible amino
acid residues (16), it has been suggested that a protein conformation change, even away from the chromophore binding site,
takes place in the Pr ⟶ Pfr phototransformation of phytochrome (32). It is possible that a conformational change in
a remote area of the protein occurs only with a causative
movement, either cooperative or noncooperative, of the chromophore. Upon the Pr ⟶ Pfr phototransformation, an additional
histidine residue becomes exposed as determined by the chemical
accessibility method (16) and NMR spectroscopy (21). The
modification of the histidine residue does not abolish the
photoreversibility of phytochrome. Significantly, a histidine
residue is present in the chromopeptide segment of phytochrome
(5). It is thus possible that this histidine residue may become accessible to its modifying agent in aqueous medium as
the chromophore rearranges around the binding crevice. The
chemical modification of this histidine residue is not likely
to affect the photoreversibility of phytochrome drastically,
although it can affect the rate of photo- and dark-reversion
of Pfr. Whether the aforementioned exposure of an additional
histidine residue(s) results from the chromopeptide or the
chromophore binding site, or arises from a conformational
change occurring at a site remote from the chromophore binding
crevice, must therefore, be ascertained by a detailed kinetic
analysis of the photo- and dark-reversion of Pfr.

Molecular Model and Implications

To accommodate the results obtained so far, we proposed a model
in which the Pr⟶Pfr phototransformation generates a hydrophobic surface on the Pfr proteins as the result of a chromophore reorientation relative to the binding crevice of the
chromophore, largely based on spectroscopic characterization of
large phytochrome (3). In intact phytochrome, the 6-kdalton
peptide is highly exposed in the Pr form, whereas it interacts
closely with the Pfr chromophore, and at the same time shields
part of the chromophore binding surface. Our preliminary
studies suggest that the molecular differences between the 118
and 124 kdalton phytochromes are quantitative in nature rather
than qualitative. For example, high-field NMR spectra of both
large and intact phytochromes are similar in the aliphatic region, suggesting that their gross conformations are similar.
Also, an additional hydrogen-tritium exchange is observed upon
Pr⟶ Pfr phototransformation in both phytochromes, but the
latter shows a lesser number of exchangeable protons and
longer exchange rates. These quantitative differences and some
of what may be regarded as qualitative differences in molecular
properties, such as the lack of ANS-induced spectral bleaching
in the intact Pfr form, may well be attributable to a preferential oligomerization of the intact protein in solution.

Fig. 3 is presented to summarize presently available results
in the form of a working model for the Pr⟶ Pfr phototransformation, modified from the previous model for large phytochrome. The model is consistant with available data described
above and with the fact that the hydrophobic pocket in the
intact Pfr is not fully exposed. Linear dichroic measurements
of the immobilized phytochrome suggest that the chromophore
indeed reorient by 32° or 148° upon Pr⟶ Pfr phototransformation (37).

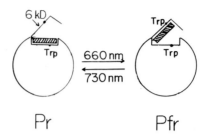

Fig. 3.   A molecular topographic model for phytochrome.
          Modified from (22).

What are the possible biological implications of the proposed
model shown in Fig. 3?   The hydrophobicity of the Pfr phytochrome generated as a result of the chromophore reorientation/
movement away from its pocket or local conformation changes
is probably responsible for the binding of large phytochrome
to ANS and liposome (26, 38) and for the oligomer formation
(17) and binding of intact phytochrome to liposome (39).

In binding of phytochrome to membrane or as-yet unidentified
receptor, both the hydrophobic surface and the chromophore-
peptide (6 kdaltons) may be involved as an anchoring device
(3,22), not unlike the interaction between signal peptide and
endoplasmic reticulum membrane.

Chloroplast movement in Mougeotia is polarotropically controlled.
Thus, red light polarized perpendicular to the long axis of
the cell is preferentially absorbed at the front and back sides
of the cell, converting dichroically oriented phytochrome (Pr)
to its Pfr form.   As Pfr builds up at the front and back of
the cell, chloroplast edges orient away from the higher Pfr
gradient (profile position). On the other hand, red light polarized parallel to the long axis of the cell is not effective
for chloroplast movement (40).

The phototransformation of phytochrome from the Pr to the Pfr

form does not seem to increase the membrane fluidity of lipo-
somes, as determined by the temperature-dependence measurements
of pyrene excimer emission (28). This observation is consistent
with the phytochrome dichroism in Mougeotia. In order to pre-
serve the dichroic orientations of Pr and Pfr in Mougeotia, it
is necessary that the phytochrome molecules in the plasmalemma
are restricted in their lateral and rotational diffusion. If
membrane fluidity changes significantly in Mougeotia plasmalemma, it is unlikely that phytochrome dichroism will be preserved
or that the dipole orientation will be changed by ca 90° upon
phototransformation. On the other hand, it is possible that
the phototransformation of Pr to Pfr generates a new binding
surface on the Pfr protein approximately perpendicular to the
binding surface plane of the Pr protein. One can thus postulate that the peripheral membrane protein Pr rotates upon its
phototransformation to the Pfr form, which then binds to the
lipid bilayer core via the newly generated hydrophobic surface.
Since both electrostatic and hydrophobic forces contribute to
the binding of phytochrome to membrane (38, 39, 41), an anisotropic rotation of the protein molecule can be envisioned.
Furthermore, a protein rotation of this sort (i.e. a rotation
of the protein molecule due to the relocation of the binding
site of the Pr and Pfr forms) can be accommodated in the chromophore reorientation/hydrophobic model of Pfr-phytochrome
(Fig. 3) by assuming that the newly generated hydrophobic surface on the Pfr protein results from the exposure and reorientation of the chromophore. Alternatively to the above
mechanism, only the chromophore reorients; the topography of
membrane binding remains fixed for the Pr- and Pfr-phytochrome.

Acknowledgements

This work was supported by the Robert A. Welch Foundation
(D-182) and National Science Foundation (PCM81-19907).

References

1. Song, P.S.: In: Advanced Plant Physiology (Edited by M.B. Wilkins), Pitman Books, London, pp. 354-379 (1984).

2. Schopfer, P.: In: Advanced Plant Physiology (Edited by M.B. Wilkins), Pitman Books, London, pp. 380-407 (1984).

3. Song, P.S., Chae, C., Gardner, J.G.: Biochim. Biophys. Acta 576, 479-495 (1979).

4. Klein, G., Ruediger, W.: Liebigs Ann. Chem. 2, 267-279 (1978).

5. Lagarias, J.C., Rapoport, H.: J. Am. Chem. Soc. 102, 4821-4828 (1980).

6. Ruediger, W.: Structure and Bonding 41, 101-141 (1980).

7. Song, P.S., Chae,Q.: Photochem. Photobiol. 30, 117-123 (1979).

8. Song, P.S.: Photochem. Photobiol. Rev. 7, 77-139 (1983).

9. Thuemmler, F., Ruediger, W.: Tetrahedron: Symp. 39, 1943-1951 (1983).

10. Song, P.S.: In: Photoreception and Sensory Transduction in Aneural Organisms (Edited by F. Lenci and G. Colombetti), Plenum Press, New York, pp. 235-240 (1980).

11. Song, P.S.: In: The Biology of Photoreceptors (Edited by D. Cosens and D.V. Prue), Cambridge University Press, Cambridge, pp. 181-206 (1983).

12. Vierstra, R.D., Quail, P.H.: Proc. Natl. Acad. Sci. USA 79, 5272-5276 (1982).

13. Hahn T.R., Kang, S.S., Song, P.S.: Biochem. Biophys. Res. Commun. 97, 1317-1323 (1980).

14. Baron, Q., Epel, B.L.: Plant Physiol. 73, 471-474 (1983).

15. Hahn T.R., Song, P.S., Quail, P.H. Vierstra, R.D.: Plant Physiol. 74, 755-758 (1984).

16. Hunt, R.E., Pratt, L.H.: Biochemistry 20, 941-945 (1981).

17. Sarkar, H.K., Moon, D.K., Song, P.S., Chang, T., Yu, H.: Biochemistry 23, 1882-1888 (1984).

18. Kim, J.H., Song, P.S.: J. Mol. Struct. 103, 67-80 (1983).

19. Haupt, W., Weisensell, M.H.: In: Light and Plant Development (Edited by H. Smith, Butterworths, London, pp. 63-74 (1976).

20. Sarkar, H.K., Song, P.S.: Biochemistry 21, 1967-1972 (1982).

21. Song, P.S., Sarkar, H.K., Tabba, H., Smith, K.M.: Biochem. Biophys. Res. Commun., 105, 279-287 (1982).

22. Song, P.S.: Annu. Rev. Biophys. Bioengin., 12, 35-68 (1983).

23. Sarkar, H.K., Song, P.S.: Biochemistry 20, 4315-4320 (1981).

24. Vierstra, R.D., Quail, P.H.: Biochemistry 22, 2498-2505 (1983).

25. Cha, T.A., Maki, A.H., Lagarias, J.C.: Biochemistry 22, 2846-2851 (1983).

26. Hahn, T.R., Song, P.S.: Biochemistry 20, 2602-2609 (1981).

27. Hahn, T.R., Quail, P.H., Sarkar, H.K., Song, P.S., Vierstra, R.D.: Manuscript to be submitted (1984).

28. Kim, I.S., Kim, E.S., Song, P.S.: Biochim. Biophys. Acta 747, 55-64 (1983).

29. Sarkar, H.K.: Ph.D. dissertation, Texas Tech University, Lubbock, TX.

30. Smith, W.O., Jr.: In: Photomorphogenesis (Encyclopedia of Plant Physiology, 16A) (Edited by W. Shropshire, Jr. and H. Mohr), Springer Verlag, Berlin, Heidelberg, New York, pp. 96-118 (1983).

31. Yokota, T., Baba, J., Konomi, K., Shimazaki, Y., Takahashi, N., Furuya, M.: Plant Cell Physiol. 23, 265-271 (1982).

32. Pratt, L.H.: Annu. Rev. Plant Physiol. 33, 557-582 (1982).

33. Pratt, L.H.: Photochem. Photobiol. 27, 81-105 (1978).

34. Hahn T.R., Chae, Q., Song, P.S.: Biochemistry 23, 1219-1224 (1984).

35. Hahn, T.R., Song, P.S.: Biochemistry 21, 1394-1399 (1982).

36. Chai, Y.G., Huh, J.W., Song, P.S.: Photochem. Photobiol. 39S, 84S (1984).

37. Sundqvist, C., Björn, L.O.: Photochem. Photobiol. 37, 69-75 (1983).

38. Kim, I.S., Song, P.S.: Biochemistry 20, 5482-5489 (1981).

39. Hong, C.B., Hahn, T.R., Song, P.S.: Photochem. Photobiol. 39S, 23S (1984).

40. Haupt, H., BioSci. 23, 289-296 (1973).

41. Furuya, M., Freer, J.H., Ellis, A., Yamamoto, K.T., Plant Cell Physiol. 22, 135-144 (1981).

Received July 18, 1984

Discussion

Fuhrhop: What does accessibility to tetranitromethane (TNM) mean? Is that supposed to be hydrophilic or hydrophobic? What is the pH? What are the assumed oxidation products?

Song: The accessibility as monitored by TNM oxidation may be defined as
a degree of the chromophore exposure relative to the chromophore pocket.
We do not have a precise picture as to the polar/hydrophobic nature of the
protein barrier of the pocket. We have not done any characterization of
the oxidation products of phytochrome and other tetrapyrroles such as bilirubin and biliverdin.

Myer: TNM reacts with a wide variety of protein groups, tyrosine, methionine, cysteine etc. What are you measuring with the TNM reaction and how
are these results related to the state of the chromophore?

Song: We measure the spectral bleaching of the chromophore upon oxidation
by TNM. Other amino acid residues, especially tyrosine, do react with TNM,
but they are oxidized >100 times slower than is the phytochrome chromophore.
For example, added tyrosine did not inhibit the chromophore oxidation by
TNM. Since the CD spectra in the far UV region are maintained under the
conditions used, we can safely assume that TNM is quite specific.

Zuber: Have you determined the number of binding sites of ANS in Pfr?
Is there a specific binding site below the chromophore? It could well be
that there are large differences between the Pr and Pfr forms (differences
in the hydrophobicity of the surface regions).

Song: We had difficulty obtaining decent Scatchard plots, but roughly
speaking Pfr binds one ANS "specifically" at the chromophore pocket vacated
by the chromophore reorientation in large phytochrome. ANS also binds "non-
specifically" at other site(s). "Specific" binding is referred to that
binding of ANS which results in spectral bleaching of the Pfr chromophore
as a result of the chromophore expelled from its pocket by ANS.

Rapoport: Is it true that none of the phytochromes - 60.000; 120.000;
124.000 - have been sequenced? Is there any special reason for this lack
of sequence data?

Song: None of the phytochromes have been sequenced. I believe two or three
groups are now involved in doing this.

Lamola: With respect to your elegant tryptophan-triplet to chromophore-
singlet energy transfer experiments: Do you know that the intrinsic tryptophan phosphorescence lifetime is the same for Pr and Pfr?

Song: This is an important question, but we have no data, as it is not
possible to remove the chromophore (and thus eliminate energy transfer).
All I can say is that ^{3}Trp lifetime of Pfr approaches that of denatured
Pr where there is minimal energy transfer.

Lamola: Do the relative intensities of tryptophan phosphorescence for
Pr and Pfr agree with the relative tryptophan phosphorescence lifetime
for Pr and Pfr?

Song: Roughly speaking, yes. This is tested with the photostationary
mixture of Pr and Pfr and we can get a reasonable agreement between the
intensities and the lifetimes.

<u>Lamola</u>:   Were the intensity data corrected for contribution from trivial (reabsorption) transfer?

<u>Song</u>:   No corrections were made since the absorbance was <0.05.

<u>Gouterman</u>:   Is the chromophore a bilirubin or a biliverdin type? Does the chromophore itself fluoresce or phosphoresce?

<u>Song</u>:   The chromophore is more like a verdin type, even though ring A is saturated, in the sense that the central methine bridge is conjugated. The chromophore itself has not been isolated, but chromopeptides are probably fluorescent. I do not recall anyone having observed phosphorescence. I think it is probably because of an extremely short lifetime of the fluorescent state, compared with intersystem crossing.

<u>Scheer</u>:   I have a puristic comment concerning the terms accessibility, which is a kinetic measure, and stability, which means thermodynamics. It is very difficult to distinguish these two, and it requires to go circles where you get to the same state once directly and once via a state in which the chromophore is fully accessible. This is not yet possible with phytochrome.

<u>Song</u>:   I think you are right. We use "accessibility" only as an operative term.

<u>Scheer</u>:   I have also a specific question. You stated that the phytochrome chromophore has essentially no absorption around 290 nm. Can you state this quantitatively, because the extended isophorcabilin has an ε of 8000.

<u>Song</u>:   We have qualitative data. There is indeed prompt fluorescence from the chromophore upon excitation at 290 nm. But the absorption is certainly much smaller than in the visible and that by Trp in the UV spectral range.

<u>Falk</u>:   A shift of 10 nm on going to the native form of Pfr may be due to any change in the vicinity of the chromophore and should not depend on a Trp residue coming near. A twisting at any single bond of the chromophore methine fragment of about 5^{0} can cause this shift and from the standpoint of energy such twisting can be achieved easily by any external influence.

<u>Song</u>:   Your comment on the 10 nm shift in intact Pfr is well taken. We would certainly keep this point in mind in interpreting the spectral shift.

<u>Falk</u>:   The intensifying of a doublet-like ^{1}H-NMR signal in the aliphatic region on going from Pr to Pfr cannot be due to a conformational change - at best it is indicating that a certain conformation already present in the Pr-form becomes more populated.

<u>Song</u>:   Your comment on the NMR peaks is well taken. At this point, we were simply trying to see any NMR differences between Pr and Pfr. We saw a noticeable difference with intact protein in the aliphatic proton resonance region, and perhaps we should not take this as a definite indication of conformational difference. We are content to say there is a subtle difference between the two phytochrome forms.

Falk: I am surprised that TNM does not oxidize the (Z)- and (E)-diastereo-
mers with different velocities.

Song: I might also add to your comment that bilirubin and biliverdin of
two entirely different structures are oxidized at the same rate.

Falk: As you think Pr is solvated within the cell fluid, whereas Pfr is
attached to a membrane, what do you think are the mechanistic implications
of action dichroism?

Song: The action dichroism in *Mougeotia* and *Protonemata* as well as the
linear dichroism of immunochemically immobilized Pr and Pfr forms carried
out by Björn and his coworkers are supportive of our model involving the
chromophore reorientation upon Pr⟶Pfr transformation. The action dichroism most likely suggests that phytochrome is membrane-bound in these systems.

Eilfeld: Can you explain with your model, why your Pfr of large phytochrome 114/118 kD is immediately bleached by ANS, Pr is not, and native
Pr and Pfr is not, only upon irradiation?

Song: The native Pfr is also preferentially bleached by ANS if [ANS]
>2 mM is used. Also, cycling in the presence of ANS leads to some bleach-
ing. However, the magnitude of bleaching is certainly much less than in
large phytochrome. I think the difference here is due to the possibilities
that 1) intact phytochrome tends to exist as larger oligomers than large
phytochrome and 2) the 6-kDa peptide interacts with the Pfr chromophore,
covering part of the chromophore crevice, which otherwise is fully exposed
for ANS binding in large phytochrome. This explanation is shown in our
model described in the text.

PHYTOCHROME, THE VISUAL PIGMENT OF PLANTS:
CHROMOPHORE STRUCTURE AND CHEMISTRY OF PHOTOCONVERSION

Wolfhart Rüdiger, Peter Eilfeld and Friedrich Thümmler
Botanisches Institut der Universität München,
Menzinger Str. 67, 8000 München 19

Introduction

The process of vision is traditionally considered to be
typically for the animal kingdom. However, plants are able
to sense both quantitative and qualitative changes in the
light conditions under which they grow. It is obvious that
photosynthetic organisms should show genetic adaption to
light. But it is nevertheless amazing to see that all major
stages of development of the individual plant, (e.g. seed
germination, later stages of vegetative growth, transition
to reproductive development, senescence) are influenced by
the environmental factor light. This phenomenon called
"photomorphogenesis" (1) requires photoreceptor pigments
which are able to transduce the light signal into the proper
physiological effects. The best known plant photoreceptor is
phytochrome (2) the chromophore of which has been known since
1969 to be a bilin, i.e. an open tetrapyrrole (3). Phytochrome has the remarkable property of photoreversibility,
i.e. it exists in the two forms Pr and Pfr which are inter-
convertible by light.

$$Pr \underset{730 \text{ nm}}{\overset{660 \text{ nm}}{\rightleftharpoons}} Pfr$$

The Pr form (absorption maximum at 667 nm) is physiologically
inactive whereas the Pfr form (absorption maximum at 730 nm)
is the physiologically active form. The different absorption

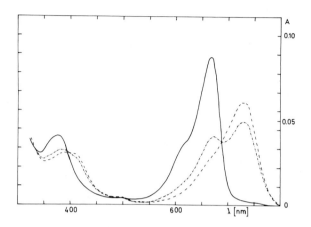

Fig. 1 Absorption spectra of native oat phytochrome in 20 mM
       potassium phosphate buffer, pH 7.8, containing 1 mM
       EDTA, at 4°C.
       (———) Pr form, (—·—·—) photoequilibrium after saturat-
       ing irradiation at 660 nm (———) calculated spectrum
       for pure Pfr assuming 80% Pfr and 20% Pr in photo-
       equilibrium (Eilfeld and Rüdiger, unpublished
       results).

spectra (Fig. 1) reflect differences in the chromophore
electronic transitions in the Pr and Pfr form, respectively.
These differences could be due either to changes in the
chromophore itself (e.g. chemical structure, conformation)
or to changes in the protein environment which in this case
should interact differently with the chromophore. A combination of both changes in the chromophore and the protein
environment could even more easily explain the observed
optical properties of Pr and Pfr. We wish to describe here
the present state of knowledge concerning the chemical
structure of the chromophore in Pr and Pfr form. Furthermore, the photoconversions Pr → Pfr and Pfr → Pr which
involve several intermediates shall be compared with the
photoconversion in the animal visual pigment rhodopsin.

The Pr chromophore

A similarity of the Pr chromophore and the chromophore of
phycocyanin was suggested (4) long before the chemical
structure of either of these chromophors was elucidated. The
elucidation of the structure of the Pr chromophore proceeded
along the same lines as that of the phycocyanin chromophore:
Chromic acid degradation (3,5,6) allowed deduction of
ß-substituents; UV-Vis spectra of the chromophore after
protein denaturation (7) or proteolysis (8) revealed the
conjugated system; cleavage of the free chromophore (9)
allowed the identification with phytochromobilin obtained by
total synthesis (10); the structure deduced from these
studies was confirmed by high-resolution ^{1}H NMR spectroscopy
of a small chromopeptide obtained by proteolysis from Pr
(11). It should be kept in mind that these results refer to
the chromophore in either denatured Pr or small chromopeptides derived therefrom. It is called Pr' chromophore (1a).
    The phycocyanin chromophore (1b) differs from 1a in having
an 18-ethyl instead of an 18-vinyl side chain.

Whereas the optical properties of denatured Pr and of
chromopeptides derived therefrom are more or less identical
(Pr' chromophore) those of native Pr are very much different
(fig. 2). The long-wavelength band of the native Pr chromophore is much sharper and has a much higher extinction
coefficient than that of the Pr' chromophore. Denaturation
or proteolysis of native Pr therefore causes a decrease in
extinction of a factor of 3 at 660 nm (12). The change in
the short-wavelength band of Pr is negligible upon denaturation or proteolysis (12). The optical properties of
native Pr are primarily determined by perturbation of the
chromophore by the protein. The absorption spectra of the
Pr' chromophore resemble those of free bile pigments in
solution. These flexible molecules can occur in a great
number of conformations of which cyclic conformations

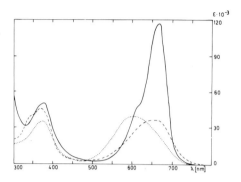

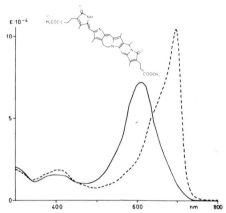

Fig.2 Absorption spectra of
chromopeptides obtained
from phytochrome (Pr or
Pfr) by pepsin digestion
measured in aqueous
formic acid. (---) pep-
tide from Pr = Pr'
chromophore cation; (·····)
peptide from Pfr = Pfr'
chromophore cation. This
is transformed into the
Pr' chromophore cation
by 10 s irradiation with
white light. For compari-
son, the absorption spec-
trum of Pr in phosphate
buffer, pH 7.8 (——) is
included

Fig.3 Absorption spectra of
isophorcabilin (struc-
ture inserted) in
methanol (——) and in
methanol/1% HCl (---)
(Scheer and Kufer,
unpublished results)

1a : R = CH=CH₂
1b : R = C₂H₅

2a : R = CH=CH₂
2b : R = C₂H₅

(see structure 1a,b)predominate in solution (13,14). The
prediction that bile pigments have more pronounced long-
wavelength absorption bands in extended than in cyclic conformations has been experimentally verified by investigation
of isophorcabilin which has a fixed extended conformation
(structure in fig. 3). The absorption spectrum of protonated
isophorcabilin (fig. 3) surprisingly matches the absorption
spectrum of the native Pr chromophore much better than that
of non-protonated isophorcabilin. Perturbation of the
chromophore by the native protein may therefore be mainly
a) stretching of the chromophore, i.e. stabilizing an
extended or semi-extended conformation and b) protonating
the chromophore, i.e. stabilizing the chromophore cation.
Of course, further effects of the protein (see fig. 6) cannot
be excluded by this consideration.

The Pfr Chromophore

As outlined in the Introduction, the difference between the
optical properties of Pr and Pfr could be due to differences
(1) of chromophore structure, (2) of chromophore conformation, (3) of the protein environment or a combination of
these effects. Differences due to effects (2) and (3) require
the native protein and therefore should disappear by unfolding or proteolysis of the protein. Differences in the
chromophore structure on the contrary should survive unfolding or proteolysis of the protein.

The absorption spectrum of the Pfr' chromophore in either
the chromopeptide (15) (fig. 2) or in denatured Pfr (not
shown) is quite different from that of native Pfr (fig. 1).
It can be deduced from these results that the protein has an
important influence upon the optical properties of the
chromophore in native Pfr (effects (2) and (3), see above).

This influence is removed by denaturation or proteolysis whereas the chemical structure remains typical for the Pfr or Pfr' chromophore. When the first results on denaturation of Pfr became known (7) it was argued that the chemical structure of the Pfr' chromophore could be an artifact produced by the denaturating agent (e.g. acid guanidinium chloride) (16). But this was excluded later when it was shown that the same chromophore (i.e. a chromophore with the same optical and chemical properties) is always obtained from native Pfr under various conditions, also without reagents like guanidinium chloride (15,17). This chromophore is refered to here as the Pfr' chromophore.

The structure of the Pfr' chromophore (2a) was demonstrated to be the 15E isomer of the Pr'chromophore (1a) by a variety of methods. Chromic acid and combined chromic acid-ammonia degradation yielded the same products from the Pfr' as from the Pr' chromophore; this means that ß side chains and covalent linkage to the peptide are identical in 1a and 2a (6). The UV-Vis spectrum of 2a (fig. 2) is characteristic for the cations of E isomers of open tetrapyrrols, especially the low extinction of the short-wavelength band and the hypsochromic position of the long-wavelength band (15,18,19). Characteristic is also the low stability of 2a towards oxidation and reduction as compared to 1a (15) and especially the easy photoisomerization of 2a to 1a (see fig. 2). This could be demonstrated also for all relevant model compounds (18-20) including chromopeptides from phycocyanin. The final prove for the E configuration and also for the localization of the E methine bridge at C-15 came from high resolution ^{1}H NMR measurements of the Pfr' chromopeptide in comparison with the E chromopeptide from phycocyanin (21,22).

These results prove that the photoconversion of phytochrome involve Z,E- (or cis, trans-) isomerization of the chromophore. However, this reaction per se cannot explain the

difference in the optical properties of native Pr and Pfr.
The absorption bands of model pigments show a hypsochromic
shift upon Z → E isomerization only if steric hindrance is
increased and therefore a twisted conformation is induced
instead of a cyclic quasi-planar conformation by the isomerization (23). In small chromopeptides from phycocyanin
and phytochrome, Z,E isomerization does not change the
position of the absorption bands at neutral pH (19). The
characteristic shift can only be detected after acidification, i.e. after formation of the chromophore cation
(see fig. 2).

Dehydration of native phytochrome by transfer into 100%
glycerol yields products with the absorption spectrum of
small chromopeptides (17). These "bleached" forms of Pr or
Pfr (free bases of Pr'and Pfr') have nearly identical
absorption spectra at neutral pH but they are distinct
compounds: after acidification, dehydrated Pfr shows the
absorption of the Pfr' chromophore cation·dehydrated Pr
that of the Pr' chromophore cation (17). Furthermore,
bleaching by dehydration is reversible: rehydration of
bleached Pfr yields native Pfr, rehydration of bleached Pr
yields native Pr (not shown). It can be concluded from these
results that the chromophore configuration (15 E in Pfr,
15 Z in Pr) survives bleaching by dehydration whereas those
influences of the protein, which make up the specific optical
properties of native Pr and Pfr are (reversibly) abolished
by removal of water.

What are the specific influences of the protein which cause
the bathochromic shift of the Pfr chromophore from 630 nm
(100% glycerol) to 730 nm (native state)? This extreme shift
cannot be produced only by changes of the chromophore conformation which is anyhow similar in native Pr and native
Pfr (24). It had been suggested that the native Pfr

chromophore is deprotonated because anion formation of model
chromophores is connected with strong bathochromic shifts (7).
This is questionable, however, because the pK_a values of (E)
model compounds (25) are higher (e.g. 13.6) than those of the
corresponding (Z) model compounds (e.g. 11.9). Charged groups
in the protein environment can, under certain conditions give
a similar bathochromic shift of the chromophore absorption
(26,27). It is interesting in this context that the shift is
larger in native phytochrome (to 730 nm, 124 kdalton)   than
in partially degraded, but fully photoreversible phytochrome
forms (to 720 nm, 118 to 60 kdalton). This will be discussed
in detail in connection with phytochrome intermediates.

Intermediates

As pointed out above, the native Pfr chromophore differs
from the native Pr chromophore in more than one respect. It
is therefore obvious that the photoconversions Pr $\rightarrow$ Pfr and
Pfr $\rightarrow$ Pr imply several steps of chromophore alteration which
should proceed in a certain order. Those steps which involve
alterations of the optical properties of Pr or Pfr chromophores are detectable by spectroscopy; such spectrally
altered phytochrome species are conventionally called intermediates. Intermediates have been detected (review in 16)
and also more recently investigated by flash photolysis
(28-31) and by low-temperature spectroscopy (32-34). In most
cases, the intermediates were characterized merely by
absorption difference spectroscopy. In order to obtain true
absorption spectra, one has to consider all possible artifacts (light scattering by freezing, temperature-dependent
absorption changes etc.) (33).

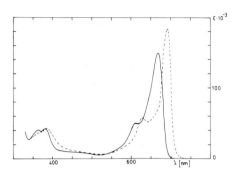

Fig. 4 Absorption spectra of lumi-R (---) and Pr (——) at
-160°C. Solvent: 100 mM potassium phosphate buffer,
pH 7.8, containing 14 mM mercaptoethanol and 5 mM
EDTA, diluted with glycerol (final conc. 66%).
The spectrum of lumi-R was calculated with ε 694
= 190 000 (33).

The first detectable intermediate on the pathway Pr ⟶ Pfr,
called lumi-R, is stable at temperatures below -120°C but
it rapidly decays at room temperature. At 4°C, only the last
part of its decay has been observed on the nanosecond time
scale in flash photolysis experiments (28). In fig. 4, its
absorption spectrum including the range 330-500 nm is
reported for the first time. Characteristic is the bathochromic shift of both chromophore absorption bands. The
oscillator strength of the long wavelength band is slightly
larger, vibrational fine structure is more pronounced than
in Pr. The corresponding changes in the short wavelength
band are only small upon transition from Pr to lumi-R.

Upon increasing the temperature, the next intermediate,
meta-Ra, is formed in increasing amounts by dark relaxation
of lumi-R. Pure meta-Ra is formed by irradiation of Pr at
-85°C to -65°C. Its absorption spectrum (not shown) can be
calculated from the absorption difference spectrum (fig. 5).
The short-wavelength band is nearly identical with that of
lumi-R whereas the unstructured long-wavelength band shows

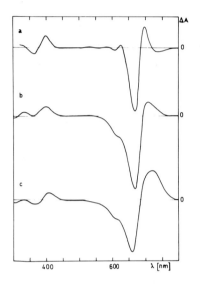

Fig. 5 Absorption difference
spectra of red irradiated
phytochrome (formation of
intermediates) minus un-
irradiated Pr. Solvent as
in fig. 4
a) Pr (114/118 kdalton)
   2 min. irradiation at
   -83°C (meta-Ra);
b) Pr (114/118 kdalton)
   15 min. irradiation at
   -27°C (meta-Rb);
c) Pr (124 kdalton)
   8 min. irradiation at
   -27°C (meta-Rc).

a hypsochromic shift and a smaller extinction coefficient
compared to lumi-R. In summary, the absorption spectrum of
meta-Ra resembles that of the free base of isophorcabilin
(see fig. 3).

The next intermediate, meta-Rb, has also been called Pbl
because its long-wavelenght band is strongly bleached in
comparison with all other intermediates. A cyclic conformation has been proposed for the chromophore in meta-Rb
whereas extended or semi-extended chromophore conformations
are assumed for all other phytochrome forms (34). A similar
"bleached" chromophore (Pr') is also found in denatured Pr,
in the chromopeptide (see fig. 2), in dehydrated, and in
chemically bleached phytochrome (17,35). We assume that the
specific chromophore protein interaction is interrupted in
all these cases, so that the chromophore can accommodate

itself to the most favorable (cyclic) conformation. Recent
low-temperature absorption measurements revealed that
meta-Rb is the main product of Pr irradiation at -40°C only
with degraded phytochrome (60 kdalton or 114-118 kdalton)
whereas native phytochrome (124 kdalton) yields another
product instead. This product for which we propose the name
meta-Rc, is characterized by an absorption maximum at
710-716 nm (calculated from the difference spectrum, see
fig. 5). It has also been observed in vivo (36). The final
product, Pfr, is formed with native phytochrome at -25°C or
higher temperatures.

On the pathway Pfr $\longrightarrow$ Pr, we observe two intermediates which
we call here lumi-F and meta-F. Lumi-F predominates at -140°C,
meta-F at -60°C. The absorption spectra of both intermediates
(not shown) are very similar. Lumi-F of native phytochrome
(124 kdalton) has the absorption maximum at 675 nm. The
absorption difference spectrum (lumi-F minus Pfr) is similar
to the corresponding difference spectrum of red-preirradiated
pea epicotyl tissue irradiated with far-red light at -196°C
(37). We find the absorption maximum of meta-F at 660 nm in
accordance with the findings of Burke et al. (34). Both
intermediates are formed from native as well as from
partially degraded phytochrome.

Comparison with rhodopsin and bacteriorhodopsin

Phytochrome does not seem to be related with rhodopsin or
bacteriorhodopsin. These chromoproteins contain the iso-
prenoid retinal as the chromophore; the former transforms
light signals finally into nerve impulses, the latter acts
as a light-driven proton pump. Phytochrome contains a bile
pigment chromophore; the exact mechanism of its biological
action is still unknown. But it is clear that phytochrome

action implies membrane effects as well as differential gene
activation. However, there are similarities on the molecular
level: all chromoproteins show dark relaxation steps after
the primary photoreaction, i.e. they form intermediates, and
all chromophores undergo cis-trans (or Z,E) isomerization
upon irradiation. The cis-trans isomerization is a good
candidate for the primary photoprocess, although other
candidates (like intramolecular proton transfer (38)) cannot
be excluded at present. Honig et al. (39) have proposed an
interesting model for bacteriorhodopsin and rhodopsin in
which Z,E isomerization is considered as the primary photoreaction and proton transfer an eventual later process. We
have applied the general considerations of these authors to
phytochrome and compared the results with the optical
properties of phytochrome intermediates.

The conclusions are shown in fig. 6. The Pr form "1" contains
the ZZZ chromophore cation in an extended conformation perturbed by a counter-anion of the protein. Lumi-R "2" contains
th ZZE chromophore cation with a more distant counter-ion.
Both "1" and "2" are similar to the isophorcabilin cation;
perturbation by the negative point charge in "1" leads to a
hypsochromic shift as compared to "2" (Scharnagl and Schneider, unpublished results). Deprotonation of "2" leads to
meta-Ra "3" which is similar to isophorcabilin (free base).
The influence of a positive charge (Y^+) near to C-10 or C-18^2
leads to a large bathochromic shift (Scharnagl and Schneider,
unpublished). This influence varies with the distance chromophore —— Y^+. It is most pronounced in native Pfr"4" (730 nm)
but less in partially degraded Pfr (720 nm) or in meta-Rc
(710-716 nm). The photochemical step from Pfr to lumi-F "5"
implies E ⟶ Z isomerization; the chromophore moves away from
Y^+, the influence of which is abolished by this means. The
optical difference to meta-F "6" is very small; this step
probably involves only accommodation of the protein into a
new conformation. The influence of the counter-ion X^- comes

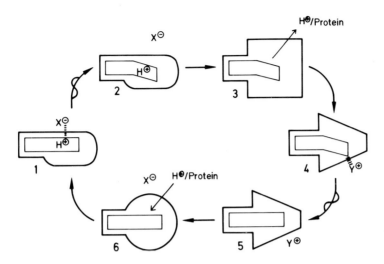

Fig. 6 Scheme of phytochrome photoconversion
1=Pr, 2=lumi-R, 3=meta-Ra, 4=Pfr, 5=lumi-F, 6=meta-F.
〜〉photoreaction; ⟶ thermal dark relaxation. ▭ ZZZ
chromophore; ◸ ZZE chromophore. $X^{\ominus}$, $Y^{\oplus}$ charged groups
of the protein interacting with the chromophore;
ıı electrostatic interaction between chromophore and
protein. The different shapes of the protein bag
reflect conformational changes of the protein.

into play only after protonation of the chromophore to the
final product Pr. Although this model contains much that is
speculation, it can explain experimental data including pH
changes during phototransformation (40) better than a model
in which only charge positions are varied (41).

Acknowledgement

We thank the Deutsche Forschungsgemeinschaft for financial
support and the NATO for a travel grant (SA 5205RG 545/83).

References

1.   Shropshire, W.jr., Mohr, H. (eds.) Photomorphogenesis.
     Encyclopedia of Plants Physiology New series Vol. 16A,B,
     Springer, Berlin, Heidelberg, New York, Tokyo (1983)

2. Rüdiger, W.: Structure and Bonding 40, 101-140 (1980)
    Rüdiger, W., Scheer, H. in Ref. 1, pp. 119-151

3. Rüdiger, W., Correll, D.L.: Liebigs Ann. Chem. 723,
    208-212 (1969)

4. Parker, M.W., Hendricks, S.B., Borthwick, H.A.: Bot. Gaz.
    111, 242-252 (1950)

5. Rüdiger, W. in: Phytochrome. Mitrakos, K., Shropshire,
    W.jr. (eds.) p. 129-141, Acad. Press, London, New York
    (1972)

6. Klein, G., Grombein, S., Rüdiger, W.: Hoppe-Seyler's
    Z. Physiol. Chem. 358, 1077-1079 (1977)

7. Grombein, S., Rüdiger, W., Zimmermann, H.: Hoppe-Seyler's
    Z. Physiol. Chem. 356, 1709-1714 (1975)

8. Fry, K.T., Mumford, F.E.: Biochem. Biophys. Res. Commun.
    45, 1466-1473 (1971)

9. Rüdiger, W., Brandlmeier, T., Blos, I., Gossauer, A.,
    Weller, J.-P.: Z. Naturforsch. 35c, 763-769 (1980)

10. Weller, J.-P., Gossauer, A.: Chem. Ber. 113, 1603-1611
    (1980)

11. Lagarias, J.C., Rapoport, H.: J. Am. Chem. Soc. 102,
    4821-4828 (1980)

12. Brandlmeier, T., Scheer, H., Rüdiger, W.: Z. Naturforsch.
    36c, 431-439 (1981)

13. Scheer, H.: Angew. Chem. 93, 230-250 (1981)

14. Braslavsky, S.E., Holzwarth, A.R., Schaffner, K.:
    Angew. Chem. 95, 670-689 (1983)

15. Thümmler, F., Brandlmeier, T., Rüdiger, W.: Z. Natur-
    forsch. 36c, 440-449 (1981)

16. Kendrick, R.E., Spruit, C.J.P.: Photochem. Photobiol. 26,
    201-244 (1977)

17. Thümmler, F., Rüdiger, W.: Physiol. Plant 60, 378-382
    (1984)

18. Kufer, W., Cmiel, E., Thümmler, F., Rüdiger, W., Schnei-
    der, S., Scheer, H.: Photochem. Photobiol. 36, 603-607
    (1982)

19. Thümmler, F., Rüdiger, W.: Tetrahedron 39, 1943-1951
    (1983)

20. Falk, H., Müller, H., Schlederer, T.: Monatsh. Chem. 111,
    159-175 (1980)

21. Rüdiger, W., Thümmler, F., Cmiel, E., Schneider, S.:
    Proc. Natl. Acad. Sci. USA 80, 6244-6248 (1983)

22. Thümmler, F., Rüdiger, W., Cmiel, E., Schneider, S.:
    Z. Naturforsch. 38c, 359-368 (1983)

23. Falk, H., Grubmayr, K., Kapl, G., Müller, N., Zrunek, U.: Monatsh. Chem. 114, 753-771 (1983)

24. Song, P.S., Chae, Q., Gardner, J.D.: Biochim. Biophys. Acta 576, 479-495 (1979)

25. Falk, H., Wolschann, P., Zrunek, U.: Monatsh. Chem. 115, 243-249 (1984)

26. Sugimoto, T., Ishikawa, K., Suzuki, H.: J. Phys. Soc. Japan 40, 258-266 (1976)

27. Falk, H., Müller, N., Purschitzky, A.: Monatsh. Chem. 115, 121-124 (1984)

28. Braslavsky, E.E., Matthews, J.J., Herbert, K.J., de Kok, J., Spruit, C.J.P., Schaffner, K.: Photochem. Photobiol. 31, 417-420 (1980)

29. Cordonnier, M.M., Mathis, P., Pratt, L.M.: Photochem. Photobiol. 34, 739-740 (1981)

30. Pratt, L.M., Inoue, Y., Furuya, M.: Photochem. Photobiol. 39, 241-246 (1984)

31. Inoue, Y., Konomik, K., Furuya, M.: Plant Cell Physiol. 23, 731-736 (1982)

32. Song, P.S., Sarkar, M.K., Kim, I.S., Poff, K.L.: Biochim. Biophys. Acta 635, 369-382 (1981)

33. Rüdiger, W., Thümmler, F.: Physiol. Plant 60, 383-388 (1984)

34. Burke, M.J., Pratt, D.C., Moscowitz, A.: Biochemistry 11, 4025-4031 (1972)

35. Eilfeld, P., Rüdiger, W.: Z. Naturforsch. 39c, in press (1984)

36. Spruit, C.J.P., Kendrick, R.E.: Photochem. Photobiol. 18, 145-152 (1973)

37. Spruit, C.J.P., Kendrick, R.E.: Photochem. Photobiol. 26, 133-138 (1977)

38. Sarkar, H.K., Song, P.S.: Biochemistry 20, 4315-4320 (1981)

39. Honig, B., Ebrey, T., Callender, R.H., Dinur, U., Ottolenghi, M.: Proc. Natl. Acad. Sci. USA 76, 2503-2507 (1979)

40. Tokutomi, S., Yamamoto, K.T., Miyoshi, Y., Furuya, M.: Photochem. Photobiol. 35, 431-433 (1982)

41. Sugimoto, T., Inoue, Y., Suzuki, H., Furuya, M.: Photochem. Photobiol. 39, 679-702 (1984)

Received July 2, 1984

Discussion

Buchler: To the non-expert the enormous bathochromic shift of the main absorption band occurring on going from Pr to Pfr remains amazing. You explained it by assuming the formation of a positive charge in the proximity of the chromophore, quoting molecular orbital calculations in favour of this interpretation. Dr. Falk yesterday reported efforts of synthesizing phytochrome analogues that did carry such positive charges and did not show these large bathochromic shifts. Therefore, it may be essential that this positive charge has a very specific orientation with respect to the chromophore plane. Could you explain the orientations that were used in the molecular orbital calculations showing these bathochromic shifts? This could perhaps be clarified by further synthetic efforts putting a positive charge to the right (or predicted) position with respect to a model chromophore, although these synthetic efforts will be hard.

Falk: Synthetic efforts directed towards this goal are in progress in my laboratory.

Eilfeld: Prof. Falk's experiment is not necessarily an argument against our model. In our model, the charge is fixed at a very certain position. In Falk's experiment, the charge can move and the absorption spectrum he observed was only an averaged one, special effects cancelled out. On the model, Prof. Schneider can give you more information.

Schneider: The calculations show that a positive charge at the right position can cause a bathochromic shift. The question which remains unanswered is, therefore, is there a positive charge which could interact with the chromophore?

Eilfeld: There is at least an indirect hint that there is a positive charge nearby the chromophore in Pfr. ANS (Aminonaphthalenesulfonic acid) which at pH 7.8 is an anion, bleaches preferentially Pfr, yielding P_{BL}. Pr is only slightly affected. This can be interpreted as a charge interaction between the positive charge on the protein and the negatively charged ANS. Therefore, ANS intercalates between the protein and the chromophore, interrupting their specific interaction.

McDonagh: Why does phycocyanobilin not undergo a photochemical $Z \longrightarrow E$ isomerization similar to that which occurs in phytochrome?

Eilfeld: The proteins of phycocyanobilin and phytochrome differ a lot. The protein of phytochrome is designed to support the photoconversion. In the case of the peptides, photoconversion Pr-peptide $\longrightarrow$ Pfr-peptide is not possible due to the missing native protein. Similarly, with phycocyanobilin there seems to be no supporting interaction between the chromophore and the protein.

Scheer: Phycocyanin is not photoreactive in its native state, but becomes reactive if it is tickled with e.g. urea, thiocyanate, or low pH, as shown by several groups. This may indicate, although there is no as sound a proof as in phytochrome, that Z,E-isomerization can start in the phycobiliproteins if the coupling with the chromophore is loosened.

Song: Having drawn a cartoon model for phytochrome, I can tell you that your model will get two responses: one, total rejection of the model because it is in cartoon form, and two, your model is taken 100 % literally. In this regard, I feel your apoprotein model shows a drastic change upon Pr⟶Pfr transformation. The work of Pratt and Butler on small phytochrome suggests that entropy changes are not as pronounced in phytochrome as in rhodopsin studied by Lamola et al.

Scheer: I agree with your comments regarding cartoons, but I should make one remark regarding the scheme shown by Dr. Eilfeld. The shape is not to mean the entire protein, but rather the cavity in which the chromophore is located, and this may change considerably as your results have shown.

Eilfeld: In my scheme only the "binding site", i.e. the protein region nearby the protein, is depicted. The complete protein is much larger. I agree with you that gross conformational changes are small, but I would say that local changes nearby the chromophore must be very dramatic which are, however, compensated by changes in other regions of the protein.

Lamola: In the system of bilirubin bound to its primary site, where it is quite clear that the primary photochemical process is twisting of one of the terminal pyrrole groups about its exocyclic double bond, the primary process is totally shut off at low temperatures where the protein environment is stiff. This appears to be in contrast to the interpretation of fast Z⟶E isomerization in phytochrome at low temperatures. Would you please comment on this.

Eilfeld: Phytochrome is a molecule specifically designed by nature to respond to light signals. The primary photoprocess occurs even at 77^0K (Pratt, Butler etc.) or at room temperature in the picosecond range. We therefore assume that there is a protein cavern in which the Z,E-isomerization can be induced by light without hindrance by the protein. Of course, further reactions cannot take place at 77^0K due to the rigidity of the protein surroundings.

Song: The critical point of Dr. Lamola's question is whether isomerization is the primary process at 77^0K, as the isomerization is inhibited in BR-albumin at low temperature. Your difference spectrum between Pr and Lumi-R shows no significant alteration in the spectral shape, suggesting that at least the conformations of the two forms are similar.

Eilfeld: In fact, Pr and lumi-R are spectrally very similar. Therefore, there are no changes in chemical structure like deprotonation. In our model, only change of configuration is involved in the primary photoprocess. As Prof. Falk has pointed out, configuration has a small influence upon the absorption spectra. This we can observe upon transition from Pr to lumi-R. Sterical requirements of such a photoisomerization $Z_{15}⟶E_{15}$ are also not very critical.

Song: The Pfr chromopeptide absorbs at 600 nm, as compared to ∿650 nm for the Pr form. This is a large shift. Do you think only the configurational isomerism and protein-chromophore interactions are involved in Pr⟶Pfr transformation? In other words, is the chemical structure of the Pfr same as that of Pr?

Eilfeld: The shift observed with the chromopeptides was explained by
Prof. Falk in terms of conformational requirements of the configurational
isomerism. Starting form native Pr and Pfr, enzymatic digest yielded chromopeptides. Their structures have been confirmed by ^{1}H-NMR. Upon enzymatic
digest, chemical changes of the chromophore are very unlikely. Denaturation
with a variety of reagents yields species with the same spectra as the
chromopeptides. But, as has been demonstrated with the model compound
isophorcabilin, the chromophore in the Pr state is protonated. Therefore,
in our model, three factors determine the spectra of native Pr and Pfr:
a) Stretching the chromophore ($\longrightarrow$extended conformation); b) Protonation
in the Pr form; c) Charged groups on the protein residing nearby the chromophore.

Falk: What about the temperature dependence of the spectra of the phototransformation intermediates?

Eilfeld: Molar extinctions as well as absorption maxima of Pr, Pfr, or
intermediates, change with temperature (e.g. Pfr λ_{max} = 732 nm at 20^{0}C,
λ_{max} = 745 nm at -140^{0}C). For comparison, it is therefore necessary to
obtain spectra at the same temperature. For this purpose, we produced intermediates by irradiation at a certain temperature (e.g. Pfr irradiated
at -65^{0}C yields meta-F). Then the samples were rapidly cooled down to
-140^{0}C and their spectra were recorded. With this procedure, no decay of
intermediates takes place and the different behaviour of intermediates
upon temperature change is taken into account.

RECENT ADVANCES IN THE PHOTOPHYSICS AND PHOTOCHEMISTRY OF SMALL, LARGE, AND NATIVE OAT PHYTOCHROMES

The Photophysical Parameters of P_r, and Kinetics and Calorimetry of the Early Stages of the $P_r \rightarrow P_{fr}$ Phototransformation

Kurt Schaffner, Silvia E. Braslavsky and Alfred R. Holzwarth
Max-Planck-Institut für Strahlenchemie
D-4330 Mülheim a. d. Ruhr

Introduction

This progress report summarizes recent communications (1-5) on our study of native (124 kDa) phytochrome from oat (6,7), and of the proteolytically degraded small (60 kDa) (8) and large (114 and 118 kDa) (6,9-12) forms of the chromoprotein (13). They have provided a consistent set of photophysical parameters for the excited singlet (S_1) state of the three phytochrome forms (1,4), and kinetic (5) and calorimetric (2,3) data for the early stages of the $P_r \rightarrow P_{fr}$ phototransformation, using picosecond time-resolved and stationary fluorescence, laser flash photolysis, and laser-induced optoacoustic spectroscopy.

Various differences between the photophysical and photochemical properties of the three phytochrome forms have been reported in the past. Thus, the Vis absorption maximum of native P_{fr} (which is identical with that of in-vivo P_{fr}) (9) is red-shifted with respect to that of the degraded P_{fr}s, whereas the P_r absorption spectrum does not change with degradation, and the A_{730}/A_{665} absorption ratio at photoequilibrium is higher for native P_{fr} (10,12). The quantum yield of the $P_r \rightarrow P_{fr}$ transformation, $\Phi_{r \rightarrow fr}$, has been claimed to increase appreciably in going from large to native P_r (12), and fluorescence yield, Φ_f (14-17), and lifetime data, τ_{meas} (16,18,19), for small and large P_r are varying and, in part, contradictory.

The comparison of the properties of small, large, and native phytochrome may eventually reveal interactions between the 6, 10 and 60 kDa polypeptide fragments (which are eliminated from native phytochrome upon degradation to the large and small forms) and the chromophore, which possibly influence some aspects of the photochromic $P_r \rightleftharpoons P_{fr}$ transformation. A comparative study of the photophysics and photochemistry of small, large and native phytochrome and a reevaluation of the data has therefore become imperative.

Results and Discussion

Fluorescence spectra, yield, and decay of P_r

The corrected emission and excitation spectra of native and (Fig. 1) and large (1) P_r at 275 K are very similar, and so are the fluorescence yields (Table) (1,4). Our value of Φ_f = 0.0029 $\pm$ 0.0004 for native P_r, after correction for impurity emission (see below), at 275 K (4), is also comparable to that reported by Hermann et al. (17) for large P_r from rye at 293 K, Φ_f = 0.0014 $\pm$ 0.0002, but it is at variance with the values reported earlier for small (Φ_f = 0.01) and large P_r ($\leq$ 0.0001) at room temperature (15,16).

The P_r fluorescence decays of all phytochromes strongly deviate from single exponential functions, as shown in Fig. 2 for pure native P_r and for the photostationary equilibrium of native P_r + P_{fr} (1,4). The relative amplitudes of the three decay components are approximate measures of the relative ground state concentration of the corresponding species. The main component with a lifetime of ca. 48 ps (Table), which compri-

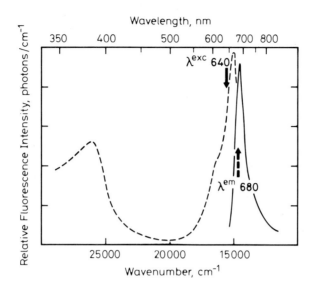

Fig. 1. Corrected fluorescence (———) and fluorescence excitation (- -) spectra of native Pr in potassium phosphate buffer solution at 275 K. Maxima: λ_{exc} = 666 nm, λ_{em} = 686 nm; I_{380}/I_{666} = 0.5 (4).

zes > 90% of the total decay amplitude, can be attributed to the S_1 state of all P_rs. The two minor components typically have lifetimes in the range of 190 ps and 1.04 ns. They are some sort of 'impurities' which are characteristic of the phytochrome preparation procedure (1,4,17,19). Their chromophores appear to be structurally related to the phytochrome tetrapyrrole chromophore.

Although the relative amplitudes of the main decay components (cf. Fig. 2: R_3) show that the purity of our preparations of small, large and native P_r is around 90% and better, about 35% of the total integrated fluorescence of, e. g., native P_r are due to the 'impurities', since the fluorescence lifetimes of these 'impurities' are so much larger than the P_r fluorescence lifetimes.

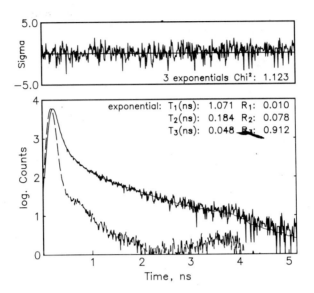

Fig. 2. Semilogarithmic plot of the fluorescence decay of pure native Pr after irradiating at 730 nm; λ_{exc} = 640 nm, λ_{em} = 680 nm. The exciting pulse (- -), the measured decay (——), and the decay function calculated from the best-fit kinetic parameters (thin line superimposed on measured decay) are shown. Inset: calculated lifetimes, $T_1 \ldots T_n$, and rel. amplitudes, $R_1 \ldots R_n$, of the decay components. Above: weighted residuals plot of the deviations of these computer-fitted parameters from the measured decay, with the chi-square value in the inset (4).

In view of the nearly identical absorption coefficients of the three P_rs (12), the radiative lifetimes, τ_{rad}, should also equal the value of τ_{rad} = 14 ns (corrected value: see Table) which has been derived (16) from the P_r oscillator strength of large P_r. Taking the measured lifetimes, τ_{meas}, fluorescence quantum yields can be calculated which are in good agreement with the 'impurity'-corrected Φ_fs. We can thus compile a set of consistent photophysical parameters for the S_1 state of each of the three phytochromes (Table).

The P_r fluorescence decays do not depend on the excitation wavelength (1,14). Also, there is no variation in τ_{meas} of native P_r in the physiological temperature range 275 - 298 K (4).

Table: Photophysical Parameters of the S_1 State of Small, Large, and Native P_r Phytochromes at 275 K

Parameter	Small	Large	Native	Lit.
Radiative lifetime, τ_{rad} (ns)	14*	14*	14*	(4,15)
Measured lifetime, τ_{meas} (ps)	45 ± 10	45 ± 10	48 ± 3	(1,4)
Fluorescence quantum yield,† uncorrected for ´impurity´ emission, $\Phi_f \cdot 10^3$	5 ± 1	3.3 ± 0.3	4.4 ± 0.8	(1,4)
corrected for ´impurity´ emission, $\Phi_f \cdot 10^3$	1.5 ± 0.3	2.0 ± 0.2	2.9 ± 0.4	(1,4)
Calculated fluorescence quantum yield, $\Phi_f \cdot 10^3$	3.2 ± 0.4	3.2 ± 0.4	3.4 ± 0.4	(1,4)
Quantum yield of the photo-transformation, $\Phi_{r \rightarrow fr}$	0.13*	0.12*		(23)
		0.08*		(6)
			0.12*	(12)

* Based on the most recent absorption coefficient (20-22). † λ_{exc} = 640 nm.

One would therefore not expect any temperature dependence of Φ_f either. The similarity in τ_{meas} of native and large P_r at 275 K suggests, furthermore, that the fluorescence of large P_r is equally indifferent within the physiological temperature range, despite a report (15) of variations in Φ_f and τ_{meas} for large P_r in this temperature range.

The radiationless processes of P_r

Since the fluorescence yield is very low, the lifetime of P_r is determined by the rate constants of the two processes which

the S_1 state undergoes preferentially: the radiationless decay
to the ground state (k_d), and the phototransformation to the
primary intermediate(s) (k_{react}), i. e., $\tau_{meas} \approx 1/(k_d + k_{react})$.

With the quantum yield for the $P_{r \rightarrow fr}$ phototransformation being
relatively small (Table), the dissipation of excitation energy
is predominantly governed by heat-producing radiationless processes. This is confirmed by laser-induced optoacoustic spectroscopy (2,3) which shows that in fact about 90% of the absorbed light energy is lost by radiationless deactivation. The
action spectrum for the heat emitted by small P_r deviates from
the absorption spectrum around 610 and 695 nm, indicating the
build-up of photoproduct(s) within 15 ns after excitation,
whereas solutions containing predominantly the P_{fr} form do not
show any significant deviations.

A comparison of the corrected action spectra of heat emission,
α (Fig. 3A), reveals the relative importance of the primary
photoreactions and the internal conversion to ground state of
excited P_r and P_{fr}. On the one hand, the difference in α
around 660 nm could be associated with a difference in the
quantum yield of the primary photoreactions in small phytochrome, with that of P_r being ca. 10% higher. This agrees with
the finding that the $P_r \rightarrow P_{fr}$ conversion is more efficient
than the reverse process in all phytochromes (4,12,23,24). On
the other hand, since the α values are of the same order of
magnitude and Φ_f of P_{fr} is at least 10 times smaller than that
of P_r, not only the photoreactions are faster for P_{fr} than for
P_r, but also the rate of internal conversion to ground state
is considerably greater for P_{fr}.

The 6, 10 and 60 kDa polypeptide fragments, which are eliminated upon degradation of the native to the large and small P_rs,
do not noticeably influence the nonradiative decay of the S_1

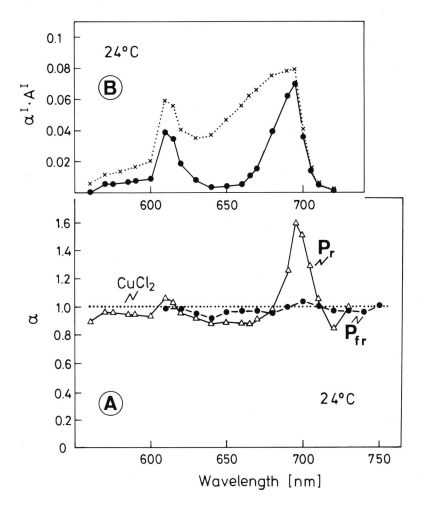

Fig. 3. A: Action spectra for heat emission (α) from small excited Pr (—△—△—) and Pfr (-●-●-), corrected for the absorbed light energy by calibration with CuCl$_2$ ($\cdots\cdots$).
B: Difference spectra of heat emission between the ´first photoproduct´ (I) in the Pr → Pfr conversion and Pr, as calculated from the Pr absorption spectrum and the corrected action spectrum of Pr (Fig. 3A), assuming fractions of α = 0.80 ($\cdot$×$\cdots$×) and 0.90 (—●—●—) dissipated of the light energy absorbed (2).

state of native P_r to ground state. The similarity in τ_{meas}, hence also in k_d, of small, large and native P_r is too close for a meaningful difference to have escaped detection. Thus, the reported difference in $\Phi_{r \to fr}$ for large and native P_r (12) results in a ca. 1.7fold change in calculated k_d when k_{react} is assumed identical for both P_rs. This change would have to be reflected in a similar change in τ_{meas}, which is definitely excluded experimentally. [Taking $\Phi_{r \to fr}$ 0.08 and τ_{meas} 45 ps for large, and $\Phi_{r \to fr}$ 0.12 and τ_{meas} 48 ps for native P_r (cf. Table), rate constants of $k_d = 2.0 \cdot 10^{10}$ and $1.8 \cdot 10^{10}$ s^{-1}, and $k_{react} = 1.8 \cdot 10^{9}$ and $2.5 \cdot 10^{9}$ s^{-1} can be calculated.]

The $P_r \to P_{fr}$ phototransformation

From which process(es) then could the difference in $\Phi_{r \to fr}$ arise, which has been claimed for the large and native phytochromes (12)? In contrast to a difference in k_d, one in k_{react} by a factor of 1.7 or more would still be within the experimental error for the lifetime of the main decay component. A more efficient thermal reversion to P_r of one of the first intermediates formed in the degraded phytochrome might in fact bring about the larger $\Phi_{r \to fr}$ value for native P_r, provided that such a thermal reversion occurs in the picosecond time range and thus has remained undetected by the nano- and microsecond flash photolysis (5,24) and optoacoustic experiments (2,3,25). In particular, the decay kinetics and activation parameters for the primary photoproducts from small and native P_r are essentially the same for periods as long as 2 ms when measured in the physiological temperature range 275 - 298 K (5).

Since the absolute amplitude of the 48 ps decay component arises specifically from excited P_r, the percentages of P_r and P_{fr} in any mixture of the two can be determined directly and without measuring several additional parameters such as absorption coefficients and initial phototransformation yields. Our values for the red light-adapted photoequilibrium concentration of P_{fr}, $P_{fr}^{660} = 0.72 - 0.77$, and for the ratio of the pho-

totransformation yields, $\Phi_{r\to fr}/\Phi_{fr\to r}$ [which is directly related to the red-light equilibrated P_{fr} concentration], are
similar for large and native phytochrome (1,4). They are in
good agreement with data reported for large phytochrome, P_{fr}^{665}
= 0.75 and $\Phi_{r\to fr}/\Phi_{fr\to r}$ = 1.0 (23), but differ from a subsequent claim that the 665 nm-equilibrated P_{fr} concentration be
higher and different for large (0.79) and native (0.86), and
that $\Phi_{r\to fr}/\Phi_{fr\to r}$ = 1.5 for native phytochrome (12).

The fact that the optoacoustic heat emission from P_r at about
610 and 695 nm apparently exceeds 100% (Fig. 3A) is attributed
to a superposition of the simultaneous emission of energy absorbed by P_r (roughly 90%) and by primary photoproduct(s) (2).
The picosecond lifetime of excited P_r allows for primary pho-
toproduct(s) to build up and absorb light within the 15 ns duration of the laser flash (cf. 26). The spectral range of the
'excess' heat emission indicates that it arises at least partially from the internal conversion of the excited intermediate (e. g., 27) termed I_{700} by Cordonnier et al. (28) and lumi-
R by Kendrick and Spruit (29). It confirms that I_{700} is indeed
a primary photoproduct of P_r, and it shows that the photochromic system $P_r \rightleftarrows I_{700}$ is already established within the first
15 ns.

Fig. 3B shows difference spectra of the heat emissions by P_r
and the primary photoproduct(s), calculated by selecting two
different heat fractions to be emitted by P_r. This spectrum
exhibits the major features of the absorption spectrum of the
first product(s) formed from P_r.

When the small Φ_f is neglected, the 'photochemical loss' in
the optoacoustic heat production can be evaluated using the
energy balance relationship (30), and the calorimetry of the
photoreactions of small P_r can be roughly estimated (2). The
internal energy content difference between P_r and the primary
photoproducts amounts to ΔE = 140 kJ/mol (cf. Fig. 4).

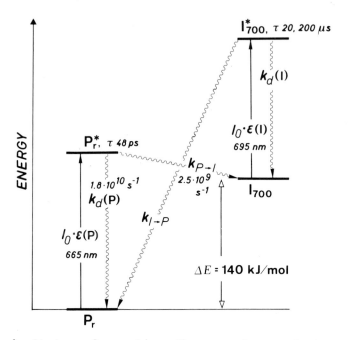

Fig. 4. State and reaction diagram of Pr and the primary photoproduct(s) I700. The rate constants are
kd(P) and kd(I) for the radiationless decays to ground
state of P$_r$ and I700, respectively, and kP→I, and kI→P
for the reactions of Pr and I700, respectively. ΔE is
the calculated internal energy content difference between Pr and I700. For the case that several primary
products are formed in parallel photoreactions (cf.
Fig. 6), the diagram must be extended accordingly. The
constants kd(I), kP→I and kI→P then represent sums of
rate constants (2,5). The values given are those determined for native (τ, kd(P), kP→I) and small (ΔE) Pr.

The time-resolved difference spectra obtained upon nanosecond
laser excitation at 275 K are quite similar for small (Fig. 5)
(5), large (24,28), and native P$_r$ (5). The decay of the 695
nm-transient maximum of small and native P$_r$ at 275 K is biexponential, with lifetimes of ca. 20 and 200 μs, relative abun-
dances of 0.4 and 0.6, respectively, and similar activation
parameters in the range 275-298 K [E$_a$ = 58 ± 5 kJ/mol, ln A =
35 ± 1 (A in 1/s units)]. Again no significant difference appears between the data of the small and large phytochromes.

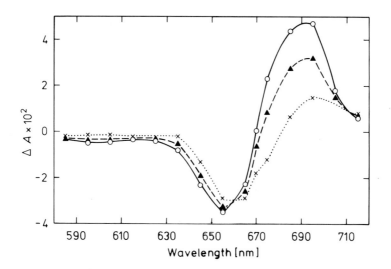

Fig. 5. Time resolved transient spectra taken immediately (—o—o—), 60 μs (-▲-▲-), and 1 ms (·×·×·)
after irradiation of small Pr with 15 ns laser
pulses; λ_{max} = 640 nm (5).

Since there is no indication that one of the two transient
components would be converted into the other, parallel paths
forming the two (I'_{700} and I''_{700}) or possibly even a third intermediate (I_{610}; see above for the optoacoustic evidence) as
the primary photoproducts of P_r appear the most likely reaction scheme (Fig. 6), rather than a sequential process.

The fact that no bleaching mirror image of the P_r absorption
has been observed around 610 - 620 nm in the flash photolysis
difference spectra (5,24,26-28,31,32) can be attributed to the
compensation of P_r absorbance in either of two ways:

(i) By a concomitant increase in absorbance of yet another
transient. The absorbance differences around 695 and 610 - 620
nm exhibit different decay kinetics (5), with the decrease at
695 nm being coupled, in small and native phytochrome, to a
corresponding very small increase at around 620 nm. They may

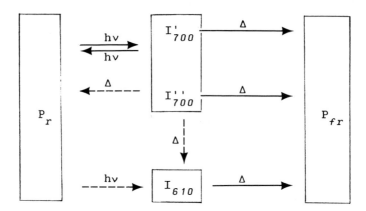

Fig. 6. Scheme of parallel primary
photoreactions of Pr (5,13).

therefore very well originate from two different types of primary photoproducts, i. e., the I_{700}s and another species absorbing at 610 nm and having a lifetime in the millisecond
range (I_{610}). In order to account for the delayed increase at
620 nm, either or both I_{700}s are to convert subsequently also
into I_{610} or to partially revert back in the dark to P_r.

(ii) By a concomitant increase in I_{700} absorbance, provided
that the decay of I_{700} is masked by its transformation into a
transient possessing a similar absorption coefficient in this
spectral region (33). Prerequisite to both interpretations is
an accidental correspondence of the absorbance in the 610 –
620 nm region of two (case i) and even three (ii) species, and
a parallel set of P_r photoreactions according to Fig. 6.

References

1. Wendler, J., Holzwarth, A.R., Braslavsky, S.E., Schaffner,
    K.: Biochim. Biophys. Acta 786, 213-221 (1984).
2. Jabben, M., Heihoff, K., Braslavsky, S.E., Schaffner, K.:
    Photochem. Photobiol. 40, in press (1984).

3. Jabben, M., Braslavsky, S.E., Schaffner, K.: J. Phys. Colloq. 44 C6, 389-396 (1983).

4. Holzwarth, A.R., Wendler, J., Ruzsicska, B., Braslavsky, S.E., Schaffner, K.: Biochim. Biophys. Acta, in press (1984).

5. Ruzsicska, B., Culshaw, S., Braslavsky, S.E., Schaffner, K.: Photochem. Photobiol., in press.

6. Vierstra, R.D., Quail, P.H.: Proc. Natl. Acad. Sci. USA 79, 5272-5276 (1982).

7. Bolton, G.W., Quail, P.H.: Planta 155, 212-217 (1982).

8. Mumford, F.E., Jenner, E.L.: Biochemistry 5, 3657-3632 (1966).

9. Vierstra, R.D., Quail, P.H.: Planta 156, 158-165 (1982).

10. Quail, P.H., Colbert, J.T., Hershey, H.P., Vierstra, R.D.: Phil. Trans. Roy. Soc. B 303, 387-402 (1983).

11. Vierstra, R.D., Quail, P.H.: Biochemistry 22, 2498-2505 (1983).

12. Vierstra, R.D., Quail, P.H.: Plant Physiol. 72, 264-267 (1983).

13. See Braslavsky, S.E.: Pure Appl. Chem. 56, in press (1984), for a review of part of these results and related studies of model chromophores.

14. Song, P.-S., Chae, Q., Lightner, D.A., Briggs, W.R., Hopkins, D.: J. Am. Chem. Soc. 95, 7892-7894 (1973).

15. Song, P.-S., Chae, Q., Briggs, W.: Photochem. Photobiol. 22, 75-76 (1975).

16. Song, P.-S., Chae, Q., Gardner, J.D.: Biochim. Biophys. Acta 576, 479-495 (1979).

17. Hermann, G., Kirchhof, B., Appenroth, K.J., Müller, E.: Biochem. Physiol. Pflanz. 178, 177-181 (1983).

18. Sarkar, H.K., Song, P.-S.: Biochemistry 20, 4315-4320 (1981).

19. Hermann, G., Müller, E., Schubert, D., Wabnitz, H., Wilhelmi, B.: Biochem. Physiol. Pflanz. 177, 687-691 (1982).

20. Brandlmeier, T., Scheer, H., Rüdiger, W.: Z. Naturforsch. 36c, 431-439 (1981).

21. Litts, J.C., Kelly, J.M., Lagarias, J.C.: J. Biol. Chem. 258, 11025-11031 (1983).

22. Roux, S.J., McEntire, K., Brown, W.E.: Photochem Photobiol. 35, 537-543 (1982).

23. Pratt, L.H.: Photochem. Photobiol. 22, 33-36 (1975).

24. Pratt, L.H., Inoue, Y., Furuya, M.: Photochem. Photobiol. 39, 241-246 (1984).

25. Heihoff, K., Braslavsky, S.E., Schaffner, K.: unpublished
    results.

26. Braslavsky, S.E., Matthews, J.I., Herbert, H.-J., de Kok,
    J., Spruit, C.J.P., Schaffner, K.: Photochem. Photobiol.
    31, 417-420 (1980).

27. Linschitz, H., Kasche, V., Butler, W.L., Siegelman, H.W.:
    J. Biol. Chem. 241, 3395-3403 (1966).

28. Cordonnier, M.M., Mathis, P., Pratt, L.H.: Photochem.
    Photobiol. 34, 733-740 (1981).

29. Kendrick, R.E., Spruit, C.J.P.: Photochem. Photobiol. 26,
    201-214 (1977).

30. Malkin, S., Cahen, D.: Photochem. Photobiol. 29, 803-813
    (1979).

31. Shimazaki, Y., Inoue, Y., Yamamoto, K.T., Furuya, M.:
    Plant Cell Physiol. 21, 1619-1625 (1980).

32. Pratt, L.H., Shimazaki, Y., Inoue, Y., Furuya, M.: Photo-
    chem. Photobiol. 36, 471-477 (1982).

33. Rüdiger, W., Thümmler, F.: Physiol. Plant. 60, in press
    (1984).

Received June 25, 1984

Discussion

<u>Song</u>:   Do you get "parallel" kinetics even with native Pr?

<u>Holzwarth</u>:   The flash photolysis kinetics are very similar both quantitative-
vely and qualitatively for the various phytochrome forms.

<u>Hoff</u>:   How do your parallel intermediates correspond to the intermediates
found by the temperature studies of Dr. Eilfeld? Is it possible that there
are more intermediates that can be distinguished spectrally and "trapped"
at different temperatures, but at room temperature are only kinetically dis-
tinguishable? Have kinetic measurements been done at lower temperature?

<u>Holzwarth</u>:   Measuring stationary spectra of trapped intermediates does not
provide kinetic information. If two intermediates have similar spectra they
can only be distinguished in kinetic spectroscopy like flash photolysis. We
have not carried out flash photolysis experiments at temperatures below 273°K.

<u>Gouterman</u>:   The "heat" measurements are done in a system that responds at
~20 μs. This is vastly larger than laser pulse so you see a very relaxed
system. Do you not lose information about any two-photon processes?

Holzwarth: The occurrence of two-photon processes can be checked on the
basis of the intensity dependence of the response. For one-photon processes the signal must be linear in the exciting light intensity. That is how
the deviations around 700 nm have been detected. The delay of ~20 μs should
not be confused with the signal formation. The 20 μs delay arises just from
the time the acoustic wave needs to travel from the irradiated volume element to the detector.

Gouterman: The "heat" measurements are hard to do in aqueous medium since
$\Delta V/\Delta T$ is small in water ($\Delta V/\Delta T$ = 0 at 4^0C for water). Then you may see a
ΔV associated with volume change of the protein or the chromophore. It is
important to work at 4^0C and separate ΔV and ΔH. This has been considered in
work by W. Parson at University of Washington.

Holzwarth: Yes, the photoacoustic signal is very low or even non-existent
in aqueous solutions around 4^0C. As you mentioned, this opens up the interesting possibility to measure volume changes of proteins. Also, another
phenomenon has to be considered, namely "electrostriction". This phenomenon
arises from the interaction of the electric field of the light pulse and
the protein, even at wavelengths where the protein does not absorb.

Lamola: This question is directed to all those involved with the phytochrome problem. The overall quantum yield Pr → Pfr is much smaller than 1.
What accounts for the inefficiency? What is the yield of the primary process or at least the yield of the earliest intermediate? Is the primary
photoprocess photoreversible? Is the sum of the quantum yields for Pr → intermediate and intermediate → Pr close to 1? These questions have relevance
for the interpretation that the primary photoreaction is a configurational
isomerization.

Holzwarth: This is indeed an important question. The first step is photoreversible, as has been reported both by the group of Lee Pratt and by our-
selves. But the photochemical back reaction occurs only at relatively high
light intensities. This is of prime importance for those carrying our flash
photolysis experiments on that system. From the fluorescence lifetime measurements we were able to calculate rate constants for radiationless decay
and photoreaction. We had to use the overall phototransformation yields for
these calculations. However, as you pointed out, it would be important to
measure directly the yield of the first intermediate. This could be done by
picosecond absorption measurements.

Song: Is it possible that your value of $[Pfr]^{660}_{\infty}$ = 0.75 is lower than the
reported value of 0.86 because the Pr fluorescence is partially quenched by
Pfr via Pr → Pfr energy transfer (and thus photoreversion of Pfr to Pr)?
Even for large phytochrome, Pratt's value of 0.75 may be a little lower
than actual value; for example, we get 0.81 by directly measuring [Pr] by
bleaching out Pfr by tetranitromethane. Bill Smith's value is also higher
than 0.75.

Holzwarth: We have worked at low concentration and with short pathlength
cuvettes. I therefore do not think that we can explain the differences by
fluorescence quenching. Our error limits are consistent with a value of up
to 0.79. I think that it is more probable that any differences arise from

the special conditions, like pH, buffer etc. that are used.

Song: Is it possible that one of the impurity components arises from protochlorophyllide?

Holzwarth: As far as I know, the fluorescence lifetime of protochloro-phyllide is longer than the values found by us for the "impurities".

Schneider: Could it be that small changes in geometry cause large varia-tions in lifetime, but only small changes in reactivity? Then a higher va-lue of [Pfr] determined by other techniques could be explained, since then only Holzwarth's technique would exclude "unnatural" phytochromes?

Holzwarth: This possibility cannot be excluded at present. In this con-nection I should like to point out that the middle component in the fluor-escence decay shows photoreversibility. We are hesitant, however, to call this portion "functional" phytochrome.

TIME-RESOLVED FLUORESCENCE DEPOLARISATION OF PHYCOCYANINS
IN DIFFERENT STATES OF AGGREGATION

Siegfried Schneider, Peter Geiselhart, Toni Mindl,
Peter Hefferle and Friedrich Dörr
Institut für Physikalische und Theoretische Chemie,
Technische Universität München, Lichtenbergstr.4
D 8046 Garching, FRG

Wolfhart John and Hugo Scheer
Botanisches Institut der Universität
Menzingerstr.67, D 8000 München, FRG

Introduction

Biliproteins are the main photosynthetic light harvesting
pigments in cyanobacteria, red algae and cryptophytes (1-4).
In the former two organisms, they are present as highly organized structures, the phycobilisomes, which are not an
integral part of the photosynthetic membrane, but rather
attached to its surface (2,3,5,6). They contain up to several
hundred biliproteins carrying up to 2000 chromophores in
a single functional unit, but also a number of so called
linker peptides which are probably responsible for the attachment and internal organization of the phycobilisomes
(7-10). Excitation energy is captured efficiently by the
phycobilisomes and transfered with a high quantum yield via
several intermediate acceptors to the chlorophyllous reaction
centers within the photosynthetic membrane. This energy
transfer has been the subject of active research over the
past decade, involving both the investigation of entire
phycobilisomes and of fragments thereof (11-19). In view

of the complex structure of the former, we have focused
mainly on a single biliprotein, C-phycocyanin (PC), which
is generally the major pigment of cyanobacterial phycobilisomes (17,19). PC has a pronounced tendency for aggregation
(4), which is also strongly influenced by the different
linker peptides (10). These linker peptides also seem to
be important in fine-tuning of the absorption of biliproteins. In this paper we want to report on the photophysical
properties of such a complex between PC and linker peptides,
viz. PC 636, which has been isolated from green-light adapted
cultures of the cyanobacterium Mastigocladus laminosus. The
time resolved fluorescence is compared to that of common
PC 618, which is free of additional linkers.

Sample Preparation

Cultures of Mastigocladus laminosus were grown as described
earlier with fluorescent tubes producing only small amounts
of red light. Monomers and trimers of PC were prepared as
reported earlier (19). PC 636 was eluted from the DEAE col-
umns between phycoerythrocyanin, a second pigment formed
in green light, and the major PC. It was dialysed against
phosphate buffer (80 mM, pH 6.0) and used without delay and
without storage in the freezer.

The chromatography of a crude extract of M. laminosus on
DEAE cellulose produces generally APC I, PC and APC II in
order of increasing salt concentration. When the cells are
grown under light of reduced intensity at $\lambda \approx 600$nm, two additional pigments are formed in small amounts. One is the well
known phycoerythrocyanin (24) eluting first from the column,
which is followed by a phycocyanin named PC 636 according
to its extremly red shifted absorption spectrum. The main
analytical difference of this pigment seems to be the pre-

sence of two colorless peptides in addition to the common
PC subunits.

It has a slightly higher buoyant density on a sucrose gradient, and dissociates upon prolonged standing or freezing
in buffer of low ionic strength. Details of the preparation
and its properties are to be published separately.

Measurements and data analysis

The fluorescence decay curves were measured using a synchronously pumped mode-locked ring dye laser (rhodamine 6G, 80
MHz repetition rate, pulse width ≤ 1 ps) in conjunction with
a repetitively working streak camera (for details see e.g.
(16)). The apparent time resolution of this system is approximately 25 ps without deconvolution procedure; it allows
measurements with low excitation intensities (10^{13} photons
per pulse and cm^2). The fluorescence decay curves measured
with the analyzer parallel, ($I_p(t)$), and orthogonal, ($I_s(t)$),
to the polarization of the exciting beam are transfered to
a minicomputer where, after proper correction for the systems
response, the expressions $I(t) = I_p(t) + 2 I_s(t)$ and
$D(t) = I_p(t) - I_s(t)$ are calculated. $I(t)$ measures the decay
of the excited state population (electronic lifetime) and
$D(t)$ the product of the former with the correlation function of the absorption and emission dipoles (17, 20). In
contrast to the anisotropy function $R(t)$ the difference
function $D(t)$ is additive and can be evaluated if more than
one emitting species is present. Lacking better information,
we approximate the correlation function by an exponential.
The best fits for both functions (I and D) are determined
under the assumption of a biexponential response function
(two emitting species) by means of a Marquardt algorithm.
Depending on the S/N ratio of the recorded fluorescence decay curves and their relative magnitude, the fit parameters

derived may be subject to considerable error. We will, therefore, discuss their trends rather than their absolute magnitude.

Results and Discussion

It is found that in all cases the decay curves can be fitted
sufficiently well as convolutions of biexponentials. The
fit parameters, e.g. the decay times (T_1,T_2 in psec)
and the relative amplitudes (A_1,A_2 in %) of the short- and
long-lived component, resp., are given in the inserts in
figure 1. The measurements were performed at three different temperatures, namely at $18^{\circ}C$ (A), at $36^{\circ}C$ (B), the
temperature the algae are grown, and at $52^{\circ}C$ (C) where irreversible thermal denaturation starts to become effective.
Partial denaturation already takes place at lower temperatures. Static measurements show a drastic loss in fluorescence yield (up to four orders of magnitude), which is much
larger than the decrease in absorption connected with a conformational change of the chromophore (21). The time-inte-
grated fluorescence intensities expressed as $A_1*T_1+A_2*T_2$
also confirm the reduction at higher temperature. It is found
as a general rule that the decrease is more pronounced in
the alpha than in the beta subunit and larger for the monomer than for the trimer. The normalized fluorescence decay
curves also show small but distinct variations with temperature. Therefore the results presented in figure 1 must be
taken as evidence for an intermediate state being present
during the process of thermal denaturation.

The alpha subunit of PC contains only one chromophore. If
it is stabilized by noncovalent interaction with the protein
to adopt only one conformation, a single exponential decay
is expected with a lifetime of 1.5 to 2.5 ns (lifetime of

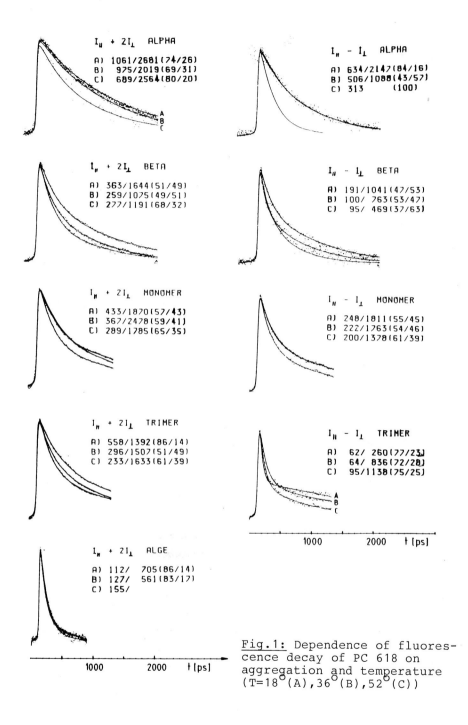

$I_\parallel + 2I_\perp$ ALPHA
A) 1061/2681(74/26)
B) 975/2019(69/31)
C) 689/2564(80/20)

$I_\parallel - I_\perp$ ALPHA
A) 634/2147(84/16)
B) 506/1088(43/57)
C) 313   (100)

$t_\parallel + 2I_\perp$ BETA
A) 363/1644(51/49)
B) 259/1075(49/51)
C) 277/1191(68/32)

$I_\parallel - I_\perp$ BETA
A) 191/1041(47/53)
B) 100/ 763(53/47)
C) 95/ 469(37/63)

$I_\parallel + 2I_\perp$ MONOMER
A) 433/1870(57/43)
B) 367/2478(59/41)
C) 289/1785(65/35)

$I_\parallel - I_\perp$ MONOMER
A) 248/1811(55/45)
B) 222/1763(54/46)
C) 200/1328(61/39)

$I_\parallel + 2I_\perp$ TRIMER
A) 558/1392(86/14)
B) 296/1507(51/49)
C) 233/1633(61/39)

$I_\parallel - I_\perp$ TRIMER
A) 62/ 260(77/23)
B) 64/ 836(72/28)
C) 95/1138(75/25)

1000   2000 t [ps]

$I_\parallel + 2I_\perp$ ALGE
A) 112/ 705(86/14)
B) 127/ 561(83/17)
C) 155/

1000   2000 t [ps]

Fig.1: Dependence of fluorescence decay of PC 618 on aggregation and temperature (T=18°(A),36°(B),52°(C))

the chromophore in a native environment). Instead, an additional short-lived component is found, whose lifetime varies with temperature between 690 and 1060 psec. A similar behaviour was verified for the alpha subunit of Spirulina platensis (17) and Anabaena variabilis (22). Since aggregation of the subunits is unlikely, one must assume at least two different sets of emitting species (chromophore-protein-arrangements). The long-lived species must be close to that in native environment, whilst the short-lived form should be closer to the denatured, less interacting species. The faster decay in the difference function D(t), furthermore, signals that the faster component is subject to a depolarization mechanism with T 1500 psec. Since no acceptor molecules are present, the depolarisation should be due to orientational relaxation of the less rigidly bound chromophores.

The beta subunit contains two chromophores in different protein environment. The respective absorption maxima are separated by about 20 nm. The stationary fluorescence spectra of both subunits are essentially equal, which indicates an efficient energy transfer from the "sensitizing" to the "fluorescing" chromophore. The energy transfer is also manifested in the fluorescence decay curves. The short-lived component (T_1 300ps) is interpreted as "leakage" fluorescence from the s-chromophore, whose lifetime is shortened due to energy transfer to the f-chromophore in the same subunit. The depolarisation time of the fast component is much shorter than that of the alpha subunit and decreases with increasing temperature from 400 to 150 psec. The longer lifetime is close to the shorter one in the alpha subunit; a lifetime of 2 ns, which would be expected for the f-chromophore in a native environment is not detected, possibly for experimental reasons. An unambiguous interpretation is presently not possible, because different subsets of chromophore-protein-arrangements cannot be excluded in view of

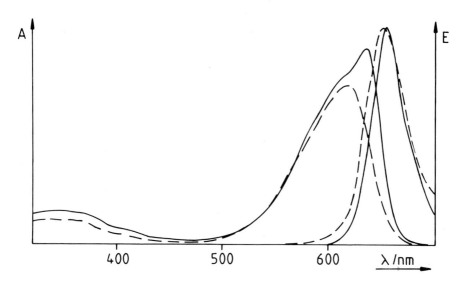

Fig.2: Absorption and emission spectra of trimers of
       PC-618 (— — —) and PC-636 (————)

the preparation sequence, which involves a denaturation-
renaturation sequence.

The isotropic decay curves of monomers and trimers are similar to each other. The short lifetime is in the range of
200-500 psec and represents the lifetime of the s-chromo-
phores, which are quenched by energy transfer. The longer
one between 1600 and 2500 psec characterizes the terminal
acceptor, i.e. the f-chromophore in the native environement.
Chromophores excited via energy transfer rather than directly
by photon absorption should emit a less polarized fluorescence. Only the short-lived leakage fluorescence is partly
polarized, but the depolarization times are moderately short.
In contrast to Spirulina platensis (17) we observe for PC
from the thermophilic algae no lengthening of the depolarization time with temperature.

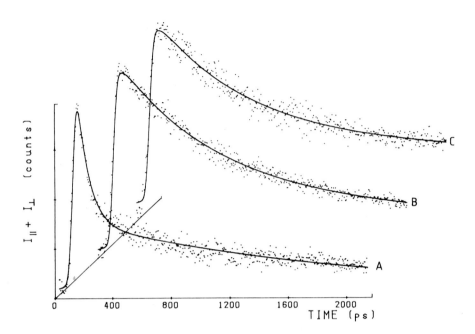

Fig.3: Fluorescence decay of PC-636 excited at 580 nm and
recorded with interference filters λ_i = 614 nm (A),
λ_i = 635 nm (B) and λ_i = 654 nm (C). Solid lines are
calculated fit curves with parameters T_1/T_2 in psec and
A_1/A_2 in % in parenthesis.
(A)   72/ 1700   (75/25)
(B)   609/ 4200   (66/34)
(C)   610/10100   (64/36)

In the intact alga, the energy is efficiently transfered
to the nonfluorescing reaction center. The observed emission
is only leakage fluorescence from PC (fast component with
lifetime of approx. 130 psec) and Allophycocyanin (slow component). Since the fraction of emission from directly excited chromophores is small, the emission is essentially
unpolarized. Furthermore, it is found that an increase in
temperature does not result in a long-lived component,
which would indicate a prevention of energy transfer.

In PC-618, the difference in excitation energy of s- and

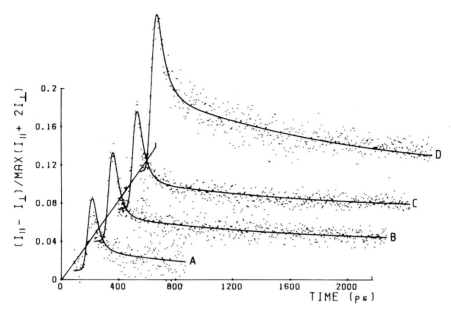

Fig.4: Calculated difference functions I_p-I_s normalized with respect to the maximum of the isotropic decay curve I(t). The fluorescence was recorded with cut-off filters, whose cut-off wavelength is 20 nm larger than the excitation wavelength. Solid lines are calculated fit curves with parameters T_1/T_2 in ps and A_1/A_2 in % in parenthesis.
(A) 35/ 760 (86/14) λ_{exc} = 610 nm
(B) 36/1130 (86/14) λ_{exc} = 600 nm
(C) 42/1165 (85/15) λ_{exc} = 590 nm
(D) 53/1290 (69/31) λ_{exc} = 580 nm

f-chromphores, resp., is small ($\approx$20 nm). Selective excitation of both chromophores is more difficult than in PC-636, in which the absorption of the f-chromophores is shifted bathochromically due to interaction with two colorless proteins (fig.2). The fluorescence decay curves of trimeric PC-636 prove to be strongly dependent on the recording wavelength (fig.3). Upon short-wavelength excitation (580 nm) the longer wavelength emission (both from s- and f-chromophores) is dominated by a fast component ($T\approx$600 psec) as in the case of PC-618. Its lifetime could be related to the energy transfer time from s- to f-chromophores.

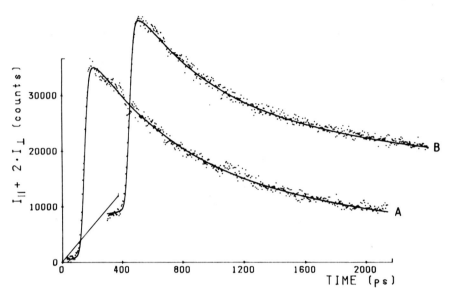

Fig.5: Isotropic fluorescence decay of trimers of PC-636.
The fluorescence was recorded with cut-off filters, whose
cut-off wavelength was 20 nm larger than the excitation wavelength. Solid lines are fit curves calculated with parameters T_1/T_2 in ps and A_1/A_2 in % in parenthesis.
(A)   610/3200 (63/37) λ_{exc} = 610 nm
(B)   535/4700 (55/45) λ_{exc} = 580 nm

The lifetime of the slow component (emission from terminal
f-chromophores) varies somewhat with observation wavelength,
a fact, which could again be indicative for a nonuniform
chromophore-protein-arrangement. If the recorded fluorescence
is restricted to near-resonant emission (614 nm), a much
faster component dominates the decay curve. We presume that
this is in part a consequence of resonant energy transfer
within the s-chromophore manifold. The latter can be verified
by inspection of the normalized difference curves displayed
in fig. 4. Upon short wavelength excitation (s-chromophore
excitation) a pronounced fraction of the short-lived emission
is polarized. The decay time of ≈ 40 ps is related to the
energy transfer time within the s-chromophore manifold
(homo-transfer). Such a transfer does not change the number

of excited molecules of this species. Accordingly, in the isotropic decay function $I=I_p+2 I_s$ no such extremely short-lived component can be detected (fig. 5).

Keeping the results for the alpha- and the beta-subunits in mind, one could of course argue that the fast component observed in the decay curves of monomers and trimers is due to a second chromophore-protein arrangement, which differs significantly from that in the native species. We believe, however, that an artefactial microheterogeneity can be excluded in the case of PC 636, because it requires much milder manipulations than those applied in the preparations of PC-subunits. Any heterogeneity is then expected to be inherent to the biliprotein.

From the large body of measurements denoted to the study of fluorescence decay and energy transfer within functionally intact phycobilisomes and its constituent aggregated biliproteins one may conclude that the transfer times are greatly reduced when the size of the aggregate, i.e. the number of chromophores is increased. It might be that in a monomeric unit the distance between the chromophores is larger than between chromophores of a trimer belonging to different monomers. This could also imply that in hexamers (and higher aggregates) the important "vertical" energy transfer between adjacent trimers is faster than the "horizontal" transfer within one trimer, which carries no energy in the direction of the reaction center (23).

Financial support by the Deutsche Forschungsgemeinschaft is gratefully acknowledged.

References

1. Scheer, H.: in "Light Reaction Path of Photosynthesis"
    (F. K. Fong, ed.), p. 7-45, Springer, Berlin 1982

2. Gantt, E.: Ann. Rev. Plant Physiol. 32,327-347 (1981)

3. Glazer, A.N.: Ann. Rev. Biochem. 52,125-157 (1983)

4. McColl, R.: Photochem.,Photobiol. 35,899-904 (1982)

5. Mörschel, E., Koller, K., Wehrmeyer, W.: Arch. Micro-
    biol. 125,43-51 (1980)

6. Wanner, G., Köst, H.-P.: Protoplasma 102,97-109 (1980)

7. Glick, R.E., Zilinska, B.A.: Plant Physiol. 69,991-997
    (1982)

8. Redlinger, T., Gantt, E.: Proc. Natl. Acad. Sci. US
    79,5542-5546 (1982)

9. Koller, K.P., Wehrmeyer, W., Moerschel, E.: Eur. J. Bio-
    chem. 91,57-63 (1978)

10. Yu, M.H., Glazer, A.N., Williams, R.C.: J. Biol. Chem.
    256,13130-13136 (1981)

11. Searle, G.F.W., Barber, J.,Porter, G.,Tredwell, C.J.:
    Biochim. Biophys. Acta 501,246-256 (1978)

12. Kobayashi, T., Degenkolb, E.O., Behrson, R, Rentzepis,
    P.M., McColl, R., Berns, D.S.:Biochemistry 18,5073-5078
    (1979)

13. Breton, J., Geacintov, N.E.: Biochim. Biophys. Acta
    594,1-32 (1980)

14. Pellegrino, F., Wong, D., Alfano, R.R., Zilinskas, B.:
    Photochem. Photobiol. 34,691-696 (1981)

15. Holzwarth, A.R., Wendler, J., Wehrmeyer, W.: Photochem.
    Photobiol. 36,479-487 (1982)

16. Hefferle, P., Nies, M., Wehrmeyer, W., Schneider, S.:
    Photobiochem. Photobiophys. 5,41-51 (1983),id.325-334

17. Hefferle, P., John, W., Scheer, H., Schneider, S.:
    Photochem. Photobiol. 39,221-232 (1984)

18. Gilbro, T., Sundstroem, A., Sundstroem, V., Holzwarth,
    A.R.: FEBS Lett. 162,64-68 (1983)

19. Hefferle, P., Geiselhart, P., Mindl, T.,Schneider,S.:
    Z. Naturforsch. 39c (1984),in print

20. Fleming, G.R., Morris, J.M., Robinson, G.W.:
    J.Chem.Phys. 17,91 (1976)

21. Scheer, H., Formanek, H., Schneider, S.:
    Photochem.Photobiol. 36,259 (1982)

22. Switalski, S.C. and Sauer, K.: Photochem.Photobiol.,
    in print
23. Scheer, H.: in "Photosynthetic Membranes: Encyclopedia
    of Plant Physiology", (Staehlin, A., Artzen, C., eds.)
    Springer, Heidelberg, in print
24. Füglistaller, P., Suter, F., Zuber, H.: Hoppe-Seyler´s
    Z. physiol. Chemie 364,691-712 (1983)

Received June 28, 1984

Discussion to this and next lecture was combined, see page 407-410.

THE INFLUENCE OF AN EXTERNAL ELECTRICAL CHARGE ON THE CIRCULAR DICHROISM OF CHROMOPEPTIDES FROM PHYCOCYANIN

Christina Scharnagl, Eliana Köst-Reyes, Siegfried Schneider

Institut für Physikalische und Theoretische Chemie
Technische Universität München, 8046 Garching, FRG

Joachim Otto
Institut für Physiologische Chemie der Universität
8000 München, FRG

Introduction

Biliproteins are important receptor pigments in photosynthesis and photomorphogenesis (1). They are characterized
by some unusual properties when compared to the free bile
pigments. Conformational changes have been proposed to explain the drastic differences in the UV-Vis absorption spectra of native and denatured biliproteins, resp. Non-covalent
protein-chromophore-interactions (e.g. hydrogen bonding,
coulombic forces to charged side groups, salt bridges etc.)
are believed to induce additional modifications of the spectroscopic properties and the photodynamic behaviour (2).
Besides of time-resolved emission spectroscopy, chiroptical
studies proved themselves very helpful to extract information
on the chromophore conformation (3-6). It is generally assumed that free bile pigments are conformationally heterogeneous and that they predominantly assume an inherently
dissymmetric cyclic-helical conformation. In a previous publication (5) we have shown that optical activity in bile

pigments can be induced not only by the whole, native protein
chain, but also by short polypeptide segments. The circular
dichroism was found to be strongly dependent on solvent
properties and pH. E.g., in aqueous buffer, a sign reversal
was observed when switching from neutral to acidic pH.
Molecular orbital calculation indicated that this change
was not necessarily due to a reversal of the chirality of
the entire chromophore, but could also be the result of
a more localized conformational change. In this contribution
we wish to report the results of an experimental and theoretical investigation referring to different phycocyanin
peptides from the cyanobacterium Spirulina geitleri, which
clearly documents the influence of external charges onto
the CD-spectra.

Materials and Methods

1) Preparation and separation of chromopeptides.
   C-Phycocyanin was extracted from Spirulina geitleri according to published procedures (7), precipitated with
   ammonium sulfate and chromatographed (DEAE-cellulose).
   After proteolytic digestion by pepsin (5) colorless peptides and free amino acids were separated by column chromatography (8). Since cleavage by pepsin is not very specific, peptides with variable length were generated for
   each of the three chromophore binding sites. A separation
   was attempted by means of isoelectric focusing using
   Servalyt (pH 3-7) as carrier ampholyte. A total of six
   colored bands was found, corresponding to pI values of
   8.96 (I), 5.55(II) 4.73(III), 4.45(IV), 4.20(V) and 3.96
   (VI). Since only the fractions III-V provided enough material for spectroscopic investigations, these fractions
   were studied in more detail. Due to a difference in solubility, each fraction could be divided further into a

methanol soluble and a methanol insoluble fraction. The
latter three fractions were completely soluble in destil-
led water. After the spectroscopic experiments were fin-
ished all urea free samples were purified again by TLC
and an amino acid analysis performed separately for the
methanol soluble and the remaining water soluble fraction.

2) CD-spectra and quantum mechanical model calculations.
CD-spectra were recorded on a dichrograph model V
(Instuments SA, Munich, FRG) equipped with a data handling
system (Leanord, Lille, France). All spectra are baseline
corrected, the measured **Δ**E values are normalized with
respect to the absorption at the red maximum. Calculations
were performed with a FORTRAN program developed and applied earlier to conformational studies of bile pigments
by Wagnière and coworkers (9-11).

Results and Discussion

The amino acid sequence of biliproteins from Spirulina
geitleri is still unknown. It is, therefore, impossible to
determine the sequence of amino acids in the chromopeptides
from the data provided by the amino acid analysis alone.
If one assumes that the sequences in the chromophore binding
regions are similar in Spirulina geitleri and Synechococcus
6301 (12) then the following sequences are likely for the
various peptides (indices 'a' refer to the methanol soluble
part, indices 'b' to the methanol insoluble remainder).

```
IIIa          Glx - Gly - Asx - Cys - Ser
IIIb    Thr - Glx - Gly - ( ) - Cys - Ser - ( ) - (Leu)

IVa           Glx - Gly - Asx - Cys - Ser - Ala - Leu
IVb                 Gly - Asx - Cys - Ser
```

```
      Va              Glx - Gly - ( ) - Cys - Ser - Ala - (Leu)
```

They seem to originate mainly from the second binding site
of the beta-subunit. This could be explained by two circumstances: Firstly, only those chromopeptides with a pI of
approximately 4.5 were used. Secondly, the peptides from
the alpha-subunit are known to be very instable; therefore,
the other fractions, which could originate from this subunit, provide only very little material. Since the CD-spectra
of figure 1 were recorded with samples containing both
fractions a and b, the explanation must include the composite
nature of the polypeptides as well as the various possibilities of chromophore conformation.

As discussed in detail in the previous paper (5) 8M urea
in tris buffer is expected to minimize the interaction between chromophore and polypeptide chain, whereas intra-chro-
mophore interaction is maximum. All three fractions behave
essentially equal (Fig. 1, top right), the CD-spectra are
comparable to those of urea denatured PC (13). In accordance
with the general understanding the chromophore should adopt
a cyclic-helical conformation. The results of model calculations for such a conformation are displayed in figure 2
(upper part; to avoid intersection of rings A and D, a torsion around the single carbon-carbon bond 14-15 is assumed).
The different optical activity in tris buffer indicates a
partial restoration of the peptide-chromophore-interaction
in water. Since at pH=7, all carboxylic acid groups are deprotonated, the interaction could be either unspecific and
coulombic in nature or be specific similar to a salt bridge.
Since especially in sample V, which most likely contained
only one species, the spectra are more complex and very weak,
partial compensation of the contributions of different chromophore configurations must be assumed. We believe that this

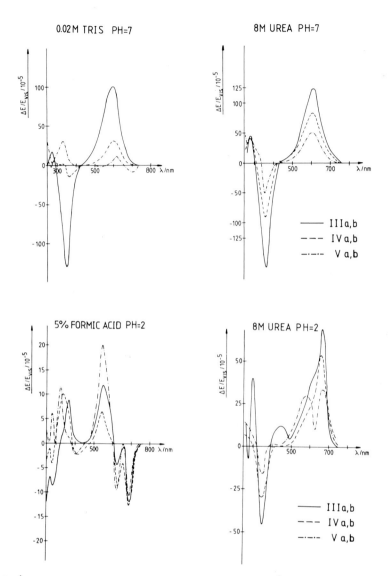

Figure 1:
Circular dichroism spectra of three fractions (III, IV, V, a,b
see text) of C-Phycocyanin chromopeptides from Spirulina geitleri: Solvent and pH-dependence

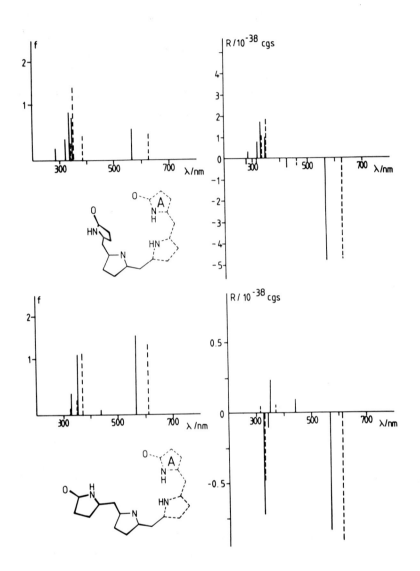

Figure 2:
Calculated oscillatory strength f and rotatory strength R for
two selected conformations under the influence of protonation:
Top: cyclic-helical geometry
bottom: semi-extended geometry
                ──────── free base,   ── ── ── cation

additional species should have a slightly more extended conformation, because the ratio of absorptivities of near UV-
band to red band decreases (not shown). Therefore, fig. 2
also shows the results of calculations for a semi-extended
form (lower part). The prerequisite for partial compensation,
existence of two conformations with opposite helicity, can
easily be understood in terms of interaction with the polypeptide.

Upon acidification of the two solutions the CD-spectra assume
a rather complex pattern, which of course can not be explained in detail unequivocally. Nevertheless, there are
a number of points which, in our opinion, are important for
the understanding of modifications of the chromophor's behaviour by interaction with the amino acids of the polypeptide chain. In the electronic absorption spectra, the well
known bathochromic shift of approx. 60 nm of the red absorption band is observed. In aqueous urea this band shows two
clear shoulders left and right of the absorption maximum,
which is indicative for several chromophore conformations.
As the model calculations demonstrate (fig. 2), the bathochromic shift caused by protonation is larger for the more
helical conformation than for the semi-extended conformation.
Furthermore we notice a drastic reduction in the absorpti-
vity ratio of near UV band to red band. Since this change
is about equal for all fractions it should be an effect intrinsic to the chromophore rather than mediated by interaction to the polypeptide chain.
The CD-spectra in urea free acidic solution appear to be
rather complicated, since there is a change in sign around
600 nm; furthermore, in the red part of the spectrum the
CD-band seems to be split into a double band, although in
the electronic absorption there is essentially no indication
for the presence of a shoulder. The position of the maximum
in CD-spectra (550 nm) coincides with the short-wavelength

slope of the red band next to the absorption minimum between
the two main absorption bands. From these facts it becomes
clear that the features of the CD-spectra can hardly be explained as being due to a superposition of the spectra of
several unperturbed protonated chromophore configurations.
Since on the other hand, the polypeptide does also change
its state of protonation, charges may be shifted in space
or their shielding could be changed with the effect that
new coulombic interaction forces become operative. In fig.3
the effect of an external electrical charge upon the CD-
spectra of a chromophore, which is held rigid in either one
of the two adopted conformations, is displayed. Besides of
an already discussed shift in transition energy (and minor
modifications in oscillatory strength) major changes in
rotatory strength are observed. It was pointed out already
earlier (5) for conformational changes that changes in oscillatory strength do not parallel change in rotatory strength.
The results in figure 3 prove this statement also for the
effect of external charges. The rotatory strength of the
red band varies under the influence of charges of either
sign. More important, however, for the weak bands calculated
in the frequency region between red and first strong UV band,
the rotatory strength is greatly enhanced and sometimes the
sign is changed compared to that of the red band. Their
oscillatory strength is still low enough that these transitions should not be detected in the electronic absorption
spectrum, but they may show up in the CD-spectrum. Since
both the magnitude of the shift and the enhancement factor
for the rotatory strength of the various transitions depend
on sign and location of the charge, charge heterogeneity
of the polypeptide can lead to complicated and strongly solvent dependent CD-spectra, especially in the region around
400 nm. This phenomenon may also explain why the sign reversal observed earlier upon removal of urea from a composite
sample of PC-peptides (5) is not observed for the samples

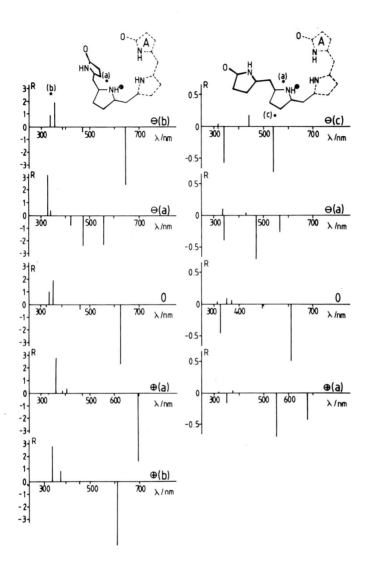

Figure 3:
Influence of an external charge on the calculated CD-spectra (rotatory strength R/10^{-38} cgs) for the protonated chromophore in two selected geometries.
Position of charges: (a) 1 Å above N (ring C)
                     (b) near ring D
                     (c) in the plane of ring C
                     (0) without charge

investigated in this work.

In their study of the phytochrome photocycle, Rüdiger and Thümmler observed an unexpected high CD-signal around 400 nm (6). Since they also assume in agreement with the model by Honig et al (14) that interaction to charges located at the protein are responsible for the observed unusual spectral shifts (and its function (2)) an investigation on the influence of external charges on the CD-spectra of phytochrome is currently performed.

Acknowledgement

We are indepted to Prof. Wagnière, Zürich, for providing to us the program for CD calculations. Furthermore, we wish to thank Prof. Rüdiger and Prof. Scheer, Munich, for helpful discussions. Free computer time by Bayerische Akademie der Wissenschaften and financial support by Deutsche Forschungsgemeinschaft is also gratefully acknowledged.

References

1.   For a review see e.g. Scheer,H.: Light Reaction Path of Photosynthesis, Fong F.K.(Ed.), Springer Verlag, Berlin 1982

2.   Thümmler,F.: Dissertation, Universität München 1984 and references there

3.   For a review see e.g. Braslavsky,S.E., Holzwarth,A.R., Schaffner,K.: Angew. Chemie 95, 670-689 (1983)

4.   Lehner,H., Scheer,H.: Z. Naturforschung 38c, 353-358 (1983)

5.   Scharnagl,C., Köst-Reyes,E., Schneider,S., Köst,H.-P., Scheer,H.: Z. Naturforsch. 38c, 951-959 (1983)

6.   Thümmler,F., Rüdiger,W.: Isr.J.Chem. 23, 195-200 (1983)

7.   Kufer,W., Scheer,H.: Hoppe-Seyler's Z. Physiol. Chem. 360, 935-956 (1979)

8.   Köst,H.-P.: Dissertation, Universität München 1974

9.   Blauer,G., Wagnière,G.: J. Am. Chem. Soc. 97, 1949-
     1954 (1975)

10. Wagnière,G., Blauer,G.: J. Am. Chem. Soc. 98, 7806-
     7810 (1976)

11. Pasternak,R., Wagnière,G.: J. Am. Chem. Soc. 101, 1662-
     1667 (1979)

12. Williams,V.P., Glazer,A.N.: J. Biol. Chem. 253, 202-211
     (1978)

13. Lehner,H., Krauss,C., Scheer,H.: Z. Naturforsch. 36b,
     735-738 (1981)

14. Honig,B., Ebrey,T., Callender,R.H., Dinur,U., Otto-
     lenghi,M.: Proc. Natl. Acad. Sci. USA 76, 2503-2507
     (1979)

Received June 28, 1984

Discussion to previous and present lecture

Woody: Would you elaborate on the way in which the charge effects on the rotational strengths are calculated? Specifically, are you using a perturbation approach or are you including the charge in the π-electron level Hamiltonian?

Schneider: The coulombic interaction is included in the Hamiltonian; in the "self-consistent-field" part of the calculation, the π-electrons are polarized. The transition dipole moments and rotatory strengths are calculated with the modified wave functions.

Woody: I have been studying the effect of electrostatic fields on the transition moment directions in the bases of the nucleic acids. Using a perturbation approach, I have found a substantial change in the transition moment directions for 9-ehtylguanine due to these effects. Have you analyzed the charge effects on the transition moment directions predicted in your calculations?

Schneider: Model calculations for merocyanine dyes including the response to an (electrical) reaction field of the solvent, showed the effect which you mentioned. Due to the polarization of the π-electrons, both the magnitude and the direction of the transition dipole moment was changed. The size of the effects depends on the geometry; I have not analyzed the data for tetrapyrroles with respect to this effect.

Song: With phytochrome, anionic detergents and ANS (an anionic, hydro-

phobic probe) bleach the visible absorption bands, apparently because an
ionic interaction between the chromophore and +-charge is disrupted, forcing the chromophore to resume a cyclic conformation. On the other hand,
neutral detergents bind at the chromophore pocket but they do not release
the interaction between the chromophore and the +-charge, thus no bleaching
occurs. Do you have any indirect evidence for the nature of charge in your
phycocyanins?

Schneider: Because different peptides show clear differences in absorption and fluorescence excitation, as well as in CD-spectra, we assume an
interaction with the amino acids which are present in addition to the smal-
lest peptide. However, as long as we do not know the amino acid sequence,
it is impossible to decide, whether the changes are caused by coulombic
interaction, hydrogen bonding, salt bridges etc. Furthermore, we cannot
distinguish presently whether the changes in CD are solely due to the polarization of the π-electrons or induced by configurational changes.

Gouterman: What is the response time of the instrument? (By this I mean
what is the shortest decay time you can determine, not the accuracy with
which a long time can be determined. Maybe these are the same?).

Schneider: The intrinsic response time of the streak camera is ~8 ps.
The excitation pulse of ~1 ps duration is, however, recorded with an apparent fwhm of ~20 ps due to jittery triggering of the streak camera and in
the interpulse distance. According to the Fourier theorem, this apparent
excitation profile can be used in a convolution procedure to determine
the decay parameters by means of a least square fit. In case of single
exponential decay, lifetimes of the order of 10 ps can be determined re-
liably. In case of nonexponential (biexponential) decay, the accuracy of
the amplitudes and decay times depends on their relative values. Sufficiently high S/N ratio assumed, the accuracy in the determination of very
short lifetimes is ~±5 ps.

Huber: What are your ideas about the strong directionality of the energy
transfer along the antenna towards the photosynthetic membrane? This is
particularly difficult to understand as the basic building block is a
symmetrical hexamer of phycocyanin. The linker peptides probably play an
important rôle.

Schneider: Even if one assumes the energy transfer rate to be equal for
a transfer up and down the stack of e.g. PC, then the possibility of transfer to APC results in an effective asymmetry in transfer probability towards the reaction center. This directionality would of course be increased
if, due to the interaction with the linker proteins, the excitation energy
of those PCs, which are closer to the core, is lowered.

Bode: Given the presence of linker proteins: due to the asymmetric location of such linkers with respect to the monomers within the trimers/hexa-
mers, also the spectral properties of these monomers within a single tri-
mer/hexamer should differ. Which impact would that have on energy transfer?

Schneider: If the interaction with the linker protein makes one chromophore to act as trap, energy back migration could be prevented in smaller

aggregates. In this case, measured transfer times would represent the time necessary for the first transfer step, while otherwise, one eventually measures a longer, "effective" transfer time determined also by the relaxation time of the chromophore, between which the energy is exchanged.

Zuber: With respect to the direction of energy migration I would like to remind you of the work of A. Glazer on the rôle of the linker polypeptides. He modulates the spectral properties (absorption, fluorescence-maxima). The effect is a series of different maxima which cause the energy to migrate to the reaction center.

Holzwarth: You find fast depolarization times in most of your aggregates which you attribute to energy transfer. Why do you not find the corresponding times in the isotropic decays and why do you not find fluorescence rise times at long detection wavelengths?

Comment (to Prof. Huber and Dr. Bode): We have made a comparison between phycobilisomes having approximately identical rod length, but different composition in phycobiliproteins. The inhomogeneously built phycobilisomes have considerably faster overall transfer (e.g., red alga *Rhodella* and *Porphyridium*) than the homogeneously built rods of e.g. *Macystis* phycobilisomes.

Schneider: An energy transfer between like species does not change the isotropic decay function which is the reason why this function is used to follow the development of the number of excited molecules with time, independently of any depolarizing mechanism.

Transfer between unlike species can show up in the isotropic decay function if the fluorescence yield is different for the two species. The inherent difference in fluorescence quantum yield can be enhanced by recording the fluorescence in narrow regions only, e.g. one where only the energy acceptor emits. In the case of PC, our experimental conditions are unfavorable, because we excite both s- and f-chromophore directly and monitor the emission over a broad wavelength range.

Sund: How did you prepare the α- and the β-subunit?

Reply by Scheer: The subunits have been separated by two different methods, involving complete unfolding with 8M urea denaturation at pH 3 or pH 7, and refolding of the separated subunits.

Sund: I would prefer to use the term polypeptide chain instead of subunit because subunit in general means that it contains more than one polypeptide chain.

Scheer: The α-subunit has a single chromophore and is monomeric under the conditions of measurement, the β-subunit has 2 chromophores and is >80% monomeric.

Sund: Is it possible to recombine the α- and β-polypeptide chains to native phycocyanin which then shows the same fluorescence decay as the original one?

Scheer: We have not done the fluorescence of the re-hybridized subunits.
This project is just under way.

Comment with regard to symmetry and linkers: Pure phycocyanin from *S. platensis* and *M. laminosus* exists either as a monomer (heterodimer), or trimer (heterohexamer); we observe larger aggregates only in the presence of linkers, which indicates to us that there is no symmetric hexamer.

BILINS AND BILIN-PROTEIN LINKAGES IN PHYCOBILIPROTEINS:
STRUCTURAL AND SPECTROSCOPIC STUDIES

Henry Rapoport

Department of Chemistry, University of California, Berkeley,
California 94720, USA

Alexander N. Glazer

Department of Microbiology and Immunology, University of
California, Berkeley, California 94720, USA

Introduction

Phycobiliproteins are a family of macromolecules which serve
as light-harvesting components of the photosynthetic systems
of cyanobacteria (blue-green algae), the red algae, and the
cryptomonads (1). The presence of covalently attached open-
chain tetrapyrrole prosthetic groups (bilins) accounts for the
ability of these proteins to absorb visible light. Biliproteins absorb light maximally between 470 and 650 nm, the region
of the spectrum which lies between the blue and far red absorption peaks of chlorophyll a. As shown in Table I, individual
biliproteins vary in the chemical nature and number of the
bilins. Four different tetrapyrroles have been identified in
cyanobacterial and red algal biliproteins--phycocyanobilin
(PCB), phycoerythrobilin (PEB), phycourobilin (PUB), and a
phycobiliviolinoid chromophore (1). A single biliprotein subunit may carry different bilins; for example, the β subunit of
R-phycocyanin carries both PCB and PEB.

The structure of a derivative of PCB, phycobiliverdin (4),

TABLE I. Bilin composition of certain biliproteins[a]

Biliprotein	Type of subunit and bilin content[b]
Allophycocyanin	α 1 PCB; β 1 PCB
C-Phycocyanin	α 1 PCB; β 2 PCB
R-Phycocyanin	α 1 PCB; β 1 PCB, 1 PEB
Phycoerythrocyanin	α 1 PXB; β 2 PCB
C-Phycoerythrin	α 2 PEB; β 3-4 PEB
B-Phycoerythrin	α 2 PEB; β 3 PEB; γ 2 PEB, 2 PUB
R-Phycoerythrin	α 2 PEB; β 2 PEB, 1 PUB; γ 1 PEB, 3 PUB

[a]For original sources of data, see refs. 1 and 2. Data on the α and β subunits of B-phycoerythrin are from ref. 3; that on R-phycoerythrin from A.V. Klotz and A. N. Glazer, unpublished results.

[b]Abbreviations are: PCB, phycocyanobilin; PEB, phycoerythrobilin; PUB, phycourobilin; PXB, a phycobiliviolinoid tetrapyrrole.

released from C-phycocyanin by boiling methanol, has been determined and confirmed by synthesis (see refs. 5 and 6 for recent reviews). The corresponding PEB derivative, released from R-, B-, and C-phycoerythrins by boiling methanol, has likewise been characterized (5, 6). The structures of the protein-bound bilins were inferred from the examination of these structures and from spectroscopic and degradation studies (4, 5). The structures of phycourobilin and of the phycobiliviolinoid chromophore attached to the α subunit of phycoerythrocyanin, have yet to be established.

The release of PCB and PEB from biliproteins by refluxing in methanol is not complete. Consequently, it cannot be assumed that the structure of all the bilins linked to C-phycocyanin and to phycoerythrins can be inferred from those of the elimination products resulting from methanolysis. Interpretation of the early work on protein-bilin linkages is hampered by the

same kind of uncertainty. In one approach, the imides released
from phycocyanin and phycoerythrin upon treatment with chromic
acid under "non-hydrolytic" and "hydrolytic" conditions were
identified and quantitated, and the results interpreted to show
the presence of two linkages between polypeptide and protein,
involving rings A and B or C of PCB or PEB (7, 8). The underlying assumption of this approach was that each bilin within a
given biliprotein was linked to the polypeptide in the same
manner. It may be noted parenthetically that a recent reinves-
tigation of chromic acid oxidation of phycocyanin and allophycocyanin failed to confirm the earlier results indicating
the presence of a linkage to rings B or C (9).

Another early approach to the bilin linkage question was to
prepare "minimal chromopeptides" by extensive proteolysis (10).
From amino acid analysis of such partially purified PCB- and
PEB-bearing peptides, as well as from chromic acid degradation
studies on such fragments, it was inferred that two linkages
were present between the bilin and peptide in all instances, a
thioether linkage to ring A and an ester linkage between a
seryl residue and the propionyl substituent of either ring B or
C of the bilin (10). However, recent numerous amino acid
sequence studies (see refs. 2 and 11 for reviews) have shown
that in allophycocyanin, C-phycocyanin, C-phycoerythrin, and
phycoerythrocyanin, the only linkage between the bilin and the
protein was through a cysteinyl residue. Moreover, many small
chromopeptides produced by either tryptic digestion or cyanogen
bromide cleavage, lacked a hydroxyamino acid.

The approach we have taken to the determination of the structures of polypeptide-linked bilins is to prepare analytically
pure chromopeptides by procedures which do not produce detectable changes in the spectroscopic properties of the bilin, and
then to determine the structure of such bilin-peptides by amino
acid sequence analysis, high resolution NMR, and mass spectrometry, and synthesis. These studies have led to several impor-

tant novel findings. First, the only linkages between the
bilin and polypeptide we have found thus far are thioether
bonds through cysteinyl residues. Second, in the β subunit
of C-phycocyanin, one PCB is linked through ring A, the second
through ring D. Third, in the α and β subunits of B-phycoerythrin, four PEB groups are attached to the polypeptides through
single thioether bonds to ring A, whereas one PEB is linked to
the β-subunit of B-phycoerythrin through two thioether bonds,
one to each of rings A and D of the bilin. Fourth, we have
isolated a tryptic peptide from the β subunit of R-phycoerythrin which carries a PUB group linked through two thioether
bonds, whereas a tryptic chromoprotein peptide derived from
the γ subunit of this protein carries a PUB attached through
a single thioether bond.

The discovery of the diversity of modes of linkage, and consequently of bilin structures, reveals an added level of
structural complexity which must be fully understood before the
spectroscopic properties of this class of proteins can be interpreted.

Amino Acid Sequence Studies on Bilin Peptides

The nomenclature α-1, α-2, β-1, etc. is used below for bilin
peptides. This designation defines the polypeptide subunit and
the position, relative to the amino terminus of the polypeptide,
that the particular bilin peptide occupies. For example, C-
phycocyanin peptide β-2 is the bilin peptide derived from the
β subunit of C-phycocyanin that carries the second PCB moiety
on that chain. Bilin-linked cysteinyl residues are indicated
by an asterisk. The discussion that follows is confined
entirely to those aspects of sequence determination that bear
on the question of bilin polypeptide linkage.

(1) PCB Peptides from <u>Synechococcus</u> 6301 C-Phycocyanin (12)

Tryptic digestion of the α subunit yielded the peptide α-1,
Cys*-Ala-Arg, in high yield. Cyanogen bromide cleavage of the
β subunit released the bilin heptapeptide β-1, Ala-Ala-Cys*-
Leu-Arg-Asp-Hsr (Hsr = homoserine), and a 37 residue bilin peptide. Thermolysin digestion of the latter peptide permitted
the isolation of peptide β-2, Ile-Thr-Gln-Gly-Asp-(Cys*, Ser)-
Ala. Each peptide was shown to have a single linkage to PCB, a
thioether bond involving a cysteinyl residue. Neither peptide
α-1 nor β-1 contain hydroxyamino acids. The bilin could be
readily removed from either peptide α-1 or β-1 by treatment
with Hg^{2+}, Ag^+, or HBr in trifluoroacetic acid. In contrast,
these treatments resulted in very slow cleavage of the bilin-
peptide linkage in β-2 (12).

(2) PEB Peptides from <u>Porphyridium</u> <u>cruentum</u> B-phycoerythrin (3)

All of the PEB attached to the α and β subunits of B-phycoerythrin was accounted for in five unique tryptic chromopeptides--

    α-1 Cys*-Tyr-Arg

    α-2 Leu-Cys*-Val-Pro-Arg

    β-1 Met-Ala-Ala-Cys*-Leu-Arg

    β-2 Met-Ser-Phe-Ala-Ala-Gly-Asp-Cys*-Thr-Ser-Leu-Ala-
         Ser-Glu-Val-Ala-Gln-Phe-Asp-Arg

    β-3 Leu-Asp-Ala-Val-Asn-Ser-Ile-Val-Ser-Asn-Ala-Ser-
         Cys*-Met-Val-Ser-Asp-Ala-Val-Ser-Gly-Met-Ile-Cys*-
         Glu-Asn-Pro-Gly-Leu-Ile-Ser-Pro-Gly-Gly-Asn-Cys-
         Tyr-Thr-Asn-Arg

In peptides α-1, α-2, β-1, and β-2, the bilin is attached
through a single thioether bond. Peptides α-1, α-2, and β-1
lack hydroxyamino acids. In peptide β-3, sequential Edman
degradation showed that the bilin was linked to the polypeptide
at two cysteinyl residues ten residues apart in the linear
sequence. This conclusion was confirmed by the isolation of
two fragments obtained by exhaustive digestion of peptide β-3
with pronase (β-3P), or with thermolysin followed by pronase

(β-3TP), whose structures are shown below.

β-3P

```
        ┌──── [PEB] ──┐
Ser-Cys            Cys-Glu-Asn-Pro-Gly
```

β-3TP

```
            ┌──────── [PEB] ────────┐
Ala-Ser-Cys          Met-Ile-Cys-Glu-Asn-Pro-Gly
```

(3)   PUB Peptides from Gastroclonium coulteri R-Phycoerythrin
      (A.V. Klotz and A. N. Glazer, unpublished results)

Determination of the sequences of two PUB peptides from this R-phycoerythrin have shown that both singly and doubly-linked bilins are present.   Peptide β-3 has the sequence

```
1               5                       10                      15
Leu-Asp-Ala-Val-Asn-Ser-Ile-Val-Cys-Asn-Ala-Ser-Cys*-Ile-Val-
                20                      25                      30
Ser-Asp-Ala-Val-Ser-Gly-Met-Ile-Cys*-Glu-Asn-Pro-Gly-(Leu)-Ile-
                35                              40
Ala-Pro-Gly-Gly-(Asn)-(Cys)-Tyr-(   )-(Asn)-Arg
```

Comparison of this sequence with that of peptide β-3 from B-phycoerythrin, which carries the doubly-linked PEB, shows a remarkable conservation of amino acid sequence. Only three differences are seen in the established sequence, Cys for Ser (at residue 9), Ile for Met (at residue 14), and Ala for Ser (at residue 31). In each case, the bilin residue is linked to two cysteines ten residues apart.

A second PUB peptide, derived from the γ subunit of the R-phycoerythrin, has the sequence--

```
Ser-Gly-Tyr-Ser-Gly-Ala-Ala-Leu-Asp-Phe-Pro-Val-Ala-Pro-Ser-
Leu-Ala-Gly-His-Tyr-Ser-Leu-Thr-Asn-Cys*-Gly-Gln-Pro-(Ser)-Gly-
Ala-(Ser)-Lys
```

The PUB is attached to this polypeptide through a single thioether bond. The sequence of this bilin peptide shows no homology to previously determined bilin peptide sequences from a
variety of biliproteins (2, 11).

Nuclear Magnetic Resonance and Mass Spectrometry Studies

(1) PCB Peptides from C-Phycocyanin

We have carried out detailed 360 and 500-MHz ^{1}H NMR studies on
two chromopeptides from the β subunit of Synechococcus sp. 6301
C-phycocyanin whose sequences are given above. The first of
these, β-1-phycocyanobiliheptapeptide, has been unambiguously
assigned the structure of a cysteinylthioether-linked phycocyanobilin. The thioether is attached to ring A at C-3' (13),
in agreement with previous postulates.

The second chromopeptide characterized from this source is β-2-
phycocyanobilioctapeptide. Cleavage of the bile pigment from
this chromopeptide results in a product bilin exhibiting different characteristics when compared with the A-ring linked,
β-1 cleavage pigment. Whereas release of the tetrapyrrole is
brought about by reflux in methanol in the β-1 case, these
conditions do not release the β-2 pigment. Treatment of the
β-2 chromopeptide with TFA/HBr is necessary to effect cleavage.
The UV/Vis spectra of the β-1 and β-2 phycocyanobilins are
similar but not identical. In addition, NMR studies on the β-
2-phycocyanobilioctapeptide show significant differences in
chromophore resonances when compared with the β-1, A-ring
linked natural product. These analyses have led to the assignment of a D-ring linked structure for the β-2 fragment via a
cysteinylthioether at C-18'.

(2)   PEB Peptides from B-Phycoerythrins

As described above, five unique phycoerythrobilin (PEB) peptides have been prepared from Porphyridium cruentum B-phycoerythrin by a combination of tryptic and thermolytic digestion
without alteration in the spectroscopic properties of the
bilin.   Four of these five chromopeptides are:

      α-1   Cys(PEB)-Tyr-Arg

      α-2   Leu-Cys(PEB)-Val-Pro-Arg

      β-1   Met-Ala-Ala-Cys(PEB)-Leu-Arg

      β-2T   Phe-Ala-Ala-Gly-Asp-Cys(PEB)-Thr-Ser

where α and β refer to the subunits from which the peptides are
derived.

High resolution ^{1}H NMR analysis of peptides α-1, α-2, β-1, and
β-2T has provided proof that all of the singly-linked PEB peptides contain a thioether bond to the 3'-position of ring A, and
strong evidence in support of a trans dihydro ring A in each of
these chromopeptides (14, 15).   The circular dichroism spectra
of the four singly-linked PEB peptides show that the configuration at C-16 is R in each instance.

The fifth bilin-containing fragment of the β subunit of
Porphyridium cruentum B-phycoerythrin produced by cleavage with
thermolysin has been shown by sequence analysis (3) to have the
structure

$$\overset{\text{(D)}}{\underset{\text{Ala-Ser-Cys}}{\ulcorner\rule{2cm}{0pt}}}\text{[PHYCOERYTHROBILIN]}\overset{\text{(A)}}{\underset{\text{Met-Ile-Cys-Glu-Asn-Pro-Gly}}{\rule{2cm}{0pt}\urcorner}}$$

Secondary ion mass spectrometry of this bilin-peptide yielded a
protonated molecular ion of 1629 mass units corresponding to
that predicted from the composition of the fragment, and indicated that the heptapeptide is linked to ring A and the tri-
peptide to ring D.   NMR spectra provided definitive evidence
for a thioether linkage at the C-3' carbon of ring A and a

second thioether linkage at C-18' of ring D of the bilin.   This
is the first documented report of a bilin linked through two
thioether linkages to a polypeptide (16).

These studies provide the first comprehensive analysis of the
structure of all the polypeptide-linked prosthetic groups on
the α and β subunits of B-phycoerythrin.

(3)   Stereochemical Studies

Complete structure elucidation of plant and cyanobacterial bile
pigments has been limited by the common practice of cleaving
the linear tetrapyrrole from its covalently bound protein (17).
This process has been a valuable tool for gross structural
determination, however, it suffers from several disadvantages.
Firstly, all direct information on the chromophore-protein
linkage, including two stereo centers, is lost.   Secondly, side
reactions leading to complex mixtures of unnatural chromophores
have been observed. Finally, the yield in the cleavage step
is frequently low.   These limitations, when combined with the
fact that complex proteins are involved containing multiple
chromophores with possible structural variations, indicate a
more controlled method for structural and especially stereochemical study is necessary.

With these problems in mind, we have developed methodology to
ascertain the nature of the native protein-tetrapyrrole
covalent bond. By mild, selective degradation of the protein
moiety without alteration of the chromophore or its attachment
site, the three chiral centers about the A ring of the bile
pigment are retained. Therefore, we are in a unique position
to study the stereochemistry at these centers.   For simplicity,
we have chosen to concentrate initially on the phycocyanin
series of bile pigments, as it contains only three chiral
centers, all contiguous about the A ring.

We have pursued these stereochemical studies by the synthesis and characterization, including the stereochemistry, of a series of 3,4-dihydropyrromethenones and 2,3-dihydrodioxobilins. High resolution ^{1}H NMR spectral analysis allows the determination of the A ring coupling constants for a series of cis and trans model compounds. From these data and correlations, the relative stereochemistry for the C-2-H and C-3-H in the A ring of phycocyanin and similar bile pigment structures has been assigned as trans (17).

Acknowledgement

This research was supported in part by the Division of Biological Energy Conversion and Conservation DOE; the National Institute of General Medical Sciences, DHHS, GM 28994; and the National Science Foundation, PCM 82-08158.

References

1.   Glazer, A.N.: in The Biochemistry of Plants (Hatch, M.D. and Boardman, N.K., eds), Vol. 8, pp 51-96, Academic Press, New York 1981.

2.   Glazer, A.N.: Biochim. Biophys. Acta 768, 29-51 (1984).

3.   Lundell, D.J., Glazer, A.N., DeLange, R.J., Brown, D.M.: J. Biol. Chem. 259, 5472-5480 (1984).

4.   Rudiger, W.: Angew. Chem., Int. Ed. Engl. 9, 473-480 (1970).

5.   Gossauer, A., Plieninger, H.: in The Porphyrins (Dolphin, D., ed.), Vol. 6, Part A, pp. 586-650, Academic Press, New York 1979.

6.   Gossauer, A.: Tetrahedron 39, 1933-1941 (1983).

7.   Rudiger, W., O'Carra, P.: Eur. J. Biochem. 7, 509-516 (1969).

8.   Rudiger, W.: Ber. Detsch. Bot. Ges. 88, 125-139 (1975).

9.   Troxler, R.F., Brown, A.S., Kost, H.-P.: Eur. J. Biochem. 87, 181-189 (1978).

10.   Killilea, S. D., O'Carra, P., Murphy, R.F.: Biochem. J. 189, 311-320 (1980).

11. Zuber, H.: in Photosynthetic Prokaryotes: Cell Differentia-
     tion and Function (Papageorgiou, G.C., and Packer, L.,
     eds.), pp 23-42, Elsevier, New York 1983.

12. Williams, V. P., Glazer, A.N.: J. Biol. Chem. 253, 202-211
     (1978).

13. Lagarias, J.C., Glazer, A.N., Rapoport, H.: J. Am. Chem.
     Soc. 101, 5030-5037 (1979).

14. Schoenleber, R.W., Leung, S.-L., Lundell, D.J., Glazer,
     A.N., Rapoport, H.: J. Am. Chem. Soc. 105, 4072-4076 (1983).

15. Schoenleber, R. W., Lundell, D.J., Glazer, A.N., Rapoport,
     H.: J. Biol. Chem. 259, 5485-5489 (1984).

16. Schoenleber, R.W.; Lundell, D.J., Glazer, A.N., Rapoport,
     H.: J. Am. Cheml Soc. 259, 5481-5484 (1984).

17. Schoenleber, R.W., Kim, Y., Rapoport, H.: J. Am. Chem. Soc.
     106, 2645-2651 (1984).

Received July 4, 1984

Discussion

Buchler: In your assignments, did you ever consider the application of
paramagnetic shift reagents?

Rapoport: We have considered the use of paramagnetic shift reagents, but
have not applied them as yet. It might be of interest to examine one of our
chromopeptides in detail with such reagents to see if additional conformational information could be obtained.

Buchler: Have you found any derivatives in which there are aromatic amino
acids in which ring current effects could help you to find out about the
vicinity of an aromatic amino acid to certain regions of the tetrapyrrole
chromophore?

Rapoport: On the basis of primary structures, there are not many aromatic amino acids in close proximity to the chromophore. We have seen no effects assignable to such ring currents.

Huber: The double linkage of some bilin chromophores to the protein puts
a constraint on the possible conformations that can be evaluated on the
basis of the known C-phycocyanin structure. The homologous peptide in C-
phycocyanin is in helical conformation. The spatial separation of the two
cysteines is therefore around 14 Å, compatible with an extended conformation of the chromophore.

Song: Biosynthetically, is the stereospecific addition of Cys-SH to a bilin chromophore enzymatic or non-enzymatic?

Rapoport: The specifics of that reaction are not known, but I would assume it is enzymatically mediated.

McDonagh: There is evidence that the prosthetic group of at least some
biliproteins is biosynthesized via ring opening of heme. This being so,
linkage of the tetrapyrrole to its apoprotein is presumably an enzymic
process. Since you observe covalent links to either one or both ends of
the tetrapyrrole, do you think that perhaps the enzyme(s) responsible is
incapable of distinguishing one end of the tetrapyrrole from the other?

Rapoport: That is one possibility. The other possibility is that the enzymes distinguish one end of the tetrapyrrole from the other and the linkage formed is deliberate, for functional reasons. I tend to favor the latter hypothesis.

Scheer: With regard to an ester bond, Schmidt and Werner Kufer in our
lab have carried out a degradation sequence involving reduction followed
by diazo-reaction, which cleaves the chromophores between the rings B and
C under conditions avoiding any possible hydrolysis. They did not find such
a bond in two different phycocyanins.

Did you see any difference in the absorption spectra of mono and doubly
linked peptides?

Rapoport: The electronic spectra are the same if the conjugate systems
are the same.

Köst: Can you think of a reason why you find PCB either A- or D-ring
linked but not both A-and D-ring linked? Also, you find PEB either A-ring
linked or A-and D-ring linked, but not only D-ring linked?

Rapoport: I have no biological rationale for these findings, but perhaps
the search is not yet over.

Blauer: Are the sequences of amino acid residues in the chromopeptides
you have shown similar to those observed in some other proteins? In myoglobin there seems to be a sequence near the heme-binding site which is
similar to some sequences found in plant chromopeptides. For this comparison, we disregard the difference between covalently and non-covalently
bound tetrapyrroles.

Rapoport: We have not done an exhaustive search for homology between the
amino acid sequences of phycobiliproteins and the available such data on
all other proteins. However, there is no homology between the tetrapyrrole
attachment site in biliproteins and heme attachment sites in cytochromes c.
Likewise, the tetrapyrrole attachment site in phytochrome is very different
from those in the biliproteins.

Schirmer et al., This Symposium, report that the arrangement of six helical
segments in the α and ß-subunits of M. laminosus C-phycocyanin closely resembles that seen in myoglobin. They comment that the chromophore attachment site in helix E in the ß-subunit of phycocyanin is homologous to the

heme attachment site (His E7) on hemoglobins. This homology is not apparent just from direct comparison of the amino acid sequences.

Falk: You extensively used the COSY-technique assignments, did you try NOESY to establish stereochemistry, especially with respect to peptide-chromophore interactions?

Rapoport: We are just beginning to use two-dimensional NOESY (nuclear Overhauser effect spectroscopy). The data are quite complex and as yet we can not reach any conclusions.

Falk: The simple subtractability of peptide from chromopeptide ^{1}H NMR spectra may indicate no interaction of chromophore with peptide fragments, what do you think?

Rapoport: For these small chromopeptides, there is clearly little effect of the peptide on the ^{1}H NMR of the chromophore.

Falk: How did you assure that on degradation of the protein, a possible serine-propionic acid side chain is not broken?

Rapoport: Our evidence clearly states that there is no ester bond between protein and chromophore. This evidence is based both on enzymatic and chemical degradation methods, throughout which such ester bonds would stay intact if they were present.

Fuhrhop: Would you please give us a rationale for the finding, that the stereochemistry of five-membered ring substituents cannot be analysed using the Karplus relationships.

Rapoport: The stereochemistry in these ring systems can not be analyzed theoretically based on the Karplus relationship because there are too many other complicating factors. As Karplus has pointed out (J. Am. Chem. Soc. 85, 2870 (1963)), vicinal coupling constants do not depend only on the dihedral angle. Thus in such rigid five membered rings with two trigonal carbons and a hetero atom, definitive stereochemical conclusions require proper model compounds. For a more detailed discussion, I refer you to our recent publication (J. Am. Soc. 106, 2645 (1984)).

STRUCTURAL ORGANIZATION OF TETRAPYRROLE PIGMENTS IN LIGHT-HAR-
VESTING PIGMENT-PROTEIN COMPLEXES

Herbert Zuber

Institut für Molekularbiologie und Biophysik, Eidg. Technische
Hochschule, ETH-Hönggerberg, CH-8093 Zürich, Switzerland

Introduction

In the primary processes of photosynthesis light energy is absorbed by a number of pigment molecules, mainly tetrapyrroles,
bound and specifically arranged in the light-harvesting or antenna pigment-protein complexes (1). Within these complexes
the antenna pigments, absorbing light of characteristic wavelength become exited and transfer the exitation energy to
another pigment molecule which in turn becomes exited. This
energy transfer (by slow or fast inductive resonance) results
in energy migration among the light-harvesting pigments and
finally to the reaction center. The whole process takes place
in tenths of a nanosecond. Of particular interest and importance in these light-harvesting pigment-protein-complexes are
structure forming and functionally active protein components
or polypeptides. They are responsible for:
a) the specific binding and arrangement of the pigment molecules (position, distance, orientation);
b) the polypeptide (lipid) environments of the pigments (functional side);
c) structural organization of the pigment-protein complexes on
   the basis of the arrangement of the polypeptides.

According to the three main groups of photosynthetic organisms

Optical Properties and Structure of Tetrapyrroles
© 1985 by Walter de Gruyter & Co., Berlin · New York

adapted in evolution to different light conditions three spectral ranges of light energy uptake and therefore of light-harvesting systems can be distinguished: 470-650 nm for phycobiliproteins (cyanobacteria, red algae), 650-700 nm for Chla/b complexes (higher plants, algae) and 720-1015 nm for BChl-com-
plexes (photosynthetic bacteria).

The object of our studies in Zürich is to provide a structural
interpretation of the light-harvesting and energy transfer
function of the antenna complexes. We concentrated on the
structural analyses (primary structure) of the phycobiliproteins from the cyanobacterium Mastigocladus laminosus (2) and
on the light-harvesting polypeptides from a series of purple
photosynthetic bacteria (3).

Phycobiliproteins and Phycobilisomes

In cyanobacteria and in eucaryotic red algae the phycobiliproteins are important light-harvesting complexes of photosystem II (2). They are found in vivo in a specific arrangement at the surface of the photosynthetic membrane forming
large, extramembrane light-harvesting antennas, the phycobilisomes. As globular proteins they are water soluble. In
the course of the evolution of cyanobacteria and red algae
various types of phycobiliproteins with different absorption
maxima were formed. On a purely spectroscopic basis one can
distinguish three main types: The allophycocyanins (absorption
650 nm), the C- or R-phycocyanins (absorption 620 nm) and the
biliproteins C-, B-, R-phycoerythrin and phycoerythrocyanin
(absorption in the green spectral range). Allophycocyanin is
an invariable constituent of all cyanobacterial and red algal
phycobilisomes. Either C- or R-phycocyanin is found in cyanobacteria and red algae. The phycoerythrins are variable in

type and amount in the various organisms, but only one type of green-light absorbing biliprotein is present in each organism. There are many types of phycobiliproteins because there is a variety of both the pigments and the polypeptides. Allophyco-cyanin and phycocyanin contain the blue phycocyanobilin and phycoerythrin the red phycoerythrobilin. Both of these types (as well as rare types like phycourobilin and phycobilins of unknown structure (phycoerythrocyanin) are open chain tetra-pyrroles and are covalently linked by a thioether bond to cysteine residues of the polypeptide chain.

As a protein family with similar function all phycobiliprote-ins have a common characteristic structure: They are composed of α- and β-polypeptide chains (subunits) with similar mole-cular weights: 16 - 20 KD. Each α- or β-polypeptide chain carries a typical number of tetrapyrrole pigment molecules. The number per polypeptide chain increases from allophycocyanin to phycocyanin, to phycoerythrin.

The α- and β-polypeptide chains can associate to form α-β-mono-mers, but normally, depending on the conditions (pH, ion con-centration, protein concentration), they aggregate to $(\alpha\beta)_2$, $(\alpha\beta)_3$, $(\alpha\beta)_4$, $(\alpha\beta)_6$ units. However, disc-shaped trimers $(\alpha\beta)_3$ and hexamers $(\alpha\beta)_6$ are preferred. These aggregation states are promoted by the specific binding of linker polypeptides (4). With the linker polypeptides specific stacks of hexamers (phycocyanin, phycoerythrin, phycoerythrocyanin) and trimers (allophycocyanin) are also found. In vivo the stacks (together with the linker polypeptides) are further aggregated to build up the phycobilisome antennas at the membrane surface. The phycobilisomes (4) as particles with a molecular weight between $7-15.10^6$D and a diameter of 30-40 nm (spaced 40-50 nm apart) are found by electron microscopy to be arranged regularly on the stroma side of the thylakoid membrane. Mainly

two types of phycobilisomes have been investigated: 1) Hemi-
discoidal phycobilisomes found in cyanobacteria (e.g. Mastigocladus laminosus (5)) and red algae (e.g. Rhodella violacea)
and 2) Hemi-ellipsoidal phycobilisomes found in red algae
(Porphyridium cruentum). The phycobilisome architecture
follows a basic principle, namely that different types of
phycobiliproteins are arranged in stacks of hexamer (trimer)
discs according to the inward pathway of heterogeneous energy
transfer. Energy migrates from the outer phycoerythrin-hexa-
mers (6 nm thick, diameter 12 nm, or phycoerythrocyanin)
via the phycocyanin-hexamers to the central allophycocyanin-
trimers (3 nm thick, diameter 11-12 nm) in the phycobilisome
core. This corresponds to an energy-transfer from short-wave
lenght absorbing donor pigments to long-wave length absorbing
acceptor pigments. Therefore the tetrapyrrole pigment molecules
(300-400 bilin molecules in hemi-discoidal phycobilisomes)
should be arranged in a specific way for energy transfer within the polypeptide aggregates. In the course of the formation
and specific aggregation to larger polypeptide assemblies
the tetrapyrrole chromophores are placed in the best possible
locations for energy transfer.

To understand the very effective mechanisms of energy absorption and transfer within the phycobilisomes, it is fundamently
important that one possesses detailed knowledge of the
positions and orientations of the pigment molecules inside
the polypeptide aggregates (hexamers, trimers, stacks of the
discs). These structural data can be obtained by primary
structure analysis and finally three-dimensional structure
analysis of the individual phycobiliproteins (α- and β-poly-
peptide chains). A few years ago, we initiated structural
analysis of three phycobiliproteins and the phycobilisomes
from the cyanobacterium Mastigocladus laminosus. The ob-
jective of the structural analysis of the Mastigocladus

laminosus phycobilisome is to be reached in three stages:
1) Determination of the primary structure of all three phycobiliproteins of the phycobilisome (allophycocyanin, C-phycocyanin, phycoerythrocyanin).
2) Determination of the three-dimensional structure of a phycobiliprotein, e.g. C-phycocyanin by X-ray diffraction
   analysis of single crystals.
3) Investigation of the specific arrangement of the phycobili-
   protein-polypeptide chains inside the phycobilisome.
The primary structures of the three phycobiliproteins already
have been determined in our laboratory (6,7,8) and the three-
dimensional structure of C-phycocyanin (X-ray analysis) has
been performed by Schirmer, Dr. Bode and Dr. Huber, Max-
Planck-Institut, Martinsried, see the following collaborative
paper (9) after preparation of suitable crystals by Dr.
W. Sidler in our laboratory.

All six polypeptide chains (3α- and 3β) have approximately the
same length (~160-170 amino acid residues). They can be
aligned on the basis of sequence homology. Comparing the polypeptide chains in pairs one finds the sequence homology to be
between 21-67% which demonstrates a functional and phylogenetic relationship. The β-chains of C-phycocyanin and
phycoerythrocyanin (171 and 172 residues, respectively), are
larger than the other poly-peptide chains, due to an insertion
of 10 amino acid residues (position 152-161). The insertion
(deletion) probably occurred in evolution as a result of the
specific development of the β-chain of C-phycocyanin (phycoerythrocyanin) which forms part of the outer phycobiliproteins.
Another insertion (deletion) region occurs between the
residues 73 and 83. The structural changes in this region are
probably related to conformational changes in the vicinity of
the chromophore binding site, residue 84. Each phycobiliprotein polypeptide chain develops a particular conformation

which probably has a specific effect on the functional properties of the chromophore. The tetrapyrrole chromophore
binding sites are functionally very important. The phycobilin
chromophore bound at the homologous cystein 84 (numbering on
the basis of α-CPC (α-PEC)) occurs in all polypeptide chains.
This chromophore at residue 84, which is also in a central
position in the primary structure, may be the most essential
pigment molecule for energy transfer in the folded α- and β-
polypeptide chain (position in the three-dimensional structure: helix E (9)). From the functional point of view, the
amino acid residues surrounding this open chain tetrapyrrole
are also of interest (active site residues). Although the
three-dimensional structure analysis ultimately will provide
more information, the primary structure data already reveal a
number of interesting aspects:

The sequence region, C-terminal to the chromophore binding
site 84 (position 84-102), is characterized by high sequence
homology (47% among all 6 polypeptide chains). This stresses
the functional importance of this active site. The following
amino acid residues found in all polypeptide chains could be
in the neighbourhood or in contact with the tetrapyrrole molecule: Arg 86, 93, Asp 87 (ion pairs), Thr 96, Tyr 97 hydrogen
bonds, Leu, Ile, Val, Ala, Tyr hydrophobic interactions. The
β-chains of the outer phycobiliproteins C-phycocyanin and
phycoerythrocyanin contain a second chromophore binding site
at cysteine 155 (153) in the vicinity of the C-terminus.
Interestingly, this position lies within the 151-160 residue
insertion region (helix H, (9)). This suggests a relation
between the amino acid insertion and the additional chromophore binding site. Both may have been incorporated during
the evolution of the β-chains of C-phycocyanin and phycoerythrocyanin to increase the number of pigment molecules,
i.e. the efficiency of energy transfer. Similar pigment and
amino acid insertions at the same and additional sites have

been found in phycoerythrin from <u>Fremyella diplosiphon</u>
(W. Sidler, E. Rüdiger and H. Zuber, paper in preparation).
The additional pigments should be located in the three-dimensional structure (9) in helix B (bilin 50/61, β-chain) and
in helix G (bilin 142, α-chain) which should be a reasonable
position for energy transfer.

On the basis of the primary structure data one can predict the
number, size and distribution of secondary structure elements
(α-helices, β-structure, β-turns), for example using the method
of Chou and Fasman. In this secondary structure prediction the
large number of α-helices in the α- and β-polypeptide chains
particularly in the N- and C-terminal region, and to a much
smaller degree some β-structure strands, were estimated (2).
A similar secondary structure pattern was noted in α-CPC and
α-PEC, or in β-CPC and β-PEC, respectively. On the other hand,
both the α- and β-chain secondary structures of allophycocyanin
are similar to the α- and β-chains of C-phycocyanin (and phycoerythrocyanin). This finding of similar secondary elements
again implies a functional and phylogenetic relationship between all three phycobiliproteins. In the three-dimensional
structure analysis of C-phycocyanin only α-helical structures
were found (see following paper (9)).

The fact that in the α- and β-polypeptide chains of all phycobiliproteins of <u>Mastigocladus laminosus</u> there are a) similar
primary structures (sequence homology, similarities in secondary structures) and b) identical binding sites for the central
tetrapyrrole 84, suggests that there is a general structural
and functional principle for these light-harvesting molecules.
The structural principle is the basis for the <u>in vivo</u> aggregation of the various polypeptides and the specific organisation
of the pigment molecules within the phycobilisome antenna.

We can postulate the following structural and functional organisation of the α- and β-polypeptides and the tetrapyrrole
molecules optimized with respect to energy transfer.

1) α-β-monomers are formed by association of α- and β-polypeptides. The pair of central α- and β-tetrapyrroles are
   fixed in a certain position, and become the basic unit for
   exciton interactions.

2) Aggregation of α-β monomers to form cyclic trimers $(\alpha\beta)_3$
   and a cyclic system of 6 central pigment molecules (arrangement of pigments for "cyclic" energy transfer). Exciton
   interactions between the pigment molecules (α-β, see (9),
   are possible and are confirmed on analysis of CD spectra.
   Binding of specific linker polypeptides influences the
   polypeptide and chromophore conformations, and may explain
   the spectral red shifts (4).

3) Aggregation of two cyclic trimers (with linker polypeptides)
   yields a cyclic hexamer disc $(\alpha\beta)_6$ polar with respect to
   the bound linker polypeptides (X). The interactions between
   the pigment molecules for inductive resonance transfer
   modulated by linker polypeptides is increased. It is likely
   that within one pigment cycle the exciton interactions are
   reasonably strong, but that exciton transfer between cycles
   occurs by a Förster-type mechanism.

4) The polar hexamer (trimer) discs $(\alpha\beta)_6$.X, core: $(\alpha\beta)_3$.X)
   are the basic units for heterogeneous energy transfer with-
   in the phycobilisome: They are arranged in a specific way
   by the bound linker polypeptides, according to their absorption maxima in stacks of defined sizes (stacks of outer
   phycobiliproteins (C-phycocyanin, phycoerythrocyanin),
   stacks of allophycocyanin in the core of the phycobilisome).

5) Finally the stacks of the outer phycobiliproteins (rods)
   and the central allophycocyanin stacks (core) are aggregated specifically to errect the whole phycobilisome antenna.
   The concentric array of the outer phycobiliprotein rods

around the allophycocyanin core focuses the heterogeneous ener-
gic transfer system in the direction to the allophycocyanin-
linker polypeptide trap (α-APC-B, APC· 16-18 KD linker· 80-100
KD linker) and finally to the reaction center within the membrane.

The sequence homology of the α- and β-polypeptide chains not
only provides insight into functionally important regions but
also into their phylogenetic relationship. During the evolution
of the phycobilisome antenna (evolution of the cyanobacteria)
the primary structure of the α- and β-polypeptides of the phycobiliprotein evolved by a process of protein differentiation.
By this process the structures of the polypeptides varied to
receive their specific functional properties within the phycobilisome, i.e. the specific polypeptide environment of the
chromophore and the particular aggregation properties. The process of protein differentiation can be followed by comparing
the primary structures of the α- and β-chains in pairs (sequence homology, sequence differences). The largest differences were found in the case of Mastigocladus laminosus between
the α-chains of C-phycocyanin, phycoerythrocyanin and the β-
chains of these phycobiliproteins including α- and β-allophycocyanin (different function). The greatest sequence homology was
found between the α- and β-chains of C-phycocyanin and phycoerythrocyanin (similar function and location of the outer phycobiliproteins). On the basis of the sequence homology (sequence differences) it is possible to establish a most likely
phylogenetic tree (phylogenetic distances are indicated as PAM
units (2)). From a one chain precursor molecule the α- and β-
polypeptide chains developed separately by repeated gene duplication. At the β-chain side the β-chain of C-phycocyanin and
the α- and β-chains of allophycocyanin, and at a later stage
the β-chain of phycoerythrocyanin, are produced. At the α-
chain side the α-chains of C-phycocyanin and phycoerythrocyanin
developed. The present day phycobilisomes of Mastigocladus

laminosus also evolved stepwise in connection with the process
of protein differentiation of the α- and β-chains (2). The precursor one-chain phycobiliprotein could form small light-harvesting aggregates. With the development of the α- and β-chains
a more effective phycobilisome appears. The formation of α-,
β-allophycocyanin and α-, β-C-phycocyanin leads to a separation
in layers of the outer C-phycocyanin and the central allophycocyanin and to an inward-directed energy transfer. With the
formation of α-, β-phycoerythrocyanin (or phycoerythrin in
other cyanobacteria or red algae) a further layer with increased efficiency for energy uptake and transfer was added.
With a hypothetically equal mutation rate the development of
the phycobilisome of cyanobacteria should be completed after
10^6-10^7 years (starting some $2.5.10^7$ years ago). After this
process the phycobiliproteins and the phycobilisomes remained
practically unchanged in structure during species differentiation of the cyanobacteria and also during incorporation of
the phycobilisome (chloroplast) into the eucaryotic cell of
the red algae (some 900 million years ago). This is revealed
by the high sequence homology (70-80%)of todays phycobiliproteins from various cyanobacteria and red algae.

Light-harvesting BChl-protein Complexes of Purple Photosynthetic Bacteria

A number of structural and functional features have been found
in the intramembrane light-harvesting BChl-protein complexes
of purple photosynthetic bacteria. Interestingly, a similar
structural principle for heterogeneous energy transfer among
cycles of pigments seems to occur. However, the polypeptides
of this hydrophobic antenna system are located in the lipid
environment of the membrane (1). In purple bacteria the

majority of the BChl-pigment-molecules are associated with the
light-harvesting complexes (10,11). Therefore the absorption
spectra of whole cells or isolated membranes (chromatophores)
are largely representative of those of the light-harvesting
complexes. The various purple bacteria display typical spectra
between about 800 and 1000 nm and are classified on the basis
of these complexes (1,12): Group I (e.g. Rs. rubrum: 870-890
nm, Rp. viridis: 1015 nm) has only one complex: B 870 (890) or
B 1015 nm respectively. Group II (e.g. Rp. sphaeroides, Rp.
capsulata, Rp. gelatinosa: 800, 850, 870 (890) nm) possesses
two complexes: B 800-850 and B 870 (890). Group III (e.g.
Chromatium vinosum: 800, 820, 850, 870 (890) nm shows three
types of complexes: B 800-820, B 800 - 850 and B 870 (890).
These complexes with different absorption maxima are, as in the
case of the phycobiliproteins, the basic units for heterogeneous energy transfer. As we also found by sequence studies
the light-harvesting BChl a-protein complexes contain two
types of small polypeptides (α-, β-polypeptides, 50-60 amino
acid residues) in a ratio 1:1. These two polypeptides are
similar to the α- and β-polypeptide chains of the phycobiliproteins, the basic structural unit of the intramembrane
light-harvesting antenna. They also bind a characteristic
number of BChl a molecules (like the phycobiliproteins):
The inner complex B 870 (890) near the reaction center binds
2 BChl a and normally 1 carotenoid molecule, the outer complexes B 800-820 and B 800-850 bind 3 BChl a and normally 1
carotenoid per polypeptide pair.

We determined the primary structures of the α- β-polypeptides
of the B 870 and B 800-850 complexes from a number of purple
photosynthetic bacteria (Rs. rubrum (13,14), Rp. sphaeroides
(15), Rp. viridis (18), Rp. capsulata (16,17), Rp. gelatinosa)
by automated or manual Edman-degradation after fragmentation
of the polypeptides to peptide fragments. As in the case of

the phycobiliproteins, the primary structure data reveal the
structural and functional organisation of the polypeptide
chains including the pigment binding site and the phylogenetic
relationship of the various α- and β-polypeptide chains (3).
The α- and β-polypeptides of each complex are only slightly
sequence homologeous (7-13%), which demonstrates a possible
early separation in evolution of the α- and β-genes (polypeptides) by gene duplication. On the basis of their higher
sequence homology (13-18%) it can be postulated that α-chains
or the β-chains of the outer B 800-850 complex evolved from
the α-chains or the β-chains respectively of the inner B 870
complex (or vice versa). By this process of protein differentiation (in analogy to the phycobiliproteins and the phycobilisomes) the light-harvesting antenna around the reaction center
was built up. Again, as in the instance of the phycobiliproteins, the α- and β-polypeptides of the individual complexes isolated from various purple bacteria have relatively
high sequence homologies (28-78%), which represents the
species differences of these polypeptides.

If one aligns the α- or β-polypeptides for maximum homology a
typical domain structure can be demonstrated (3,14,15). A
polar, charged N-terminal domain (α: 12-19 residues, β: 20-22
residues), a hydrophobic central domain (α: 21 residues,
β: 23 residues) and a polar, charged C-terminal domain (α:
19-25 residues, β: 3-10 residues).

The most homologeous amino acid residues are the His residues.
One type is found in all polypeptide chains within the hydrophobic domain. A second conserved His residue is present in
all β-chains. According to the BChl-protein-complexes of the
green bacterium prosthecochloris aestuarii (19) these conserved His residues are probably the main binding sites (via
a Mg-atom) for BChl a molecules. This again would be a

situation similar to the phycobiliproteins: A central pigment binding site exists in both α- and β-polypeptide chains and an additional site for one pigment in the β-chains.

The domain structure of the light-harvesting polypeptides indicates a specific organization of these polypeptides within the membrane (3). It is reasonable to assume that the hydrophobic domain is located within the membrane (in the hydrocarbon tail region), while the N- and C-terminal domain should be found at the polar head region or at the membrane surface. Assuming a thickness of the photosynthetic membrane between 40-60 Å (30-40 Å hydrocarbon tail region, 10-20 Å polar head region) the hydrophobic domain of the polypeptides would fit only in the α-helical form and only transmembrane oriented (helix length: 31-34 Å) in the membrane. IR-spectroscopic data support this hypothesis (20). In this case the polar and charged cluster regions would be located in the polar head region. The transmembrane orientation of the light-harvesting polypeptides is also supported by digestion experiments with proteolytic enzymes (21). In intact vesicles with the cytoplasmatic side outside only the N-terminal parts of the α- and β-polypeptides exposed at the cytoplasmatic membrane surface could be split off (with proteinase K: α: 6 residues, β: 16 residues, with trypsin, chymotrypsin, staph. proteinase only α: 3-7 residues, collaboration with Dr. Wiemken, Dr. Bachofen, Universität Zürich).

On the basis of this transmembrane model of the light-harvesting polypeptides we can postulate the following:
1) The central BChl a molecules bound at the periplasmatic side of the hydrophobic domains of the α- β-polypeptides are probably exciton-couples (either within the (α-β) polypeptide pair or between two pairs, again on the basis of CD-spectra).

2) The BChl a molecule bound to the second conserved His residue at the β-chain is probably the monomeric BChl a (800 nm
   absorption).

3) The carotenoid binding site and regulatory sites respectively could be in the N-terminal domain at the hydrophobic
   cluster regions and at the acid cluster regions.

One next considers the likely molecular organization of the
α- and β-polypeptides and their BChl a molecules within the
BChl-protein-complexes. The α- and β-light-harvesting polypeptides are responsible for the specific binding and arrangement of the BChl pair within the BChl-protein complexes in a
lipid environment. This arrangement is based, as in the phycobiliproteins, on the aggregation of the α- and β-polypeptides
(α-helices) within these complexes. The two transmembrane α-
helices aggregate by polypeptide-polypeptide interactions at
the N- and C-terminal domains near and at the membrane
surface and at the α-helical hydrophobic domain via the amino
acid side chains. The amino-acid-side-chain interactions of
the two helices are dependent on the distribution of the side
chains at the helix surface. In the right-handed α-helices
(5-6 turns) four right-handed rows of side chains are possible.
Twisting of the two helices yields a tight packing of the α-β-
polypeptide helix pair (transmebrane superhelix), whereby part
of the side chains interact. The residual side-chains on the
surface of the superhelix could further interact to form larger
arrays of α-β-polypeptide helix pairs, which should correspond
to the BChl-protein complexes. Besides the charged amino acid
side chains, the side chains with polar residues are the types
best suited for precise polypeptide-polypeptide interactions.
In all light-harvesting polypeptides studied the polar residues are situated at the N-terminal (cytoplasmatic) half of
the α-helices. Further in this N-terminal α-helix region the
polar residues of all α- or β-polypeptides are concentrated

on particular regions at the α-helix surface. Accordingly,
these polar regions should be hypothetically together with
the hydrophobic side chains the polypeptide-polypeptide interaction sites for the aggregation of the α-, β-polypeptides and
of the (α-β) polypeptide pairs to erect the light harvesting
BChl-protein-complexes. Taking into account the repetition of
3-4 interacting sites in the β-chain and of 2-3 sites in the
α-chain one finds as a result a cyclic arrangement of the (α-β)
polypeptide pair (i.e. cyclic hexamers), similar to the situation with the phycobiliproteins. In this tightly packed polypeptide arrangement, the BChl a molecules are all oriented
outside and lie in a polypeptide-lipid environment. This environment should be the basis for the various spectral properties of the light-harvesting complexes. In this system
pairs of BChl a molecules should be strongly excitonically
coupled (distance 10-15 Å), while the distances between the
pairs are greater (20-30 Å) and the resulting coupling weaker.
Therefore the cyclic arrays (e.g. hexamers) of pigment molecules could have excitonic behavior similar to that postulated
in the case of the phycobiliproteins.

To form the whole light-harvesting antenna the cyclic arrays
of the BChl-protein complexes (B 870, B 800-850) aggregate
further within the membrane yielding a two dimensional array
of e.g. light-harvesting hexameric cycles (B 800-850) in the
neighbourhood of the reaction center. By this arrangement a
system for heterogeneous energy transfer is set up by com-
bining the inner B 870 complex with the outer B 800-850 complex and focusing the energy migration in the direction of
the reaction center. The BChl molecules outside and therefore
between the complexes could form "channels" for energy
migration. In this concentric arrangement the various light-
harvesting complexes hypothetically could interact by the
N- and C-terminal polar and charged domains, folded at the

membrane surface. The conformational state of these folded regions could (depending on the light conditions) influence the polypeptide- i.e. the BChl a-distances and therefore the efficiency of the energy transfer.

In conclusion, it can be postulated that the intramembrane and extramembrane light-harvesting antennas of purple photosynthetic bacteria and cyanobacteria, respectively, are organized in a similar way by specific aggregation of pairs of polypeptides carrying the pigment molecules to form large arrays of polypeptides and pigment molecules for heterogeneous energy transfer of excitons to the reaction center. Most likely, this specific organization is a general structural principle for light-harvesting in all photosynthetic organisms.

Acknowledgments

This work was supported by the Swiss National Science Foundation (project 3.190.-0.77, 3.534-0.79, 3.286-0.82) and by the ETH Zürich.

References

1. Clayton, R.K., Sistrom, W.R.: The Photosynthetic Bacteria Plenum Press, New York, London (1978).

2. Zuber, H.: In: Photosynthetic Prokaryotes: Cell Differentiation and Function, Papageorgiou, C.G., Packer, L. (eds), Elsevier Science Publishing Co., New York, Amsterdam, Oxford, 1983, pp 23-42.

3. Zuber, H.: IVth Int. Symposium on Photosynthetic Prokaryotes, Bombannes, France, 1982. Abstracts C 40.

4. Glazer, A.: Biochim. Biophys. Acta 768, 29-51 (1984).

5. Nies, M., Wehrmeyer, W.: Arch. Microbiol. 129, 374-379 (1981).

6. Frank, G., Sidler, W., Widmer, H., Zuber, H.: Hoppe Seyler's Z. Physiol. Chem. 359, 1491-1507 (1978).

7. Sidler, W., Gysi, J., Isker, E., Zuber, H.: Hoppe Seyler's Z. Physiol. Chem. 362, 611-628 (1981).

8. Füglistaller, P., Suter, F., Zuber, H.: Hoppe Seyler's Z. Physiol. Chem. 364, 691-712 (1983).

9. Schirmer, T., Bode, W., Huber, R., Sidler, W., Zuber, H.: Proceedings of the Int. Symposium on "Optical Properties and Structure of Tetrapyrroles", Walter de Gruyter, 1984, pp. ...

10. Cogdell, R.J., Thornber, J.P.,: Febs Letters 122, 1-8 (1980).

11. Cogdell, R.J., Valentine, J.: Photochem. Photobiol. 38, 769-772 (1983).

12. Thornber, J.P., Cogdell, R.J., Pierson, B.K., Seftor, R.E.B.: J. Cellular Biochem. 23, 159-169 (1983).

13. Brunisholz, R.A., Cuendet, P.A., Theiler, R., Zuber, H.: Febs Letters 129, 150-154 (1981).

14. Brunisholz, R.A., Suter F., Zuber, H.: Hoppe Seyler's Z. Physiol. Chem. 365, 675-688 (1984).

15. Theiler, R., Suter, F., Wiemken, K., Zuber, H.: Hoppe Seyler's Z. Physiol. Chem. 365, 703-719 (1984).

16. Tadros, M. Suter, F., Drews, G., Zuber, H.: Eur. J. Biochem. 129, 533-536 (1983).

17. Tadros, M., Suter, F., Seydewitz, H.H., Witt, J., Zuber, H., Drews, G.: Eur. J. Biochem. 138, 209-212 (1984).

18. Brunisholz, R.A., Jay, F., Suter, F., Zuber H.: Hoppe Seyler's Z. Physiol. Chem., submitted.

19. Fenna, R.E., Matthews, B.W.: Nature 258, 533-557 (1975).

20. Theiler, R., Zuber, H.: Hoppe Seyler's Z. Physiol. Chem. 365, 721-729 (1984).

21. Brunisholz, R.A., Wiemken, V., Suter, F., Bachofen, R., Zuber, H.: Hoppe Seyler's Z. Physiol. Chem. 365, 689-701 (1984).

Received July 27, 1984

Discussion

<u>Hoff</u>:   The circular arrays of B 800/850 antenna protein you propose seem to make energy transfer between reaction centers difficult. Is there a way to organize the B 800/850 and B 890 complexes differently, while retaining the stoichiometry and binding sites, for example in a more or less linear array such that the B 890 complexes are linked?

<u>Zuber</u>:   No. We tried very hard to find a different way of organization of the B 800/850 and B 870 complex. The cyclic arrangement seems to be the most satisfying system known to fulfill the experimental data (structural, spectral). This also concerns the energy transfer between the reaction centers. In the model proposed the reaction centers are all connected via the light-harvesting complexes.

<u>Holzwarth</u>:   I think Dr. Zuber's model allows for energy transfer between reaction centers in a lake model if one accepts this transfer to be mediated by the antenna. I am not aware of a photosynthetic system where it is proven that energy transfer between reaction centers occurs directly.

<u>Huber</u>:   The highly symmetrical circular arrangement of the membrane-bound light-harvesting complexes is unlikely in view of the very asymmetric shape of the reaction center of Rp. viridis whose crystal structure is known.

<u>Zuber</u>:   A highly symmetrical cyclic (circular) arrangement of the polypeptide components was found by electron microscopy in the light-harvesting complex B 1015 of Rp. viridis. Our model of the cyclic arrangement of the α- and β-polypeptide pairs corresponds to this data. Therefore, it seems that we are confronted in the particular case of the B 1015 complex/reaction center of Rp. viridis (and probably of the light-harvesting complexes/reaction center of other purple bacteria) with a system which combines symmetric and asymmetric units. In my opinion, this is not impossible, particularly if one assumes or takes into account a) similar binding sites of the L- and M-subunits with respect to the light-harvesting polypeptides and b) the two-fold symmetry of the reaction center chlorophylls.

<u>Scheer</u>:   Cogdell obtained evidence which would, to my feeling, conflict with your placing the chromophores at the outside of the purple bacterial antenna complexes. He has shown that the characteristic long-wavelength shift is only observed if you have large aggregates like hexamers $(\alpha\beta)_6$. This spectrum disappears before dissociation if the antenna is treated with SDS. Since at least part of the long-wavelength shift is due to aggregation, it seems that the chromophores are brought close together in the hexamer, e.g. they are in the center rather than at the periphery of the hexamer.

<u>Zuber</u>:   In fact, the chromophores (BChla) in our hexamer model (of B 800/850 complexes) are close together a) in BChla pairs for exciton interactions (12-15 Å distances) and b) for energy transfer between the α-β-BChl pairs (distances ~20 Å). (I am not aware of the findings of R. Cogdell that the red shift is only observed in hexamers.) There is no conflict with the data of Cogdell and I do not understand your question. The characteristic long-wavelength shift should be possible in our hexamer $(\alpha$-$\beta)_6$ model, even

with the BChl molecules outside oriented. By forming the hexamers the C-terminal part of the polypeptide chains, in addition to the α-helical side chains, would also interact with the BChl molecules in a specific way to cause the spectral red shift.

Buchler: You suggested a histidine in a BChl-protein complex as a binding site for the BChl molecule via a Mg-N bond. Are there other chlorophylls in which a N(His)-Mg(Chl)-coordination has been shown? (From the viewpoint of the coordination chemist a Mg-N-coordination is not that necessary as the well-known Fe-N-coordination in heme proteins.)

Zuber: Possible histidine-BChl interactions have been shown in the case of the three-dimensional structure analysis (Fenna and Matthews) of the BChla-protein complex of the green photosynthetic bacterium *Prostheco-chloris aestuarii* (7 BChl-histidine interaction sites via Mg-atoms per subunit).

Schneider: Do you think that the linker protein interacts directly with the chromophore to change its properties (e.g. excitation energy) or is this interaction mediated by the protein segment close to the binding site?

Zuber: Both (direct and indirect) types of interactions are possible. However, it is too early (without knowledge of structural details) to make any statement on these interactions.

Sund: Do the polypeptide chains of the different phycocyanin molecules contain free sulfhydryl groups?

Zuber: Yes, the α- and β-polypeptides of the various phycobiliproteins contain one to two cysteine residues (not bound to tetrapyrroles) which could form free sulfhydryl groups.

Gouterman: How do the structures of the phycobiliprotein antennae relate to the bacteriochlorophyll antennae and the reaction site? Does one feed energy to another?

Zuber: The phycobilisomes (with the various phycobiliproteins) are bound as extra-membrane antennae to the thylakoid membrane surface. At these binding sites they probably interact (directly or indirectly via other poly-peptides by polypeptide-polypeptide interaction) with the Chla-protein complexes within the membrane. Energy transfer takes place between the phyco-bilin molecules of allophycocyanin and the Chla molecules and finally the Chla molecules of the reaction centers (mainly photosystem II).

THE CRYSTAL AND MOLECULAR STRUCTURE OF C-PHYCOCYANIN FROM MASTIGOCLADUS LAMINOSUS AND ITS IMPLICATIONS FOR FUNCTION AND EVOLUTION

Tilman Schirmer, Wolfram Bode, Robert Huber
Max-Planck-Institut für Biochemie
D-8033 Martinsried

Walter Sidler and Herbert Zuber
Institut für Molekularbiologie und Biophysik, ETH-Hönggerberg
CH-8093 Zürich

Summary

C-Phycocyanin is a component of the light-harvesting organelles in cyanobacteria. The protein from Mastigocladus
laminosus forms well ordered crystals. These have been
studied by crystallographic methods and an electron density
map at 3 Å resolution has been determined based on phases from
multiple isomorphous replacement. The phases have been
improved by solvent flattening and incorporation of partial
model information. An atomic model on the basis of the known
chemical amino acid sequence has been built, which is currently
refined but already at this stage allows to draw a number of
interesting conclusions about the function and evolution of
C-phycocyanin.

The unit building up the crystal structure is a trimer of ($\alpha\beta$)
units. The trimer resembles a hollow disk with an outer
diameter of 110 Å, an inner diameter of 35 Å and a thickness
of 30 Å. The trimer is located on a crystallographic triad.
The trimer is known to be the predominant molecular species in
solution and probably plays also an important role in the

phyco-bilisome organelles.

The α and ß units are structurally similar and related by an
approximate diad perpendicular to the molecular triad. The
molecular conformation is characterized by predominance of
α-helices and the absence of ß-sheet structures. There are 8
α-helices in both the α and ß units, which we designate
X1,X2,A,B,E,F,G,H. The spatial arrangement of helices A,B,E,
F,G and H is closely similar to the arrangement of the helical
segments A,B,E,F,G,H in the hemoglobins.

Helices X1 and X2 have no counterpart in the hemoglobin fold,
but represent an N-terminal extension. Helices X1 and X2 of
the α-subunit are in extensive contact with helices A,E, and
F of the ß subunit and vice versa. This interaction is of
basic importance for the α-ß association. Interaction of the
α-ß units within the trimer is less extensive.

The two tetrapyrrole pigments of the ß-subunit have well
defined electron densities and are in extended conformation.
One chromophore is attached to a cysteine residue of helix E,
the other chromophore to helix H. The chromophore attachment
site on helix E is homologous to the heme attachment site
(His E7) in the hemoglobins, emphasizing the evolutionary
relationship documented by the similar polypeptide chain folds.
The tetrapyrrole attached to helix E of the α subunit is less
well defined. Further crystallographic refinement is required
in this case.

The tetrapyrrole pigments bound to helix E of the α and ß
subunits are relatively close in space within the trimer.
The tetrapyrrole bound to the helix H of the ß subunits is
isolated. There is evidence of an extensive interaction of the
chromophores in the intact bilisome organelles. Apart from
the interaction within the trimer there is the possibility of
stacking interactions of the 30 Å thick trimers.

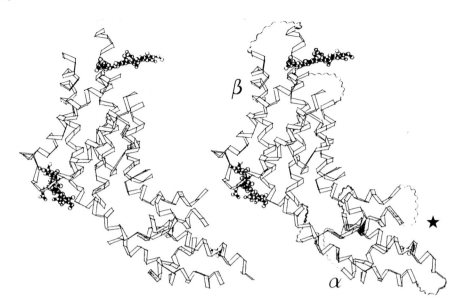

Fig. 1 Preliminary model of the C-phycocyanin monomer (αß).
The polypeptide chains are symbolized by ribbons connecting
the Cα-atoms (dark: α-chain, bright: ß-chain). The two chromophores of the ß-subunit are shown as "balls and sticks". The
model is not complete:〜〜missing loops,★:position of the chromophor of the α-subunit. View along the crystallographic
c-axis.

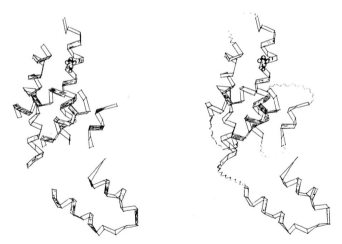

Fig. 2 C-phycocyanin α-chain, rotated around the local two-
fold axis in the position of the ß-chain for comparison.

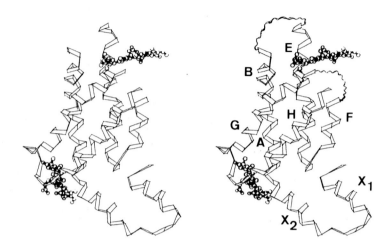

Fig. 3 C-phycocyanin ß-chain, same orientation as in Fig. 1.
The helices are labelled by the capital letters X_1, X_2, A, B, E,
F, G and H (A, B, E, F, G and H referring to the globin nomen-
clature).

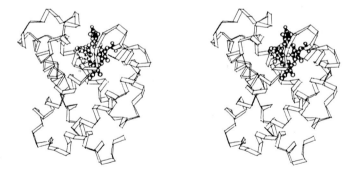

Fig. 4 Sperm whale myoglobin, aligned with the ß-chain of
C-phycocyanin. The haem group and Histidin E7 are shown as
"balls-and-sticks".

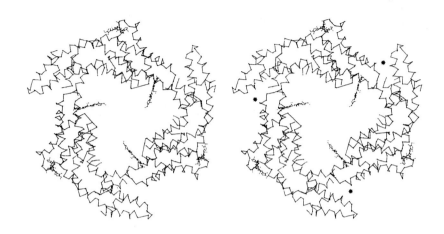

Fig. 5 C-phycocyanin trimer $(\alpha\beta)_3$ of the crystal.
The threefold crystallographic axis is running perpendicular
to the drawing plane through the center of the trimer.

All figures were prepared by a computer program written by
A.M. Lesk and K. D. Hardman.

Received July 4, 1984

Discussion

Gouterman: What is the percent homology between phycocyanin and myoglobin?

Schirmer: The overall sequence homology is low, but, comparing only the
main helical parts of the molecules (A,B,E,F,G and H), it turns out that
18 % of the myoglobin residues are identical either with the α or β chain
of phycocyanin.

Section IV
Chlorophylls

Chairmen: M. Gouterman and A.A. Lamola

OPTICAL SPECTRA OF PHOTOSYNTHETIC REACTION CENTERS. THEORETICAL AND
EXPERIMENTAL ASPECTS

A.J. Hoff

Department of Biophysics, Huygens Laboratory of the State University,
P.O. Box 9504, 2300 RA Leiden, The Netherlands

Introduction

In this paper we will discuss interactions between those pigments in photosynthetic organisms that are engaged in the primary photoprocesses. Such
interactions play a central role in the conversion of light into chemical
and electrical free energy, firstly because they probably are essential for
the photoredox or charge separation reaction and secondly because they allow trapping of excitation energy. For readers not familiar with photosynthesis, first a brief discussion of the electron transport chains, including
the photoredox reactions, is given. The attention will then be focussed on
the bacterial photosystem. The make-up of the bacterial photoreaction center
is introduced and tentative assignments of electron donors and acceptors are
given. Subsequently, the optical investigations (absorbance, linear and circular dichroism) that have led to assignments are reviewed. Then, a recently
developed technique for the recording of triplet-minus-singlet absorbance
difference spectra will be introduced, and it will be shown that this technique is of considerable help in unraveling absorbance difference spectra.
Notably, as discussed in the last section, this technique which makes use
of absorbance-detected magnetic resonance (ADMR) of triplet states in zero
magnetic field, has allowed to discriminate between competing interpretations
of redox absorbance difference spectra and has contributed new insight into
the structure of the primary electron donors in photosynthesis.

Photosynthetic electron transport

Photosynthesis in plants and some bacteria is the complex series of photo-

Optical Properties and Structure of Tetrapyrroles
© 1985 by Walter de Gruyter & Co., Berlin · New York

physical and photochemical reactions by which the photon energy of visible
and near-infrared light is converted into cell metabolites. The reaction
can be roughly divided into three rather different processes: i) photon absorption and excitation transfer, ii) trapping and charge separation, iii)
dark electron transfer and biochemical metabolic reactions. The light is
captured by light-harvesting pigment-protein complexes (LH-PPC) and the
electronic excitation is transferred to the reaction center complex (RC-PPC
or RC), a specialized PPC consisting of light-harvesting pigments, electron
donor(s) and electron acceptors. In the RC, the electronic excitation energy
is trapped and rapidly (less than 10 ps (1,2)) and with high quantum yield
(1.02 ± 0.04 (3)) converted into the chemical free energy of two separated
charges by the reaction

$$D \; A \xrightarrow{\;h\nu\;} D^{*}A \longrightarrow D^{+}A^{-} \tag{1}$$

where D is the primary electron donor and A the first electron acceptor.
The charge separation thus produced drives a series of dark electron transfer reactions, ultimately producing the chemical (electrical) potential
gradients necessary for the functioning of the cell's metabolism:

$$\ldots D_2 D_1^{+} A_1^{-} A_2 \ldots \;
\begin{array}{c} \nearrow \; \ldots D_2 D_1^{+} A_1 A_2^{-} \ldots \; \searrow \\[4pt] \\[4pt] \searrow \; \ldots D_2^{+} D_1 A_1^{-} A_2 \ldots \; \nearrow \end{array}
\; \ldots D_2^{+} D_1 A_1 A_2^{-} \ldots \to \ldots \tag{2}$$

The array of D's and A's form the electron transport chain, ETC. In plants
and in the blue-green algae or cyanobacteria, the ETC comprises two photoreactions in two separate RC systems, called photosystem 1 (PS 1) and photosystem 2 (PS 2), and a number of dark reactions between electron carriers
(Fig. 1). The ultimate electron donor of PS 2 is water, which is oxidized
via a series of intermediates by the primary electron donor of PS 2, P680.
(The number in this and following labels refers to the peak of a bleaching
in the photo-induced redox absorbance difference spectrum; see below).
Through a large number of electron carriers (PS 1 and PS 2 are not physically connected) the electron on A_1, A_2 ... of PS 2 is transported to the donor
side (D_n, D_{n-1}, D_1) of PS 1, i.e. it is utilized to reset the photooxidized

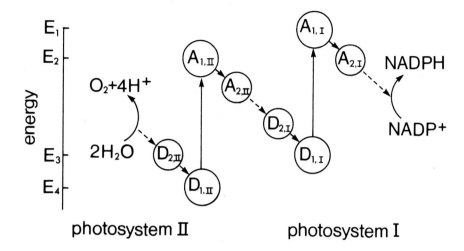

Fig. 1. The photosynthetic electron transport chain in plants. E_1 - E_4: redox energies of the primary donors and acceptors. Energies of other redox components not drawn to scale.

part of PS 1 to ready it for the next photoreaction. Obviously, there is · not a sharp distinction between the A part of PS 2 and the D part of PS 1, and several electrons and positive charges can be stored in the ETC between H_2O, P680, and P700, the primary donor of PS 1.

The photoreaction of PS 1 generates via a number of intermediates a highly reductive reduced ferredoxin that is capable to reduce in an enzymatic reaction $NADP^+$ to NADPH. The latter molecule is used, along with ATP, in the Calvin-Benson cycle of reactions to fix CO_2 to glucose. ATP is produced by the enzyme ATPase at several steps along the ETC between P680 and P700, utilizing the electrical and pH gradients produced across the membrane by the photo- and subsequent dark reactions.

Several families of bacteria are also capable of photosynthesis. They are usually divided into the purple photosynthetic bacteria (Rhodospirillaceae) and the green photosynthetic bacteria (Chlorobiaceae and Chloroflexaceae). The Rhodospirillaceae, and as recently has become apparent also the Chloroflexaceae, have similar photosystems with only minor variations. They consist of one photoreaction, which functions in a cyclic ETC:

$$D_n \ldots D_1 A_1 \ldots A_n \xrightarrow{\ h\nu\ } D_n \ldots D_1^+ A_1^- \ldots A_n \rightarrow \qquad (3)$$

The electrical and pH gradients produced across the membrane by reaction (3) generate via ATPase ATP which is partly used to reduce NAD^+ to NADH. The latter molecule catalyzes a number of metabolic reactions analogous to the Calvin-Benson cycle in plants. The Chlorobiaceae have an ETC that is quite different from that of the purple bacteria, and is rather similar to that of PS 1 of plants. It comprises one photoreaction, that generates a highly reductive ferredoxin capable of reducing NAD^+ directly. In all probability, the ETC of the Chlorobiaceae is also cyclic.

Photosynthetic reaction centers

The reaction center is a PPC comprising the real photoreactor, which is capable to carry out reaction (1). In addition to D_1 and one or more A's, the RC-PPC houses a variable number of accessory pigments. Most of our insight in the build-up of photosynthetic RC has come from the study of RC-PPC isolated from photosynthetic bacteria, notably that of the carotenoid-less mutant of the purple bacterium Rhodopseudomonas sphaeroides R-26 (4). This RC-PPC can be prepared in highly purified form and as it can be regarded as a paradigm for all other RC-PPC we will discuss it here in some detail. It consists of three subunits, two of which house the RC-ETC complex that consists of four bacteriochlorophyll a molecules, two bacteriopheophytin a molecules and two ubiquinones connected by a high-spin ferrous ion.

Two of the BChl a molecules are believed to form a pair that together make up the primary donor P865 (the wavelength varies somewhat from species to species). One of the two remaining BChl a's that both absorb near 800 nm and are usually labelled B800, is thought to act as first acceptor (B), possibly in a charge transfer complex with P865. One of the two BPh a's most probably acts as 'secondary' acceptor (H); the two quinones are labelled Q_A (the tertiary or first 'stable' acceptor) and Q_B (the next 'stable' acceptor) and are both strongly complexed to the Fe^{2+} ion. According to this (proposed but not unambiguously proven) scheme, the photoredox reaction and early transfer reactions are written (5):

$$(\text{BChl } \underline{a})_2 \text{ BChl } \underline{a} \text{ BPh } \underline{a} \text{ } Q_A \text{Fe}^{2+} Q_B \xrightarrow[1.5 \text{ ps}]{h\nu} P^+ B^- H \text{ } Q_A \text{Fe}^{2+} Q_B \xrightarrow{5 \text{ ps}}$$
$$ P B H$$

$$P^+ B \text{ } H^- Q_A \text{Fe}^{2+} Q_B \xrightarrow{200 \text{ ps}} P^+ B \text{ } H \text{ } Q_A^- \text{Fe}^{2+} Q_B \xrightarrow{200 \text{ } \mu s} P^+ B \text{ } H \text{ } Q_A \text{Fe}^{2+} Q_B^- \qquad (4)$$

Q_B is a two-electron acceptor and is in the intact photosynthetic membrane connected to a complex of electron carriers via the so-called quinone pool (comprised of Q_B and 0 to ~ 12 quinones).

The RC-PPC of the other purple photosynthetic bacteria resemble much that of Rps. sphaeroides R-26. That of the green bacteria has not yet been studied in as great a detail, although progress has been made (6). The RC-PPC of PS 1 and PS 2 have not yet been isolated in a form devoid of LH-PPC. This has considerably hampered their study by optical means, due to the strong absorbancies of non-photoreactive pigments. It is now generally thought that the primary donor of PS 1 and PS 2 are monomeric or dimeric Chl $\underline{a}$ molecules; the 'multiplicity' is controversial. In PS 1, the first and second acceptors may be Chl $\underline{a}$ molecules. In PS 2 the first acceptor is thought to be a Pheo $\underline{a}$ molecule, the second and third plastoquinone molecules complexed to a Fe^{2+} ion (7).

Optical Spectroscopy

An important tool to elucidate reaction center composition is absorbance spectroscopy, and the study of light-induced changes in the absorbance. As the latter changes may be reproduced by chemical oxidation or reduction of ETC components, we will generally call the difference spectrum obtained by subtraction of the dark, or untreated, absorbance spectrum a redox-ΔA spectrum. By way of example the absorbance spectra and the redox-ΔA spectra of an isolated bacterial reaction center are displayed in Fig. 2. The assignment of the bands in the bacterial absorbance spectrum is, although generally received, not unequivocal. It rests mainly on the belief that a band close to an absorbance band of a particular pigment in vitro must be due to the same pigment in vivo. The RC bands ascribed to BPh $\underline{a}$ are indeed

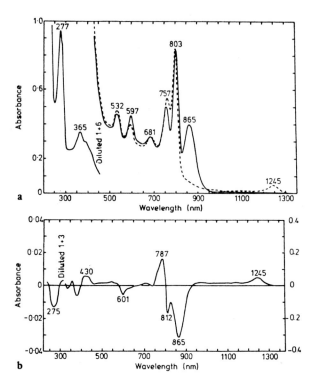

Fig. 2. a) Absorbance spectrum of reaction centers of <u>Rps</u>. <u>sphaeroides</u> R-26.
——, dark, ---, under illumination at 800 nm. b) Redox-ΔA spectrum light-minus-dark constructed from a). From (30).

close to the Q_y and Q_x bands of BPh <u>a</u> in vivo. The other bands, and in particular those at 800 and 865 nm, are far removed from the absorbance of
either BPh <u>a</u> or BChl <u>a</u> in vitro. These shifts are usually attributed to
electrostatic (excitonic) interaction between transition dipole moments in
the singlet excited state. Exciton theory, as developed by Davidov (8) and
applied by Kasha (9) and Tinoco (10), gives the band positions and oscillator strengths of two identical interacting pigments with transition dipole moment μ and originally degenerate energy levels E as

$$E_a = E + J, \; E_b = E - J \tag{5a}$$

$$J = (5.04 \; \mu^2/R^3)/(\cos \alpha - 3 \cos \beta_1 \cos \beta_2) \tag{5b}$$

$$A_a = \mu^2(1 + \cos \alpha), \; A_b = \mu^2(1 - \cos \alpha) \tag{5c}$$

where J is in cm^{-1}, μ is in Debye, R is the distance between the pigments
in nm, α is the angle between the two dipoles and the β's are the angles
between the dipoles and the distance vector.

According to this theory, two BChl a molecules may give rise to two bands,
one blue shifted and one red shifted with respect to the original absorbance
peak. As the in vitro Q_y band of BChl a is at 780 nm, it stands to reason to
regard the 870 nm band (which has about twice the monomeric molar extinction)
as the red shifted component of a dimeric BChl a exciton. The 600 nm band in
the RC is generally ascribed to the sum of the Q_x components of all four
BChl a pigments, and the 780 nm band to the two BPh a's, so that the blue
shifted (higher energy) exciton components of the dimers must be forbidden,
i.e. α = $β_1$ = $β_2$ ~ 0, and the BChl a pigments are parallel and lie roughly
head-to-tail. The 800 nm band, which also has oscillator strength roughly
corresponding to two monomers, could either be due to a similar dimer with
lesser oscillator strength (e.g. because of larger R) or the 20 nm shift
with respect to the in vitro absorbance is environmental, and the two B800
BChl a pigments show little interaction. The observation of a shoulder at
810 nm at 1.5 K (11) supports the latter interpretation.

Although the above band assignments of the RC absorbance spectrum do account
for all the bands and the relative oscillator strength in the visible and
near-infrared region, it suffers from several inconsistencies: i) the 870 nm
band is much broader than the in vitro band, whereas exciton theory would
predict a narrowing of a factor √2 (12), ii) an exciton-induced shift of
almost 100 nm is difficult to account for if reasonable distances and dipole strengths are used in the calculation of J. Moreover, it is clear that
all six pigments lie close together, and it is certainly not excluded that
several are excitonically coupled with each other. Then the picture becomes
much more complex and it is hard to reach at an unambiguous band assignment.

The steady-state redox-ΔA spectrum may help in the assignment, as photooxidation of the primary donor, together with photoreduction of one of the
'stable' acceptors, leads to bleaching of the respective absorbance bands
and shifts of the bands of accessory pigments due to the photo-induced
electrical charges (Stark effect). This is clearly seen in Fig. 2 where it

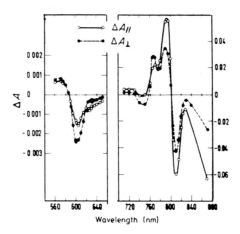

Fig. 3. Redox-ΔA light-minus-dark spectra of reaction centers of Rps. sphaeroides R-26 for parallel and perpendicular polarization of the analyzing beam with respect to the polarization of excitation light at 900 nm. From (31) (left) and (32) (right).

is shown that the 865 nm band is almost completely bleached, whereas the 800 nm band shows a blue shift and the 760 nm band an increase in oscillator strength. At 1245 nm a weak band appears. The interpretation of the redox-ΔA spectrum, however, is less simple than suggested. In fact, the apparent shift of the 800 nm band is made up of more than one component as evidenced by the redox-ΔA spectrum taken for different directions of polarization in a photoselection experiment (Fig. 3). For a simple shift, one would expect the positive and negative lobe to have equal intensity for both directions of polarization; this is clearly not the case. The spectra of Fig. 3 were interpreted as a band shift plus a bleaching of a low-intensity absorbance band peaking at 805 nm and polarized perpendicular to the 870 nm transition (which shows uniform polarization across the band). This was taken to support the hypothesis of Vermeglio and Clayton (13) that the 805 and 865 nm bands are the exciton partners of an excitonically coupled dimer transition. This proposition implies that the band center is at 835 nm, and that $2 J \approx 860$ cm^{-1}. The interpretation of the photoselection experiment is not unambiguous, however. Shuvalov et al. (14) have interpreted the spectra of Fig. 3 as a superposition of two band shifts around 800 nm, without a bleaching at 805 nm. In their view the 865 nm band is a single allowed low-energy exciton band of the dimeric primary donor, the higher-energy blue

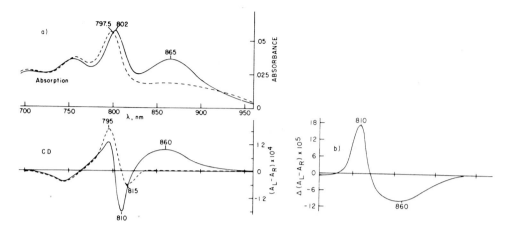

Fig. 4. a) Circular dichroism spectrum of reaction centers of R. rubrum
under mildly reducing (——) and oxidizing (- - -) conditions. From (33).
b) ΔCD spectrum light-minus-dark.

shifted partner being forbidden. This model does not allow the determination
of the band center and the exciton interaction between the two BChl a's of
P865.

Both interpretations of the ΔA spectrum of Rps. sphaeroides give good fits,
and account well for the absorbance spectrum. Obviously, they imply quite
different geometric configurations of P865 and different interactions between
the B800, P865 pigments. We will show later on that the recording of triplet-
minus-singlet absorbance difference spectra does allow to discriminate between the two interpretations.

It should be noted that neither of the interpretations explains the large
width of the P865 band. The two-exciton band interpretation additionally
does not explain the difference in width of the two exciton partners.

A representative CD spectrum of the bacterial reaction centers is displayed
in Fig. 4, together with that obtained under illumination, and the resulting
ΔCD spectrum. The long-wavelength absorption gives rise to a single positive
band, that at 800 nm to a CD band that is approximately conservative, i.e.
equal rotational strength in the positive and negative lobe. A conservative-
ly split band is characteristic of an excitonically coupled dimer (10), the

difference in peak wavelength (here about 15 nm) reflects the exciton coupling 2 J. The intensity of the CD signals is much larger than for monomeric BChl a in vitro, possibly due to exciton interaction (10). The fact that the 865 nm band is totally non-conservative indicates that either the exciton splitting is much larger than the band width, or that it is composed of a single band or of two overlapping bands of non-interacting pigments. In the latter case, the enhanced intensity must be due to interactions with the environment, evidence for which has been found in certain pigment-protein complexes (15).

The ΔCD difference spectrum shows a single broad band at 865 nm and a fairly strong, much narrower band of opposite sign at 810 nm. The appearing non-conservative band at 810 nm cannot be due to simple changes in rotational strength of the purported 800 nm dimer, as then the changes should reflect its non-conservative CD band. Thus, either rearrangement of pigment configuration occurs or the band could be the CD signal of one of the monomers of dimeric P865$^+$, if the positive charge is localized on an optical time scale.

Many of the above conclusions have been supported by optical studies of the BChl b containing photosynthetic bacterium Rps. viridis. Owing to an ethylidene group instead of an ethyl group at carbon 4, the infrared absorption band of BChl b is shifted to lower energy. As a consequence the absorbance spectrum of RC of Rps. viridis is much better resolved than that of the BChl a containing RC, and it is easier to discriminate between the various features of the ΔA spectrum, as shifts, bleachings, etc. We will postpone a discussion of the optical spectra of Rps. viridis to hereafter, where we will compare the redox-ΔA spectrum with the triplet-minus-singlet ΔA spectrum.

Optically Detected Magnetic Resonance of the Triplet State
The triplet state of the primary donor

Reaction (2) describes electron transport events in reaction centers that are 'open', i.e. capable of the photoredox reaction and stabilization of the photo-induced charges. One may, however, treat RC in such a way that forward electron transport from A_1 to A_2 is blocked, e.g. by chemical reduction of A_2, or by its extraction. Then, the photoreaction reads

$$D_1 A_1 A_2^- \xrightarrow{\ h\nu\ } D_1^* A_1 A_2^- \longrightarrow D_1^+ A_1^- A_2^- \xrightarrow{\ k_T\ } {}^3 D_1 A_1 A_2^- \ . \tag{6}$$

$$k_S \qquad\qquad k_S'$$

The radical pair $D^+ A_1^-$ now decays by recombination to either the singlet excited or singlet ground state, or to the triplet state ${}^3D$. Reaction (6) has
been observed in all photosystems and has been theoretically described in a
number of publications (reviewed in (16)). The triplet state ${}^3D$ can be observed optically and by electron spin resonance methods. Its characteristics,
as the zero field splitting parameters and the molecular decay rates, may
serve as a probe of the structure of D, although it now appears that there
are too many parameters influencing the triplet characteristics to extract
unambiguous information from them concerning the geometry of the molecules
making up the dimer (see for a discussion (17)). Moreover, there is good
reason to assume that at least on an optical time scale, the triplet state
is localized (see below). For all chlorophyllous triplet states, the optical
absorbance spectrum is rather featureless, with little absorbance in the
visible and near-infrared, and therefore formation of ${}^3D$ leads to bleaching
of the absorbance bands of D in these wavelength regions. If D consists of
more than one chlorophyllous pigment, and if the triplet is localized on one
of the pigments, a monomer absorbance band might appear. In addition, the
presence of ${}^3D$ exerts an influence on the absorbance spectrum of the other
pigments in the reaction center, because singlet-triplet electrostatic interactions generally are different from singlet-singlet interactions. This results in band shifts of pigments adjacent to D.

From the above it follows that the triplet-minus-singlet absorbance difference (T - S) spectrum has with the redox-ΔA spectrum in common the bleaching
of the near-infrared band of D. It will generally differ from the redox-ΔA
spectrum in the shifts of adjacent pigments, the appearance of monomer absorption and the absorbance due to ${}^3D$ instead of D^+. Thus, the T - S spectrum
is a very useful additional probe of the reaction center. Until recently, a
T - S spectrum could only be recorded with flash absorbance spectroscopy,

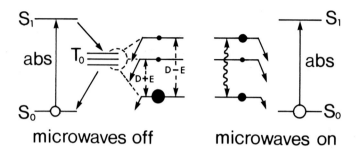

Fig. 5. Principle of absorbance-detected magnetic resonance (ADMR). Filled circles denote relative equilibrium populations of the triplet sublevels, open circles that of the singlet ground state. Corrugated arrow denotes microwave transition.

with broad band detection and inherent weak signal-to-noise ratio. In 1982, a new technique was presented, which makes use of the paramagnetic properties of the triplet state. This method, which will be briefly described in the next section, has permitted the recording of high-resolution T - S spectra of various RC preparations with high sensitivity.

Triplet-minus-singlet spectra recorded with absorbance-detected magnetic resonance (ADMR)

The technique of ADMR rests on the fact that in general the populating probabilities and population decay rates of the three triplet sublevels are different, so that under continuous illumination at temperatures where spin-lattice relaxation is sufficiently slowed, a steady-state is reached in which the population of the three triplet substates is different (Fig. 5). Irradiating the sample with microwaves resonant between two of the three sublevels will then transfer population from one substate to another, resulting in a new overall equilibrium concentration of the triplet state: Σn_i(microwaves on) $\neq \Sigma n_i$(microwaves off), where n_i represents the fractional steady-state population of the i'th sublevel. Since the sum of singlet ground state and triplet concentration equals unity (the concentration of excited singlet states can be neglected under the usual conditions of illumination), application of microwaves leads to a change in the singlet ground state concentration. Subtraction of the absorbance spectrum (microwaves on) from the spectrum (microwaves off) then yields the triplet-minus-singlet (T - S) absorbance difference spectrum (18). To improve the signal-to-noise ratio the microwaves are amplitude-modulated, and the absorbance spectrum detected with lock-in detection. In this way, the detection band width is considerably decreased compared to flash spectroscopy and with a good spectrometer changes in absorbance of 10^{-5} can be detected.

Triplet-minus-singlet spectra of bacterial reaction centers

In Fig. 6a and b (top traces) T - S spectra of reaction centers of the photosynthetic bacteria Rps. sphaeroides R-26 and C. vinosum are displayed. The spectra are composed of the bleaching of P865 (absorbing near 890 nm at

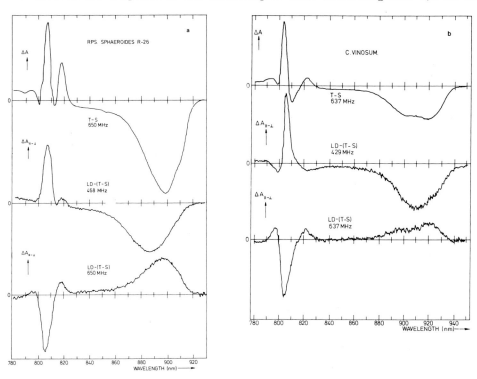

Fig. 6. Triplet-minus-singlet absorbance difference spectra of reaction centers of a) Rps. sphaeroides, b) C. vinosum. Top traces: unpolarized T - S spectra. Middle and bottom traces: linear dichroic T - S spectra obtained with selection by polarized microwaves. Parallel and perpendicular indicate the direction of polarization of the transmitted light with respect to the direction of polarization of the resonant microwaves. The small shift in peak wavelength of the near-infrared bleaching and LD band is the result of site-selection.

1.5 K) and of a positive band near 800 nm that is accompanied by less intense features resembling band shifts (one to the red, one to the blue).
Similar, somewhat better resolved spectra have been obtained for Rps. viridis (18). For all cases, the appearing band at 800 nm has been interpreted
as being due to a monomeric constituent of P865 whose absorption is shifted
to 800 nm because the singlet-singlet exciton interaction coupling the two
BChl a's in the primary donor is broken by the formation of a triplet that
on an optical time scale is localized on one BChl molecule (18,19).

The shoulder of the bleachings at 890 nm in Rps. sphaeroides corresponds to
the shoulder found for the longest-wavelength bleaching of Rps. viridis in
the low temperature absorbance spectrum (20,21) and in its T - S spectrum
(22). The shoulder in the absorbance spectrum has been attributed to charge
transfer interaction between the primary donor and an adjacent acceptor,
possibly B (20,21). Also the shoulder in the T - S spectrum of Rps. viridis
(22) has been discussed in terms of charge transfer interaction. It is remarkable, though, that the bleaching of C. vinosum is clearly split, with
most of the intensity in the longest wavelength peak. We will later see that
in T - S spectra monitored at somewhat different microwave frequency, the
longest-wavelength peak may even predominate.

The assignment of the smaller bands in the 800 nm region to band shifts is
considerably strengthened by the recordings of polarized T - S spectra displayed in Fig. 6a and b (middle and bottom traces). These spectra display
the difference in T - S absorbance for directions of the electrical vector
of the monitoring light beam parallel and perpendicular to one of the in-
plane spin axes of the triplet state (23). The intensity and sign of the
absorbance bands and shifts depend on the angle that the optical transition
makes with the spin axis. It is clearly seen that the appearing band at
800 nm is not part of a band shift. Its width can now be more accurately
defined; it is about 7 nm. Furthermore, it is quite clear that there is no
significant bleaching corresponding to the postulated exciton partner of
P865 at about 805 nm (13). Thus, we must conclude that either P865 is a
dimer composed of two BChl molecules of which the Q_y transition moments are
nearly parallel and approximately lying head-to-tail, or that the shift to
865 nm is almost entirely due to environmental influences (see below).

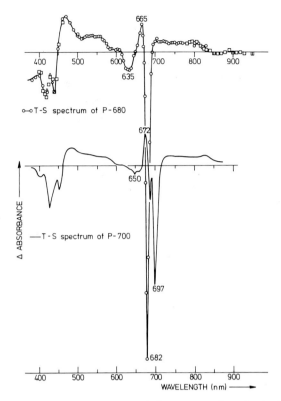

Fig. 7. T - S spectra of photosystem 2 (top trace) and of photosystem 1 particles (lower trace). From (24).

Triplet-minus-singlet spectra of photosystem 1 and 2

Fig. 7 shows the T - S spectrum of PS 1 and PS 2 particles (24). Apart from the shift in wavelength of the major bleaching (at 697 nm for PS 1, at 682 nm for PS 2), the spectra are quite similar. Notably, both spectra show a pronounced positive band to the blue of the bleaching. This band is not ob- served in the T - S spectrum of monomeric Chl a (25), but it is present in the T - S spectrum of a covalently bonded Chl a dimer (26). The latter band has been attributed to monomer absorption of the dimer in which the triplet state is localized on one of the constituent pigments. The T - S spectra strengthen the notion that the primary donor in both photosystems is a dimeric Chl a molecule.

The T - S spectrum of PS 1 shows a small feature in the lower energy flank
of the major bleaching. This band is attributed to a shift of a nearby Chl <u>a</u>,
possibly (one of) the early acceptor(s).

Site-selected T - S spectra

Fig. 8a,b shows the shape and position of the longest-wavelength bleaching
(inverted for ease of comparison with the absorbance spectrum) of <u>Rps</u>.
<u>sphaeroides</u> R-26 (27) and of <u>Rps. viridis</u> (22) as a function of resonant

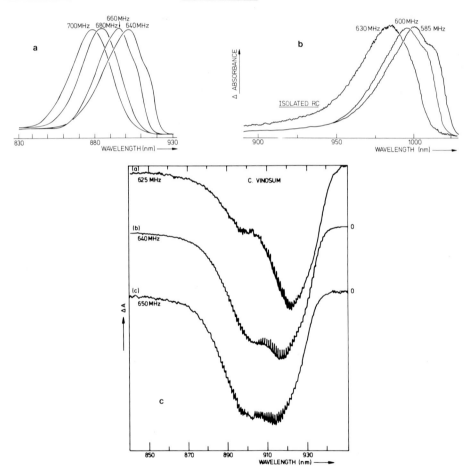

Fig. 8. Normalized long-wavelength bleachings as monitored by ADMR at 1.2 K
as a function of resonant microwave frequency within the |D| + |E| trans-
ition of reaction centers of a) <u>Rps. sphaeroides</u> R-26, b) <u>Rps. viridis</u>,
c) <u>C. vinosum</u>. Traces a) and b) are inverted. From (27) (a) and (22) (b).

microwave frequency <u>within</u> one of the (inhomogeneously broadened) ADMR transitions. It is quite clearly seen that both position and shape change rather
drastically. Note that the spectra are normalized; if the non-normalized
spectra are taken, the envelope reproduces the lineshape of the absorbance
spectrum. Thus, selection of triplet states with certain zero field splitting
parameters (a so-called 'site' (27)) by the microwave frequency results in a
considerably narrowed long-wavelength absorbance band. This means, that this
band as observed in a regular absorbance spectrum is <u>inhomogeneously</u> <u>broaden-
ed</u>. This explains at least partly why this band is so much broader than expected for an exciton band.

As discussed above, the shoulder that is pronounced at certain microwave
frequency might be due to charge transfer interaction. However, in Fig. 8c,
similar experiments on <u>C. vinosum</u> show that in this organism the shoulder is
transformed into a band, that even predominates at certain frequencies. Although this does not invalidate the charge transfer explanation, it does
make it less probable.

Tentative Explanation of the Optical Spectrum of Bacterial RC

Assembling all the available optical data we arrive at the following picture:
  i) The longest-wavelength absorbance of bacterial reaction centers is in-
     homogeneously broadened, and may even be composed of two distinct, part-
     ly overlapping bands. These bands then cannot be exciton components, as
     the polarization (as measured with LD, CD and LD-ADMR) of the two bands
     is practically the same.
 ii) Formation of ^{3}P865 leads to the appearance of a narrow absorbance around
     800 nm in BChl <u>a</u> containing bacteria. The oscillator strength of this
     band is much less than expected for a single monomeric BChl <u>a</u>, its width
     is somewhat smaller than that of the components of the 800 nm band in
     the absorbance spectrum (11).
iii) The T - S spectrum shows two shifts in opposite direction of pigments
     absorbing near 800 nm. These pigments in all probability are the two
     accessory BChl <u>a</u> molecules.

The observations i) - iii) can be explained in (at least) two ways. One, explanation A, is that P865 is composed of a strongly excitonically coupled dimer which has one strongly allowed, red shifted, transition. The shoulder or split in this band is due to charge transfer interactions. Formation of P^+ or ^3P abolishes or weakens the exciton interaction, and a monomer band appears close to the center band wavelength at 800 nm. The low intensity may be a result of redistribution of oscillator strength in a system of two or three non-degenerately coupled pigments, possibly involving the $S_2 \leftarrow S_o$ transitions (28).

According to explanation B, the 865 nm band is composed of two weakly inter-acting BChl a molecules. If P865 is oxidized, the hole is distributed over the two pigments; in the triplet state the triplet is localized on one pigment. The shift from the in vitro BChl a absorption to longer wavelength is caused by environmental interactions, possibly electrical point charges on protein residues (29). The appearing monomer band at 800 nm is due to redistribution of oscillator strength resulting in reorientation of two excitonically coupled accessory BChl a's, possibly involving P865 and one of the BPh a's as well, and two band shifts.

Both explanations have their pro's and con's, but presently this author feels inclined towards the, admittedly rather tentative and sketchy, ex-planation B. Hopefully, further experimentation will give some more clues to the strength of the coupling within P865 and the make-up of the optical spectrum of bacterial RC.

References

1. Shuvalov, V.A., Klevanik, A.V.: FEBS Lett. 160, 51-55 (1983).
2. Shuvalov, V.A., Klevanik, A.V., Kryukov, P.G., Ke, B.: FEBS Lett. 107, 313-316 (1979).
3. Wraight, C.A., Clayton, R.K.: Biochim. Biophys. Acta 333, 246-260 (1973).
4. Feher, G., Okamura, M.Y.: In 'The Photosynthetic Bacteria' (Clayton, R.K., Sistrom, W.R., eds.), pp. 349-386. Plenum Press, New York 1978
5. Parson, W.W.: Ann. Rev. Biophys. Bioengin. 11, 57-80 (1982).

6. Amesz, J., Knaff, D.B.: In 'Environmental Microbiology of Anaerobes' (Zehnder, A.J.B., ed.). John Wiley, New York 1985.

7. Klimov, V.V., Dolan, E., Shaw, E.R., Ke, B.: Proc. Natl. Acad. Sci. U.S.A. 77, 7227-7231 (1980).

8. Davidov, A.S.: Zhur. Eksptl. Theoret. Fiz. 18, 210-218 (1948).

9. Kasha, M.: Rev. Mod. Phys. 31, 162-169 (1959).

10. Tinoco, I.: J. Chem. Phys. 33, 1332-1338 (1960).

11. Feher, G.: Photochem. Photobiol. 14, 373-387 (1971).

12. Hemenger, R.P.: J. Chem. Phys. 67, 262-264 (1977).

13. Vermeglio, A., Clayton, R.K.: Biochim. Biophys. Acta 449, 500-515 (1976).

14. Shuvalov, V.A., Asadov, A.A., Krakhmaleva, I.N.: FEBS Lett. 76, 240-245 (1977).

15. Pearlstein, R.M.: In 'Excitons' (Rashba, E.I., Sturge, M.D., eds.), pp. 735-770. North-Holland Publ. Cy., Amsterdam 1982

16. Hoff, A.J.: Q. Rev. Biophys. 14, 599-665 (1981).

17. Hoff, A.J.: In 'Triplet State ODMR Spectroscopy' (Clarke, R.H., ed.), pp. 367-425. John Wiley, Inc., New York 1982.

18. Den Blanken, H.J., Hoff, A.J.: Biochim. Biophys. Acta 681, 365-374 (1982).

19. Shuvalov, V.A., Parson, W.W.: Biochim. Biophys. Acta 638, 50-59 (1981).

20. Vermeglio, A., Paillotin, G.: Biochim. Biophys. Acta 681, 32-40 (1982).

21. Maslov, V.G., Klevanik, A.V., Ismailov, M.A., Shuvalov, V.A.: Doklady Akad. Nauk SSSR 269, 1217-1221 (1983).

22. Den Blanken, H.J., Jongenelis, A.P.J.M., Hoff, A.J.: Biochim. Biophys. Acta 725, 472-482 (1983).

23. Den Blanken, H.J., Meiburg, R.F., Hoff, A.J.: Chem. Phys. Lett. 105, 336-342 (1984).

24. Den Blanken, H.J., Hoff, A.J., Jongenelis, A.P.J.M., Diner, B.A.: FEBS Lett. 157, 21-27 (1983).

25. Linschitz, H., Sarkanen, K.: J. Am. Chem. Soc. 80, 4826-4832 (1958).

26. Perisiamy, N., Linschitz, H.: J. Am. Chem. Soc. 101, 1056-1057 (1979).

27. Den Blanken, H.J., Hoff, A.J.: Chem. Phys. Lett. 98, 255-262 (1983).

28. Parson, W., Holten, D., Kirmaier, C., Scherz, A., Woodbury, N.: In 'Advances in Photosynthesis Research' (Sybesma, C., ed.), Vol. I, pp. 187-194. M. Nijhoff/W. Junk, The Hague 1984.

29. Eccles, J., Honig, B.: Proc. Natl. Acad. Sci. U.S.A. 80, 4959-4962 (1983).

30. Reed, D.W.: J. Biol. Chem. 244, 4936-4941 (1969).

31. Rafferty, C.N., Clayton, R.K.: Biochim. Biophys. Acta 546, 189-206 (1979).

32. Vermeglio, A., Breton, J., Paillotin, G., Cogdell, R.J.: Biochim. Biophys. Acta 501, 514-530 (1978).

33. Philipsson, K.D., Sauer, K.: Biochemistry 12, 535-539 (1973).

Received July 18, 1984

Discussion

Scheer: You assign the width of the P870 band to inhomogeneous broadening.
Could this be due to the loosely bound cytochrome, which is in somewhat
different position?

Hoff: No, because it is almost as broad in _Rp. viridis_, in which the
cytochrome is tightly bound.

Scheer: Can you offer an explanation to this?

Hoff: I think that the inhomogeneous broadening reflects slight changes
in local dimer geometry. Such changes will equally affect the exciton shift
in the optical spectrum and the zero field parameters, i.e. the ADMR resonance line, which is also inhomogeneously broadened. It is then not surprising that there is a correlation between the exact microwave frequency
and the shift in ADMR detected long-wavelength absorbance peak (Fig. 8).

Fuhrhop: If you turn one of the BChl-molecules in the dimer around by
180°, this should have a large effect on the coupling of electronic transitions in the dimer. Why did you then say that the X-ray dimer is very
close to your earlier proposal (turned by 180°)?

Hoff: As the dipole transition moment is a bidirectional vector (it represents an oscillating charge distribution) a 180° turn does not make any
difference to the coupling.

Fuhrhop: Do you have any idea from your data about the relative hydrophobicity of the environments of the BChl pair, the reactive BChl and the
quinone?

Hoff: No.

Holzwarth: Why is the long wavelength band of the bacterial reaction center so much more inhomogeneously broadened as compared to the other bands?

Hoff: If we accept that the long-wavelength absorbance is shifted because
of exciton interaction in the P865 dimer, then the inhomogeneous broadening
is about 10 % of this shift. The other bands are much less shifted because
of exciton interaction, not more than 10 to 30 nm. If those shifts are
largely due to interaction with P865, then a change of 10 % in this interaction (due to re-orientation of the P865 transition moment) results in
changes of only a few nm. If the bands of pigments other than P865 are
shifted because of interactions among the pigments themselves, or because
of interaction with the protein matrix, then the changes would be even less.

Woody: Were all of these experiments done on isolated reaction centers?

Could the inhomogeneous broadening result from minor modifications introduced by the isolation procedures?

Hoff: All triplet-minus-singlet difference spectra of the photosynthetic
bacteria have been recorded for isolated reaction centers and for chromato-
phore preparations. They show equally inhomogeneously broadened long wavelength absorbance for both types of preparations, excluding effects due to
reaction center isolation procedures.

Buchler: At the end of the talk you mentioned that possibly the plant
chlorophylls also form some kind of special dimer. If this were so, the
oxidized system [(Chl)(Chl$^+$)] should show such a near infrared band as is
shown by the reaction centers of e.g. *Rps. sphaeroides* R-26 (Fig. 2 of your
paper) when they are illuminated, or our lanthanoid bisporphyrinates, e.g.
Eu(OEP)$_2$ (see paper of Buchler and Knoff, this conference), where the radical nature of one of the porphyrin rings has been demonstrated. What is
the near infrared absorption of illuminated photoactive chlorophyll centers?

Hoff: There is an appearing band near 820 nm for both PS 1 and PS 2 reaction centers upon photooxidation. This band corresponds to a Chl a$^+$ cation absorption band. To my knowledge, for the plant photosystems no longer-wavelength absorption band corresponding to the 1250 band for *Rps.
sphaeroides* has been found. In bacteria the band is an appearing band, i.e.
if it is a charge-transfer band, it must be due to charge-transfer within
the cation of P865. Note, that the ab-initio calculations of Petke et al.
[(Photochem. Photobiol. 31, 243 (1980)] predict a measurable absorption for
Chl a$^+$ near 850 nm and for BChl a$^+$ near 950 nm; there appears to be no
appreciable oscillator strength in the 1250 region for monomeric BChl a$^+$.

CIRCULAR DICHROISM OF CHLOROPHYLLS IN PROTEIN COMPLEXES FROM CHLOROPLASTS OF HIGHER PLANTS: A CRITICAL ASSESSMENT

R.P.F. Gregory

Department of Biochemistry, The Medical School, University
of Manchester, M13 9PT, U.K.

Introduction

Chlorophyll performs three principal functions: absorption
of light, transmission of energy, and photochemical action.
In the thylakoid membrane of green plants there are (on a
consensus view) two photoreactions, carried out by specialized
chlorophyll molecules (or dimeric pairs). For each of these
photochemical centres there are several hundred other
chlorophyll molecules which contribute by absorbing energy
and transmitting it towards the photochemical centres. Most
of the chlorophyll, probably all of it, is attached to
protein, by links which are weak enough to be broken by
organic solvents, but strong enough to hold the chlorophyll
in what appear to be specific arrangements. The same
consensus opinion holds that the thylakoid membrane in the
functioning chloroplast contains three multi-subunit chlorophyll-protein complexes: photosystems I and II (PSI, PSII)
and a light-harvesting or antenna complex (LHC). Physical
interaction and adhesion between these three particles is
supposed to be variable and under the control of the photosynthetic system.

When attempts are made to extract the complexes from the
thylakoid, the products in general are found to be more
numerous than the three predicted above, and are assumed to
have undergone loss or alteration of particular components,

and to show in addition combinations or oligomers that may or
may not have pre-existed in the functioning thylakoid. Three
major goals of the researches are therefore (1) to describe
each chlorophyll-containing polypeptide, (2) to allocate each
component to either PSI, PSII or LHC and (3) to account for
the role of each chlorophyll-peptide in terms of the chloro-
phyll-functions of absorption, transmission and photochemistry.

Circular Dichroism (CD) is one of several optical methods for
describing chlorophyll. (The others, no less important,
include deconvolution of absorption spectra, fluorescence
spectroscopy and linear dichroism.) CD is a measurement of
the difference in absorption of left- and right-handed
circularly-polarised light, an essential condition for this
being some asymmetry in the chromophore or its environment.
Chlorophylls a and b have a relatively weak CD because the
asymmetric centres are somewhat remote from the chromophoric
group. This weak CD can in theory be greatly enhanced, in
principle, by (1) strong asymmetric association with, for
example, protein, (2) twisting of the planar chlorophyll
molecule to create asymmetry and (3) the formation of
asymmetric dimers, trimers, etc. with consequent exciton
splitting. Exciton-splitting produces 'conservative' CD
spectra with equal positive and negative areas: effects on
single molecules are non-conservative. In the foregoing
account, it is clear that CD is an expression of short-range
effects, so that the CD of multi-subunit chlorophyll-protein
complexes could be expected to be substantially equal to the
sum of the CD of the individual chlorophyll-carrying
components. By extension the same should be true of
thylakoid fragments which are functionally intact in
photochemistry and electron transport. In fact indications
of serious large-scale CD interactions are only seen when
the membranes exist that are able to aggregate or stack,

and show obvious light-scattering; this little-understood
phenomenon has been discussed (1). In summary, the CD
spectrum of a chlorophyll-protein should be specific for that
protein; it may provide information as to the organisation
and environment of the pigment molecules, and it may enable
the protein to be identified in larger complexes including
the functioning membrane.

An account now follows of various chlorophyll-protein
preparations and the application of CD. All the preparations
considered here involved surface-active agents.

Sodium dodecyl sulphate (SDS). When first employed the
conditions were harsh; only two chlorophyll-protein
complexes were obtained (2) known as CP1 and CP2. CP1 was
identified with PSI from its content of P700 (in a
preparation from a blue green alga) and its absence from a
PSI-deficient mutant of Scenedesmus (see the review 3);
subsequently the presence of P700 in CP1 has been well
established even though the electron acceptors are usually
substantially destroyed (e.g. ref. 4). The CD spectrum
(5, 6) in the Qy region showed an almost conservative
effect, positive at 677nm and negative at 690nm. This CD
signal was well to the red-side of the absorption peak at
676nm, so that only a minority of the chlorophyll was
implicated. CP2 on the other hand was eventually identified
with LHC on the basis of its low chlorophyll a/b ratio
(approx. unity), its presence in a PSII-deficient mutant of
Scenedesmus and its absence from the chlorina mutant of
barley (Hordeum vulgare L.). The CD in the Qy region showed
a simple spectrum (5,6) interpreted as a split-exciton of
chlorophyll a (+ve at 670nm) and chlorophyll b (-ve at 651nm)
which corresponded closely with the two peaks in the
absorption spectrum. The CD could be regarded as
conservative if the baseline was depressed (indicated by

features at 630 and 680nm), as might be produced by less-
organised chlorophylls. This work had an obvious success
in that the CD of CP2 could be recognised in unresolved
systems, and there was the strong indication that chlorophylls
a and b were close enough to interact. The 690nm negative
CD of CP1 could be seen as a shoulder on unresolved systems
(6), so that the method offered a useful diagnostic method
and a quantitative analysis. However, the failures were that
under these conditions up to half the total chlorophyll
appeared detached from protein ("free pigment"); there was no
indication of any complex representing PSII, and the most
prominent feature of membrane fragments, -ve CD at 680nm
was not accounted for.

In a comparative study of CP2, very similar CD was found in
six species (7) which was consistent with a structural model
for this complex; the major input to the model was
fluorescence and infra-red data, and the model was not
particularly sensitive to the CD spectrum.

Under more gentle conditions (for review see 8) the quantity
of free pigment was almost entirely avoided, but many more
bands were now formed. CP1 was accompanied by CP1a and CP1b,
which had more chlorophyll a with respect to chlorophyll b.
CP2 was now represented by three (or possibly four) bands,
LHCP 1-3. A new band, CPa, appeared which was provisionally
identified as a PSII derivative, on the grounds that it is
present in LHC-less mutants, and absent from PSII-less
mutants (8). CPa has not yet been examined by CD; the
families of CP1 and LHC resemble CP1 and CP2.

Triton X-100 has been extensively used both for PSI and LHC
extraction. Being non-ionic it is not useful for electrophoretic separation, and the two techniques of column
chromatography and density-gradient centrifugation have

yielded a set of PSI derivatives with chla/P700 values
ranging from 110 to approx. 15 (8, 9). There is some
variation in the CD spectra reported for PSI-derived
particles: the major feature has been found at 672nm (+ve),
688 (-ve) (10) and 667 (+ve), 680 (-ve) (11), while a minor
feature probably representing chlorophyll b was found at
654nm (+ve) (10), absent (11) and 650 (-ve) (author and Prof.
M. Brody, unpublished). It is now agreed that chlorophyll
b is able to pass energy to P700 and is an integral part of
PSI (12). This fraction of chlb is believed to belong to
chlorophyll (a+b)-peptides of 20-25 kDa in Chlamydomonas
(CPO see ref. 12), which have been resolved by Haworth et al.
(11) and Lam et al. (14). The CD of the "CP1-65" material
(prepared by means of Triton X-100) showed strong -ve CD
at 650 with apparently the shoulder due to CP1 at 685-690.
However the material resolved by Haworth et al. (11) as the
'antenna fraction, LHCP-I' showed a -ve maximum at 685, and
this was confirmed by Lam et al. (14), whose material showed
-ve CD at 648. Haworth's LHCP-I contained a 20 kDa peptide,
and they identified it with CPO; Lam resolved two peptide
fractions 22 to 23kDa (LHCP-Ia) and 20kDa, LHCP-Ib. The
present state of the CD relationship between PSI, CP1,
LHCP-I (a,b) and the -ve CD at 680nm is confusing.

Triton X-100 extracts yield an LHC preparation that can be
obtained in a pure form by prepitation with Mg^{++} ions (15).
This material (before precipitation) has a prominent -ve CD
at 680nm, superimposed on the CP2 spectrum (16). There are
known to be at least two peptide components in this LHCP.
(The CP2-related LHCP and its oligomers LHCP1-3 have been
labelled LHCP-II so as to distinguish them from the
LHCP-I (a,b) material related to PSI.)

Deriphat-160 (disodium N-dodecyl β-imidoproprionate) is a
zwitterionic detergent used in the gels on which SDS-extracts

of thylakoids were resolved (17) and studied by means of CD
(18). Apart from the slowest band which had very similar
CD to CP1, the other bands contained a major -ve CD component
at 680nm. It was tentatively concluded that a semi-
conservative CD (+ve 669nm, -ve 680nm) would account for the
CD at 680nm lost by the use of SDS alone, and furthermore
that this 669/680 CD could belong to the missing PSII or CPa
material.

Digitonin-deoxycholate. Digitonin preserves all photochemical
activities, but the extracts were hard to resolve until the
introduction of deoxycholate in PAGE buffer. In this way
Picaud et al. (19) resolved spinach chloroplasts into many
bands, three of which could be identified with PSI, a PSII-
LHC mixture and LHC. The CD of the PSI fraction reported by
Picaud (19) showed a strongly -ve feature at approx. 683nm;
the PS2 material had a major -ve CD at 680nm. The LHC
fraction had only a small -ve CD in this region, which
could be in fact the red-end of a baseline depression caused
by chlorophyll disorganisation. Evidence will be given
below to suggest that -ve CD at 680 is a proper characteristic of PS2 and LHC fractions.

In the present study, the method of Picaud et al. is applied
to pea and the chlorine (chlorophyll b-less) mutant of barley.
It has been shown that in the mutant barley there is no LHC
(hence no LHCP-II) but an apoprotein has been detected by
Machold (20) that could represent the antenna of PSI (LHCP-I),
although it would of course have no chlorophyll b. The
barley mutant is a simpler system and should have a simpler
pattern of complexes. It is particularly to be hoped that
the PSII-band will be both photochemically active and
(unlike the spinach preparation) uncontaminated by LHC.

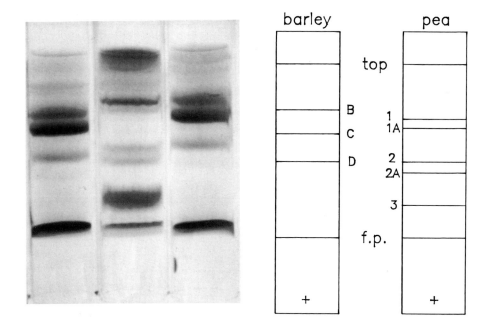

Fig. 1.   Photograph and Interpretative Drawing of Deoxycholate-
Polyacrylamide Electrophoretic Gels Comparing Peas with
Chlorophyll-b-less Barley.

The central tube contains pea extract, the outer tubes contain
barley extract.   The bands are chlorophyll complexes and are
unstained.   Conditions were as described in ref. 19.   The
diameter of the tubes was 1.8cm and the sample was loaded at
the top with the +ve voltage downwards.   The bands B, C and
D of the barley refer to Fig. 2, and the numbering of the pea
bands (1, 1A, 2, 2A and 3) is compatible with ref. 19.

Methods

CD spectra were obtained in a Cary 61 CD spectropolarimeter

modified by the substitution of an elasto-optic modulator (21)

in place of the Pockels cell. A Brookdeal 9503 lock-in ampli-

fier operating at 25kHz was substituted for the original system.

Data was digitized at 1nm-intervals and plotted after subtrac-

tion of the baseline.   The conditions were chosen to keep the

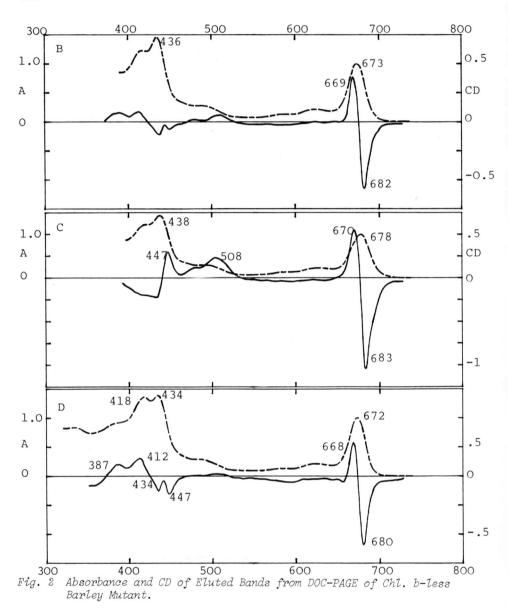

Fig. 2 *Absorbance and CD of Eluted Bands from DOC-PAGE of Chl. b-less
Barley Mutant.*

All spectra are normalized to unit absorbance at the red
maximum. Absorbance is the broken line, CD the continuous
line. In descending order the plots represent bands B, C
and D (see Fig. 1). The CD ordinate is in units of $(A_L - A_R)$
x 10^3. There was no change when the bands were eluted from
the gel.

noise-envelope less than 3% of the maximum signal-excursion.
The wavelength scale was checked by means of the 577.0 and
559.1nm mercury-emission lines.

Delayed-light emission was measured in a Becquerel
phosphoroscope operated at 6000-15000 rev/min, synchronised
with a photographer's xenon-flash lamp. Sedimentation was
studied in a Beckman L8-70 ultracentrifuge fitted with a uv
scanner, which was modified by substitution of a violet
filter (Kodak OVl) so as to isolate the 435nm mercury
emission line and hence specifically observe the sedimentation of chlorophyll-complexes only. Photosystem-I was
measured in a Rank-Chappell oxygen-electrode by means of
the Mehler reaction with 1mM methyl viologen, 0.2mM
dichlorophenolindophenol and 2mM sodium isoascorbate in
0.05M phosphate buffer at pH 7.6. Protein-chlorophyll
complexes were prepared from pea (Pisum sativum) leaves
grown in a greenhouse for three weeks. The method of Picaud
et al. (19) was used, and the appearance of the gels was
very similar to these authors' picture. The same procedure
was applied to the chlorophyll b-less mutant of barley
(Hordeum vulgare L. var. Chlorina 2).

Results

The appearance of the gels from the two species is shown in
Fig. 1. The mutant barley has no LHC band, but far more
free pigment has been formed. On the other hand, less
chlorophyll has remained at the top of the barley gels. The
positions of the bands do not correspond identically with
those of the peas. The main bands labelled B, C and D were
extracted for study, and five bands, 1, 2, 3 (19) plus la
and 2a were taken from the pea gels.

The absorption and CD spectra of the three barley bands are shown in Fig. 2. It is apparent that band B contains features of both band C and band D. Delayed light emission was found in both B and D, and it is interpreted on the basis that it indicates the presence of a functional PSII reaction centre. Photosystem I activity was present in B and C. Hence B is a complex of C and D, where C is PSI and D is PSII. The sedimentation coefficients (in the deoxycholate glycine-tris buffer prescribed for the electrophoresis, 19) were: B, 19.1S; C, 15.5S and D, 11S. These values confirm the hypothesis that B is a 1:1 complex of C and D. In the pea preparation, delayed light was observed in P1A, P2 and P2A. This is in accord with the CD data, which shows (Fig. 3) that PSII material is present in bands P1A, P2 and P2A, PSI material in bands P1 and P1A, and LHC material in bands 1A, 2, 2A and 3.

A strong -ve CD signal at approx. 680nm is a major feature of all the spectra, both from pea and barley. There is a perceptible and significant red-shift of 2-3nm in PSI material with respect to PSII, and there is an indication of the -ve CD of chlorophyll b at 650nm in the pea band-1, as reported (19).

The -ve CD at 680nm was not stable in any of the preparations, even in the dark at 4°C although the delayed light emission lasted apparently unchanged. The decay of the 680nm CD signal was most rapid in the LHC prep (band 3 of the pea) and this signal diminished by 50% in some 8 days.

Discussion

It is clear that the method of Picaud is indeed a valuable means for isolating small active complexes representing PSI, PSII and LHC. It is also clear that all possible combinations

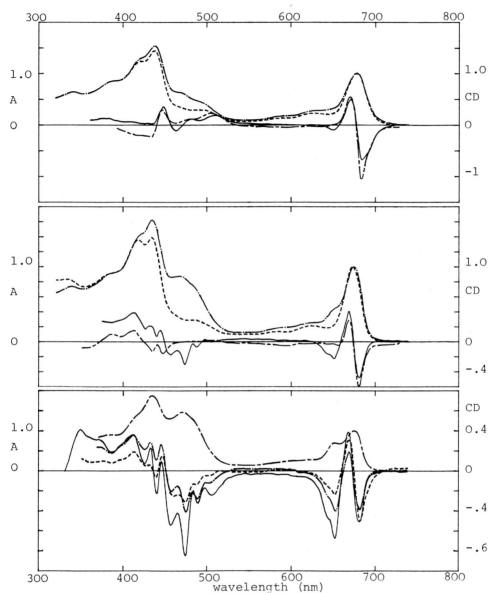

Fig. 3 Comparison of Eluted Bands from Pea Thylakoids

Top: Band 1 (Absorbance, chained line; CD, solid line) with band C of barley (as in Fig. 2) (Absorbance, dotted; CD, chained)

Middle: Band 2 (Absorbance, chained; CD, solid) with band D of barley (Absorbance, dotted; CD chained).

Bottom: Band 3 (Absorbance, chained; CD, solid) with band 1A (CD, dotted) and band 2A (CD, chained); (Absorbance not shown). Other details as Fig. 2.

of these complexes exist, and possibly more with different
proportions, since there were many minor bands that could not
be studied. These super-complexes may or may not have pre-
existed in the thylakoid. The CD of the active PSII (band D)
from barley is a good match with the prediction (18).
Judging by the blue region of the CD spectra, Picaud
achieved a lesser degree of contamination of PSII by LHC
than is reported here for pea; the CD spectrum of band 2 of
Picaud (19) is a good match with the mutant barley band D,
while only the small chl b band makes any difference in the
Qy region. Nevertheless, the spectra in this study are more
conservative, suggesting that there has been less disorganisation of chlorophyll.

There appears to be no evidence of additional bands
representing interactions between PSI, PSII and LHC in the
super-complexes, although the complexity of the spectra need
further quantitative analysis.

There are shoulders on the PSI bands in each case, but given
the evidence of LHCP-1 (11, 14) it would be wise to avoid
stating that they are definitely due to CP1.

The occurrence of -ve CD at 680nm in all the bands was a
surprise. It remains to be seen whether further resolutions
of PSI, PSII and LHC into their subunits will also separate
the 680nm CD.

Acknowledgements

The author is grateful to the Royal Society for a grant to
improve the spectropolarimeter. Ultracentrifuge data were
carried out by Dr. A. Wilkinson, and processed by
Miss S. Morris and Miss J. Smith who also measured the PSI-
activity. The help of the UMRCC Computer Graphics Unit is
also gratefully acknowledged.

References

1.   Brecht, E., Demeter, S., Faludi-Daniel, A.   Photo-
        biochem. Photobiophys. *3*, 153-7 (1981).

2.   Thornber, J.P., Gregory, R.P.F., Smith, C., Bailey, J.L.
        Biochemistry *6*, 391-396 (1967).

3.   Thornber, J.P. Annu. Rev. Plt. Physiol. *26*, 127-158
        (1975).

4.   Setif, P., Acker, S., Lagoutte, B., Duranton, J.
        Proc. Int. Congr. Photosynthesis, 5th *3*, 503-11 (1981).

5.   Gregory, R.P.F., Raps, S., Thornber, J.P., Bertsch, W.F.
        Proc. Int. Congr. Photosynthesis, 2nd, 1503-8 (1972).

6.   Scott, B., Gregory, R.P.F. Biochem. J. *149*, 341-7
        (1975).

7.   Shepanski, J.F., Knox, R.S. Isr. J. Chem. *21*, 325-31
        (1981).

8.   Anderson, J.M., Barrett, J., Thorne, S.W. Proc. Int.
        Congr. Photosynthesis, 5th, *3*, 301-15 (1981).

9.   Ikegami, I., Ke, B. Biochim. Biophys. Acta *764*, 70-9
        (1984).

10.   Philipson, K.D., Sato, V.L., Sauer, K. Biochemistry *11*
        4591-5 (1972).

11.   Haworth, P., Watson, J.L., Arntzen, C.J. Biochim.
        Biophys. Acta *724*, 151-8 (1983).

12.   Anderson, J.M., Brown, J.S., Lam, E., Malkin, R.
        Photochem. Photobiol. *38*, 205-10 (1983).

13.   Wollman, F.-A., Bennoun, P. Biochim. Biophys. Acta
        680, 352-60 (1982).

14.   Lam, E., Oritz, W., Malkin, R. FEBS Lett. *168*, 10-14
        (1984).

15.   Burke, J.J., Ditto, C.L., Arntzen, C.J. Arch.
        Biochem. Biophys. *187*, 252-63 (1978).

16.   Brecht, E., Adler, K., Faludi-Daniel, A. Biochem.
        Physiol. Pflanzen *179*, 81-94 (1984).

17.   Markwell, J.P., Thornber, J.P., Boggs, R.T. Proc.
        Natl. Acad. Sci. USA., *76*, 1233-5 (1979).

18.   Gregory, R.P.F., Borbely, G., Demeter, S., Faludi-
        Daniel, A. Biochem. J. *202*, 25-9 (1982).

19.   Picaud, A., Acker, S., Duranton, J. Photosynthesis
        Res., *3* 202-13 (1982).

20.   Machold, O.   Biochem. Physiol. Pflanz. *176*, 805-27
       (1981).

21.   Breeze, R.H., Ke, B.   Analyt. Biochem. *50*, 281-303
       (1972).

Received July 5, 1984

Discussion

Scheer: What is the ratio of P680 to additional chlorophyll in your
best PSII preparation?

Gregory: I have no data myself. Ref. 19 gave one reaction centre per
80-90 chlorophylls(but they had some LHC contamination).

Dörnemann: Did you ever try instead of your Chl b-less mutant the in-
termittant light system as described by George Akoyunoglou for peas and
by our lab for algae which give also LHCP-free green plants which can
then be converted to fully green plants by continuous light?

Gregory: No. I look forward to trying it.

Dörnemann: Where does your evidence for two LHCP's come from?

Gregory: Both ref. 11 and ref. 14 found chl (a+b)-peptides that were
not the same as the well-known LHCP.

Woody: I object on semantic grounds to the use of the term "exciton"
for interactions between non-identical chromophores. Coupled oscillator
interactions operate between non-degenerate excited states, but the term
"exciton" refers to the special case of essentially identical chromophores.

Gregory: Agreed.

Woody: You mention the use of rapid chromatography in the presence of
SDS to achieve a partial dissociation of protein-Chl complexes under mild
conditions. Isn't the denaturing effect of SDS quite rapid in general?

Gregory: Yes: in general the effect of SDS proceeds rapidly to an extent governed by the particular protein and the conditions. However, the
loss of the 680(-) CD in chloroplast extracts is relatively slow (approx.
3h). In this case a low concentration (0.05%) was used. The 15.5S (C)
material dissociated to 8S and 12S bands, with some preservation of both
PSI activity and 680(-) CD.

OPTICAL PROPERTIES AND STRUCTURE OF CHLOROPHYLL RC I

Dieter Dörnemann and Horst Senger

Fachbereich Biologie/Botanik der Philipps-Universität Marburg

Lahnberge, D-3550 Marburg

Introduction

A new chlorophyll, designated chlorophyll RC I (Chl RC I),
has meanwhile been isolated by the authors from all kind of
photosynthetic organisms, as higher plants, green algae, red
algae, cyanobacteria and the chlorophyll b containing prokaryote Prochloron spec. Hence, its ubiquity in chlorophyll containing photosynthetically active organisms seems to be proven.
Its absorption - and fluorescence characteristics as well as
the molecular weight and structure are different from all up
to now known chlorophylls. Since Kok's discovery of P-700 and
its identification as the reaction centre of PS I (Kok, 1961)
it had been assumed that Chl a bound in a particular way or to
a specific protein plays the role of the PS I-reaction centre
chromophore. The enrichment of Chl RC I in PS I particles, the
fact that the formation of both reaction centres and Chl RC I
in the pigment mutant C-2A' of Scenedesmus obliquus is inhibited by chloramphenicol and last not least the identity of
its molar ratio in methanol extracts with that of P-700 in
vivo in all investigated organisms suggest a close relationship to the reaction centre of PS I or even the identity of
Chl RC I with the chromophore of P-700 (Dörnemann and Senger
1981a, b, 1982, 1984). In the current paper we summarize the
optical properties and the results clarifying the structure
of Chl RC I.

Materials and Methods

Chlorophyll RC I was isolated and purified from the mutant
C-6E of the green alga Scenedesmus obliquus as described
earlier (Senger and Dörnemann, 1982). Preparations from the
red algae Porpyridium cruentum and Cyanidium caldarium were

Optical Properties and Structure of Tetrapyrroles
© 1985 by Walter de Gruyter & Co., Berlin · New York

performed in the same way. Isolation of Chl RC I from the blue
green alga Anacystis nidulans was performed in cooperation with
Dr. Tetzuya Katoh, University of Kyoto, Japan, in the same
way after the enrichment of photosynthetic membranes from
broken cells by differential centrifugation. Cells of the
Chl b containing prokaryote Prochloron spec. (Lewin, 1983)
which was not yet successfully grown in laboratory culture,
were obtained from Dr. R.A. Lewin, La Jolla, Calif. As the
organism only grows as symbiotic host in dedemnid ascidians,
cells were harvested by squeezing the algae from the cloacal
atria of Lissoclinum patella which was collected at Palau,
Western Caroline Islands. After washing at pH 8 the cells were
freeze-dried and shipped. The dry algal powder was extracted
with methanol and further preparation of Chl RC I performed
as usual.

For extreme purification of the chromophore a solvent system
for HPLC was developed using a 4 mm RP-18 column, particle
size 10 μm. The system consists of acetonitrile/methanol
(75/25; v/v)superimposed by a multilinear gradient of water
content starting with 25% H_2O down to 5% after 10 min. The
water content is deminished to 0% within 30 min. followed by
a 10 min. phase of regeneration which increases the water content of the solvent again up to 25% to reach the starting
point again. A similar system was developed for separation on
a semi-preperative column of 16 mm $\emptyset$.

For neutron activation analysis which was performed in the
Forschungsreaktor of the University at Mainz and kindly supported by Dr. Trautmann and co-workers, samples were sealed under
vacuum in quartz ampules and exposed to a neutron flux of
4×10^{12} n/sec cm^2 for 30 min. to form ^{38}Cl, if present in the
molecule. For zero-standard an empty ampule (dummy) was taken,
back ground was determined by measuring a sample containing
20.8 μg methylpheophorbide a. Calibration standards were samples
of NaCl (1,5 μg) and 20-Cl-methylpheophorbide a (22,4 μg),
Chl RC I ampule content was 30 μg.

^{1}H-NMR-spectra were recorded on a 400 MHz Brucker instrument by Dr. Berger and co-workers in the Fachbereich Chemie of the University at Marburg. The solvent was d_6-acetone with a deuterium content of more than 99.96%.

Fluorescence lifetime measurements were done by Dr. Holzwarth and co-workers at the Max-Planck-Institut für Strahlenchemie at Mülheim/Ruhr.

Fluorescence spectra and fluorescence yield were measured in acetone with a Shimadzu RF 502 spectrofluorometer automatically corrected against a rhodamine b standard. Samples were always excited in their blue absorption maximum using a spectral band-width of 15 nm. Emission band-width was 3 nm.

Absorption spectra were recorded in acetone with a Shimadzu MPS 5000 or a Kontron Uvikon 820 spectrophotometer, the light path was 1 cm.

Molecular weights were determined by plasma descorption mass spectrometry (PDMS) which was kindly performed by Drs. Danigel, Junclas and Schmidt at the department of Nuclear Chemistry, Fachbereich Physikalische Chemie, University of Marburg. Ionisation is achieved by exposing the sample under high vacuum to a ^{252}Cf-source. Fission products of ^{212}Cf, e.g. ^{105}Tc (10MeV) pass through the solid sample film depositing an energy of approx. 10^4 eV/nm. Time of fission and time of flight in a 10 kV electric field are measured and the time of flight correlated to mass.

All other measurements were performed as described earlier (Dörnemann and Senger, 1982).

To test the incorporation of ^{36}Cl into Chl RC I the Scenedesmus mutant C-6E was grown for 5 days in the presence of ^{36}Cl-NaCl after cells were previously cultivated under chlorine deficiency for 5 days.

Results

During the past 4 years Chl RC I has been extracted from a
great variety of photosynthetic organisms covering most
groups of Chl a containing organisms. Absorption spectra of
all isolated Chls RC I were identical and showed a shift to
longer wavelength of about 4 nm in the blue maximum and of
about 8 nm in the red maximum compared to Chl a. Spectra of
Chl RC I from spinach and the Scenedesmus mutant C-6E and of
Chl a in acetone are shown in Fig.1. According shifts could
be shown in other solvents, too. Under acid conditions the
central Mg-atom was replaced by two protons and the pheophytin RC I was formed. The blue and red absorption maxima of
pheophytin RC I were again shifted 4 and 8 nm respectively
demonstrating that the changed absorption characteristic is
due to another change in the molecule than the central atom.
The absorption spectra of pheophytin RC I and pheophytin a
are given in Fig. 2.

Chlorophylls are also well characterized by their fluorescence
emission. Changes of the absorption maxima should also result
in the change of the fluorescence emission. This is clearly
demonstrated by the fluorescence emission spectra of Chl RC I
and its pheophytin in Fig. 3. Both emission maxima are shifted
to the red region at the same degree as their absorption
maxima again demonstrating that the alteration of the molecule
must be located in the conjugated TT-electron-system of the
chromophore and not in any of the side chains.

Together with the first isolation of Chl RC I the question concerning the physiological significance of this chromophore
arose. As Chl RC I was first isolated from pigment mutant C-6E
of Scenedesmus which has by genetic defect only photosystem I
(Bishop, 1971) it was evident that the new chlorophyll is part
of PS I. Comparison of the in vivo ratio of P-700 to Chl a and
the ratio of Chl RC I and Chl a in the methanol extract gave
evidence that there is a 1:1 molar relation between P-700 and
Chl RC I. Subsequently the investigations were extended to

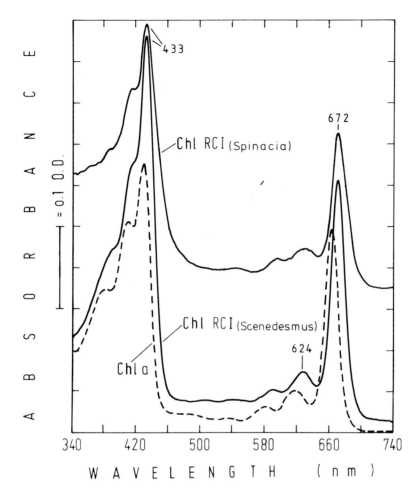

<u>Fig. 1</u>: Absorption spectra of Chl RC I from <u>Scenedesmus</u> mutant
C-6E and <u>Spinacia</u> in acetone compared to Chl <u>a</u> at
room temp. (r.t.). Baselines were arbitrarily adjusted
to avoid overlap ($E_{740}=0$)

PS I-particles from the same organism highly enriched on a
succrose density gradient. Again a 1:1 molar ratio of Chl
RC I in extract to P-700 <u>in vivo</u> was found. Chl RC I was also
detected in a 1:1 ratio to P-700 in PS I particles of spinach
as well as in the cyanobacteria <u>Anacystis</u> <u>nidulans</u> and <u>Anabaena</u>
<u>variabilis</u>. Lately the presence of Chl RC I in red algae like
<u>Porphyridium</u> <u>cruentum</u> and <u>Cyanidium</u> <u>caldarium</u> was proven, the

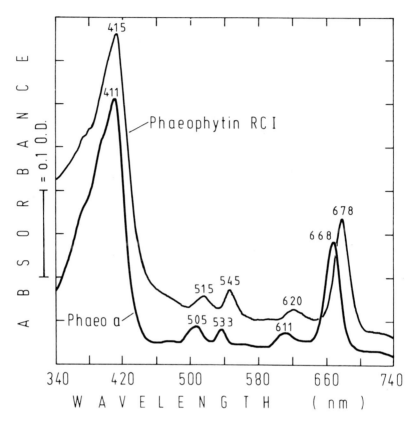

Fig. 2: Absorption spectra of pheophytin RC I and pheophytin
         a in acetone at r.t. Baselines were arbitrarily ad-
         justed to avoid overlap ($E_{740}=0$).

ratio of Chl RC I to P-700 again being 1:1. To complete the
list of chlorophyll containing organisms the only known
Chl b-containing prokaryote Prochloron spec. was investigated.
The ratio of Chl a to P-700 was determined to be about 180:1
(Withers et al., 1978). The ratio of Chl a to Chl RC I in the
methanol extract was around 140-160:1. The results of the
Chl RC I contents of the various organisms are compiled in
Tab. 1. The fact that the ratio of P-700 to Chl RC I in all
cases is close to or identical with 1 suggests that Chl RC I
is a component of the reaction centre I or even the chromo-
phore of P-700 itself. This assumption is supported by an

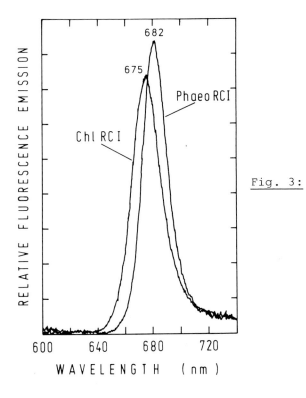

Fig. 3: Fluorescence emission spectra of Chl RC I and pheophytin RC I in acetone at r.t. Excitation: 433 (415) nm.

experiment performed with the yellow pigment mutant C-2A' of Scenedesmus which only can form chlorophyll in light (Senger and Bishop, 1971). When this organism is greened in the presence of chloramphenicol no reaction centers are formed and no P-700 activity can be measured. Under these conditions also Chl RC I was undetectable whereas a control culture without chloramphenicol showed an in vivo Chl a: P-700 ratio of 300-500:1 and the methanol extract contained about 1 Chl RC I per 400 Chl a molecules. Thus the assumption that Chl RC I is the chromophore of P-700 seems to be very likely.

A first hint concerning the possible structural change in the molecule was given by the fact that the absorption data of methylpheophorbide RC I, especially the interchanged intensities of the two typical pheophytin peaks in the region between 500 and 550 nm, correspond to the data of 20-Cl-methylpheophor-

MATERIAL	PIGMENTS	RATIO
Mutant C-6E	Chl a/Chl RCI (extract)	89
	Chl a/P_{700} (in vivo	80
Sucrose density gradient PS I particles (C-6E)	Chl a/Chl RCI (extract)	28
	Chl a/P_{700} (in vivo)	28
Mutant C-2A' after 7h greening −CAP	Chl a/Chl RCI (extract)	400
	Chl a/P_{700} (in vivo)	300–500
+CAP	Chl a/Chl RCI (extract)	∞
	Chl a/P_{700} (in vivo)	∞
Spinach PSI particles	Chl a/Chl RCI (extract)	70–90
	Chl a/P_{700} (in vivo)	70–90
Anacystis nidulans photosynthetic membranes	Chl a/Chl RCI (extract)	85
	Chl a/P_{700} (in vivo)	80
Anabaena variabilis photosynthetic membranes	Chl a/Chl RCI (extract)	86
	Chl a/P_{700} (in vivo)	90
Cyanidium caldarium	Chl a/Chl RCI (extract)	450
	Chl a/P_{700} (in vivo)	430
Porphyridium cruentum	Chl a/Chl RCI (extract)	500
	Chl a/P_{700} (in vivo)	470
Prochloron spec.	Chl a/Chl RCI (extract)	140–160
	Chl a/P_{700} (in vivo)	180

Ratios of Chl a/RCI in acetone extract and of Chl a/P_{700}
in vivo (buffer) in different organisms

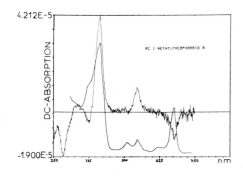

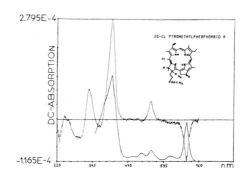

Fig. 4: Absorption- and CD-spectra of methylpheo-phorbide a, 20-Cl-pe-pro-methylpheophorbide a and methylpheophor-bide RC I. Spectra were kindly performed by Dr. Scheer, Munich.

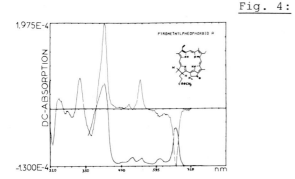

bide a as well as the CD spectra, both kindly provided by Dr. H. Scheer, University of Munich. Fig. 4 clearly demonstrates the similarity of 20-Cl-methylpheophorbide and methylpheophor-bide RC I characteristics.

As the three protons of the methin bridges between the pyrrol rings appear at very low field in the ^{1}H-NMR-spectrum a sub-stitution at C-20, which is the proton of the δ-methin bridge should be easily detectable by ^{1}H-NMR spectroscopy. Fig. 5 shows the ^{1}H-NMR-spectra of Chl a and Chl RC I and methyl-pheophorbide RC I. The spectrum of Chl a shows the ß, α and δ-proton at 9,8225, 9,5330 and 8,9175 ppm respectively, whereas in the spectra of Chl RC I and its methylpheophorbide a peak in the region around 8.9 is missing clearly indicating the substitution at the 20-position by a group or an atom giving no ^{1}H-NMR signal, like chlorine.

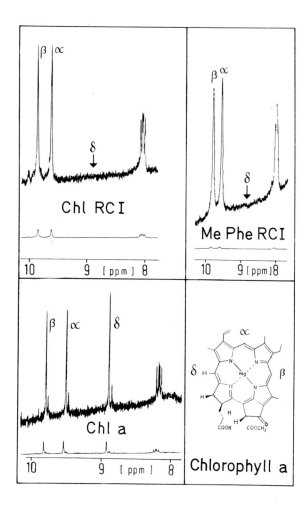

<u>Fig. 5:</u>   ^{1}H-NMR spectra of Chl <u>a</u>, Chl RC I and methylpheophor-
bide RC I in d_6-acetone. Spectra were recorded with
a Brucker 400 MHz-instrument.

This result is in agreement with the data of Scheer et al.
(1984) who reports a substitution at C-20 by a non ^{1}H-NMR-
active substituent in the methylpheophorbide of Chl RC I.

He finds an additional substitution in the C-13 position.
From the molecular weight of 656-658 and other data Scheer
concludes that the new substituents are chlorine at C-20 and
a hydroxyl group at C-13[2].

Structural changes in the chlorophyll molecule should be expressed in its molecular weight. Therefore mass determinations
were carried out with plasma desorption mass spectroscopy
(PDMS). A great advantage of this method is the fact that
even molecules of molecular weights more than 3.5 kD usually non volatile undergo the solid to gas phase transition
as M^+-ions. The molecular weight for Chl RC I determined by
PDMS was found to be 943 $\pm$ 1 mass units. This mass increase
of 50 mass units fits very well with the proposed substitution
of two protons of Chl a by a chlorine atom and a hydroxyl
group. A PDMS-diagramm with an extended M^+-peak region is
shown in Fig. 6. Our earlier reports showing a mass increase
of 35 mass units plus the addition of one molecule of water
have to be reinterpreted. It could clearly be shown that not
chlorine and water are responsible for the mass increase of
50, but the substitution of two protons by chlorine and a
hydroxyl group. These results are almost identical to those
measured by Scheer for the mass difference of the methylpheophorbides of Chl a and Chl RC I, which was determined to be
50, too (Scheer, 1984).

For the final proof of chlorine in Chl RC I a direct method

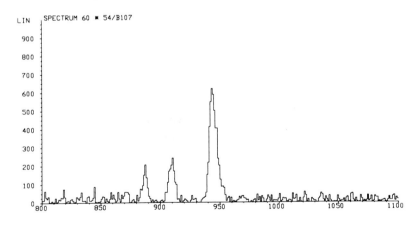

Fig. 6: PDMS-spectrum of enriched Chl RC I (only extended
M^+-peak region) showing also M^+-peaks of Chl a-
adduct and pheo a-adduct.

for determination of Cl at the ng scale was found to be the
neutron activation analysis. To avoid uncontrolled alterations
of the molecule,chlorophyll samples were first transformed
into their more stable methylpheophorbides. The results of
the neutron activation experiments are shown in Fig. 7. For
calibration several standards were used. The zero standard
was obtained from irradiating an empty quartz ampule (dummy),
chlorine background was determined from a methylpheophorbide
a sample. Chlorine containing samples were NaCl and 20-Cl-
methylpheophorbide a. As all chlorine samples contained approximately the same amount of chlorine the intensity of the sig-

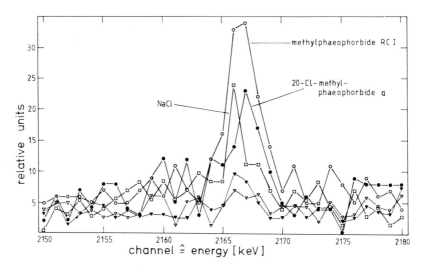

Fig. 7:   Neutron activation analysis data: dummy ▼——▼ ,
          methylpheophorbide a ▽——▽ , NaCl □——□, 20-Cl-
          pyromethylpheophorbide a o——o , methylphorphorbide
          RC I ●——● .

nal could also be evaluated semi-quantitatively. The result
demonstrates clearly that one atom of chlorine per molecule
of chromophore is present in Chl RC I. The determinations
are confirming the results and conclusions from ^{1}H-NMR and
PDMS.

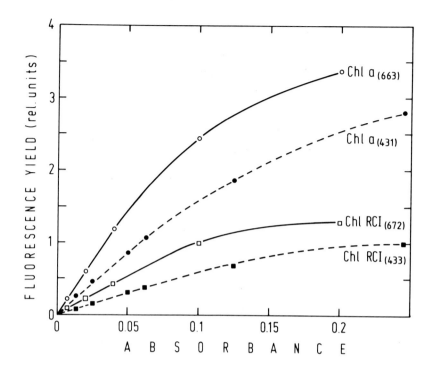

Fig. 8:   Fluorescence yield of Chl RC I and Chl a at various
          concentrations in acetone. Samples were adjusted to
          equal absorbance at their blue and red absorption
          maxima.

During isolation and fluorescence spectroscopy it became obvious that the fluorescence yield of Chl RC I was lower than
that of Chl a. This observation could be quantified by ad-
justing samples of Chl a and Chl RC I to identical absorption
in their blue and red absorption maxima respectively. It turned
out that in both cases the fluorescence yield of Chl RC I is
about 3 times (2.67 and 2.85 fold) lower than that of Chl a
(Fig. 8). This low fluorescence yield is consistent with the
low fluorescence yield of P-700 in vivo (Papageorgiou, 1975),
but one has also to consider that protein binding and environmental factors play a great role in vivo. For these measurements it was assumed that the absorption coefficients of Chl
RC I and Chl a are very similar.

Sample	fluorescence lifetime (ns)
Chl a	4.30
Pheo a	4.49
20-Cl-pyro-methylpheo a (mix.)	2.51
Chl RC I (mix.)	1.85
Chl RC I (u)	1.49
Chl RC I (l)	1.52
Pheo RC I (mix.)	1.50

Tab. 2 Fluorescence lifetimes of chlorophylls and derivatives (performed by Dr. Holzwarth, MPI Mülheim)

(mix.) = mixture of the two diastereomeres

(u)   = upper band

(l)   = lower band   of the isolated diastereomeres

From theoretical calculation and under the assumption of a similar ε the low fluorescence yield has to result in a shorter fluorescence life time of Chl RC I. This could be demonstrated by picosecond fluorescence spectroscopy which was kindly performed by Dr. Holzwarth and co-workers at the Max-Planck-Institut at Mülheim/Ruhr. Fluorescence lifetimes of different chlorophyll derivatives are given in Tab. 2.

It has to be mentioned that both, the 20-Cl-derivatives of Chl a and Chl RC I can be separated into 2 diasteromeres depending on the chlorine atome being below or above the nearly planery porphyrin system. In the case of Chl RC I lifetimes of both diasteromeres and of the mixture were measured. All halogenated chlorophyll derivatives showed shorter lifetimes than Chl a and derivatives.

Since the presence of chlorine in biological molecules is
not very common one might consider that possibly chlorine
is not incorporated into the chromophore in vivo, but that
Chl RC I represents a reactive form which is halogenated
when the chromophore is removed from the protein. This could
be ruled out by the following experiment: Pellets of cells
of mutant C-6E and greened mutant C-2A' were mixed with

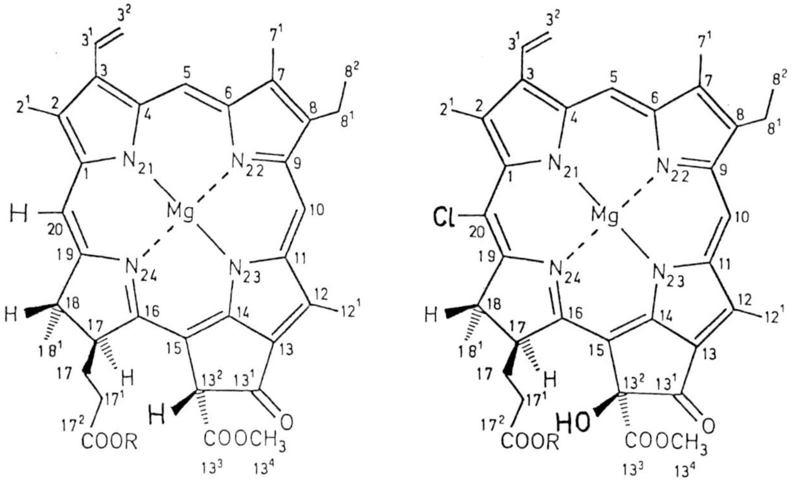

CHLOROPHYLL a

CHLOROPHYLL RCI
(13^2-hydroxy-20-cnloro-chlorophyll a)

Fig. 9:   Structural formulae of Chl a and Chl RC I showing
          the substitution of protons at C-20 and C-13^2 by
          chlorine and a hydroxyl group.

Na^{36}Cl before the extraction with hot methanol. If the chlorine
were incorporated during extraction labelling should be detectable in the Chl RC I fraction.But there was no radioactivity
found in the extracts. Thus it has to be concluded that the
chlorine is really incorporated into Chl RC I in vivo.

As conclusion from the presented data the newly discovered
Chl RC I has to be termed 20-Chloro-13^2-hydroxy-chlorophyll a
identically proposed by Scheer et al. (1984). The structural
formula in comparison to Chl a is given in Fig. 9.

Acknowledgements

The authors wish to thank Drs. Scheer, Munich, Danigel, Junclas, Schmidt and Berger, Marburg and Holzwarth, Mülheim
as well as all their co-workers for their help and cooperation.
We thank Mrs. Schreiber and Krieger for typing the manuscript
and for skillful technical assistance Mrs. Koss and Dörr. The
contribution was kindly supported by the Deutsche Forschungsgemeinschaft.

References

1. Bishop, N.I.: In Methods in Enzymology (A. San Pietro,ed.),
    Vol. 23, 130-173, Academic Press, New York (1971).

2. Dörnemann, D., Senger, H.: Proc. Vth Int. Congr. on Pho-
    tosynthesis (G. Akoyunoglou, ed.), Vol. V, 223-231, Ba-
    laban Int. Science Serv., Philadelphia, Pa. (1981a).

3. Dörnemann, D., Senger, H.: FEBS Lett. 126(2),323-327 (1981b).

4. Dörnemann, D., Senger, H.: Photochem. Photobiol. 35,
    821-826 (1982).

5. Dörnemann, D., Senger, H.: Proc. VI Int. Congr. on Photo-
    synthesis (C. Sybesma, ed.), Vol. II,2, 77-80, Nijhoff/
    Junk Publ., The Hague, The Netherlands (1984).

6. Kok, B.: Biochim. Biophys. Acta 48, 527-533 (1961).

7. Lewin, R.A.: Ann. Microbiol. (Inst. Pasteur) 134 B,
    37-41 (1983).

8. Papageorgiou, G.: In Bioenergetics of Photosynthesis
    (Govindjee, ed.), 319-371 (1975).

9. Scheer, H., Wieschoff, H., Schaefer, W., Cmiel, E.,
    Nitsche, B., Schiebel, H.-M., Schulten, H.-R.: Proc.
    VI Int. Congr. on Photosynthesis (C. Sybesma, ed.), Vol.
    II,2, 81-84, Nijhoff/Junk Publ., The Hague, The Nether-
    lands (1984).

10. Senger, H., Bishop, N.I.: Plant & Cell Physiol. 13,
    633-649 (1972).

11. Withers, N.W., Alberte, R.S., Lewin, R.A., Thomber, J.P.,
    Britton, G., Goodwin, T.W.: Proc. Natl. Acad. Sci. USA,
    75(5), 2301-2305 (1978).

Received July 2, 1984

Discussion

<u>Hoff</u>:   Have the magnetic properties (g-value) of the doublet (cation, anion) states and triplet state of Chla RC I been measured? One would expect that the presence of Cl would considerably change the g-values of the cation and anion and the decay rates of the triplet state; this would be a means to ascertain whether Chla RC I is the primary donor (P700) or not.

<u>Dörnemann</u>:   No, not yet. The amount of sample required was too large. But the experiment is planned and possibly we could do it together.

<u>Buchler</u>:   Dr. Scheer mentioned that a high resolution mass spectrum was obtained. Was this with or without the phytyl side chain?

<u>Scheer</u>:   These measurements were done with the methylpheophorbide of RC I.

<u>Buchler</u> (Question to Dr. Dörnemann): Have you tried to obtain a field ion desorption mass spectrum of Chl RC I?

<u>Dörnemann</u>:   Yes, we did, but you do not get an M^+-ion of RC I by this method. This is also known from other chlorophylls.

<u>Gregory</u>:   Some 25 years ago, a crucial property of the chlorophyll molecule was the ionizable H-atom in ring V - which was the basis of the 'phase test'. What are the properties of the 13^2-OH group?

<u>Dörnemann</u>:   The phase test was performed at a time, when the structure was not yet known, only to test the isolated chromophore. It did not work, but at this time the result was not regarded to be important. Today, of course, it is.
Special properties of the 13^2-OH group were not yet investigated.

<u>Rapoport</u>:   How do you know that the chlorine atom is at C-20?

Reply by <u>Scheer</u>:   The product of this type was first prepared by H. Fischer and coworkers, and the structure deduced by Woodward. The chlorination is regiospecific next to the reduced ring. It can be seen in the NMR by characteristic shifts of the neighboring group signals. The other alternative (OH at C-20) would produce a completely different type of spectrum, since it is known to tautomerize to oxophlorins.

LONG-WAVELENGTH ABSORBING FORMS OF BACTERIOCHLOROPHYLLS

II. Structural requirements for formation in Triton X-1oo micelles and in aqueous methanol and acetone

Hugo Scheer, Berndt Paulke and Jörg Gottstein

Botanisches Institut der Universität München,
D-8ooo München

Introduction

The bacterial photosynthetic apparatus contains several bacteriochlorophyll (Bchl) proteins of well defined functions and spectroscopic properties (1,2). In all these chromoproteins, the electronic spectra are distinctly different from that of Bchl in solution. The near infrared (Qy) absorption maxima are shifted by up to 23onm (35oocm^{-1}) to the red and increased in intensity, the visible (Qx) and/or the near UV (Soret) band(s) are red shifted to a lesser degree and usually decreased in intensity. Even more pronounced changes are observed in the circular dichroism (cd) spectra. They are more complex in the Bchl proteins, and their anisotropies are increased by up to two orders of magnitude as compared to free Bchlš.

Similar, although less dramatic changes are known for the chlorophylls of green plants, where they have mainly been attributed to aggregation (see 3 for leading references). In solution, coordinative unsaturation of the central magnesium and the presence of several carbonyl donor groups has been demonstrated as the main driving force for

---

Abbrevitations: Bchl $\underline{a}$gg = bacteriochlorophyll, type ($\underline{a}$ or $\underline{b}$) and esterfying alkohols (p=Pphytol or gg = geranylgeranol) are given as suffix and subscript, respectively. cd = circular dichroism, nir = near infrared, vis = visible. uv = ultraviolet,

aggregation from a series of detailed experiments. Similar aggregates
have been proposed to be present in chlorophyll proteins. It is not yet
clear, however, to which extent the mechanisms for aggregation identified
in solution are also relevant for the interactions of chlorophyll in
proteins.

Much less work has been carried out with Bchl (4-11) and it has already
been pointed out by Katz et al (8), that its interactions are more
complex to do the presence of an additional donor, e.g. the 3-acetyl
group. Two laboratories have recently taken up the subject and studied
the properties of Bchl and related pigments in detergent solution (9-11).
The outset of these studies was the observation, that the most commonly
used detergent for the isolation of Bchl proteins were capable by themselves to form complexes with Bchl, which had many of the characteristic
spectral properties of the former. A more detailed investigation of these
complex may be helpful in understanding the spectra of Bchl proteins.
It has also a practical aspect, because detergent effects may obscure
the interpretation of experiments with Bchl proteins (see e.g. 12).

Gottstein and Scheer (9) reported on three distinct complexes of Bchl
in Triton X-1oo, which were suggested to contain a minimum of one, two
and three strongly interacting pigment molecules per micelle, respectively.
Scherz and Parson (10) investigated complexes with LDAO which are also
believed to contain only a small number of strongly interacting molecules.
They developed a theory which related the spectral changes to exciton
interaction and hyperchromism (11). One of the most remarkable results
of the latter authors was the observation, that the complexes are not
only formed by Bchl, but also its Mg-free derivative bacteriopheophytin
(Bphe). These and earlier similar observations of Krasnovskii et al.
(5) with Bchl films strongly questioned the dominant role of the central
Mg for aggregation in the detergent complexes and suggested a different
type of interaction. Since the pigment enviroment in a micelle is
probably more similar to that in the protein, than is a homogenous
solution, these aggregation mechanisms should also be considered in Bchl
proteins. Here, we wish to report further results on long-wavelength
forms of Bchl, which indicate an important function of the long-chain

terpenoid alcohol in the aggregation of Bchl in micelles.

MATERIALS AND METHODS

Bchl esterfied with phytol and geranylgeranol was isolated from
Rhodopseudomonas spheroides and Rhodospirillum rubrum, respectively,
by the method of Svec (13, modified by dioxan precipitation of the
crude extract) and chromatographed on DEAE-cellulose (14). Bphe was
prepared by demetalation of Bchl with 6% HCl. Triton X-1oo was obtained
from Serva, Heidelberg, and LDAO from Bayrol, München. All solvents
were reagent grade. Uv - vis - nir absorption spectra were recorded with
a model DMR 22 spectrophotometer (Zeiss, Oberkochen), cd spectra with
a Dichrograph V equipped with a red sensitive photomultiplier (ISA -
Yvon-Jobin, München). The spectra are not corrected for scattering.
Controls in a scattering insensitive photometer gave shifts $\leq$ 1onm.

RESULTS AND DISCUSSION

Absorption Spectra of Triton X-1oo complexes

The Triton X-1oo complexes have been routinely prepared by dissolution
of Bchl in a buffer (tris-HCl, 1omM, pH 8.o) containing o.1% (v/v)
Triton X-1oo (fig. 1). Like in the previous study (9), the absorptions
in the nir are composed of up to four peaks (λmax 775, 83o, 86o and
92o nm) (fig.1). Two of them (83o, 92o nm) are always present in a
fairly constant ratio (2:1, corrected for absorption of B86o and B77o).
This suggests, that the two absorption bands belong to a single species,
and that there are two different long - wavelength absorbing complexes,
B86o and B83o/93o, which absorb at 86o nm and 83o/92o nm, respectively.
The Bchl/micelle ratio in these complexes has been estimated to approximately two and three, respectively, which agrees with the minimum number
of strongly interacting pigment molecules derived from the cd spectra
(fig.2). Whereas complexes absorbing around 86o nm are common to all

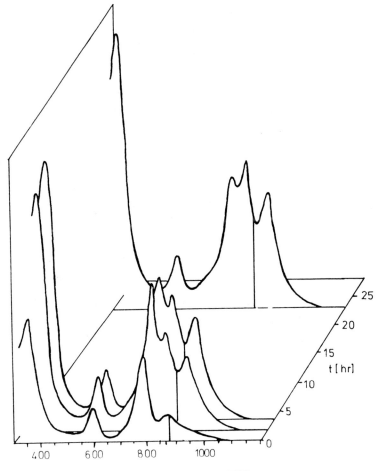

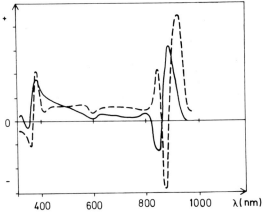

Fig. 1: Absorption spectra
of Bchl agg during the dissolution in tris buffer
(10 mM, pH 8.0) containing
1% Triton X-100

Fig. 2: Circular dichroism
spectra of the sample shown
in Fig. 1 at t=0 (———) and
3 hrs (----).

previous studies, the further development is different in different
preparations. B830/ 930nm type complexes have been observed also by
Komen (4), whereas forms absorbing mainly at 830 or 800 nm without
any absorption maxima $\geqslant$ 860 nm have been reported by others (8, 15).
It is not yet clear, if the 830 nm absorption in B830/930 orginates
from one of the latter, or is part of a more complex system as
suggested by the constant band ratio.

In both complexes, but in particular the 83o/92o nm one, the Soret and
the Qx absorption (fig.1) and cd bands are reduced in intensity relative
to the nir bands. During the time - course of dissolution, the absorptions
around 37o and 59o nm remain rather constant, even though the intensity
of the bands $\geqslant$ 8oo nm increase several fold during this process. Scherz
and Parson (1o) have succeded in the preparation of a B85o type complex
which is almost free of monomer, which clearly shows this hyperchromic
effect without the necessity for curve resolution. This effect is one
of the characteristics not only for the long - wavelength absorbing
forms in vitro, but also for many Bchl proteins.

Concentration dependence of complex formation with Triton X-1oo.

Concentration effects have been studied varying the pigment/detergent
ratios. Increasing pigment concentrations promoted the formation of the
B830/930 nm complex, which is agreement with the assignment of a larger
aggregation number to this complex. Increasing the detergent concentration, reduced vice versa the amounts of these complexes and promoted
monomer formation (λmax $\approx$ 77o nm, fig.3). At Triton concentrations
$\geqslant$ 1%, the red shifted complexes are no longer observed. Complexes
formed at o.1% Triton X-1oo are likewise gradually destroyed by the
addition of increasing amounts of the detergent. These results can be
explained by a random and rapid distribution of the Bchl molecules on
the detergent micelles present, and support the idea (9, 11) that each
aggregate comprises only a small number of Bchl molecules in a single
micelle.

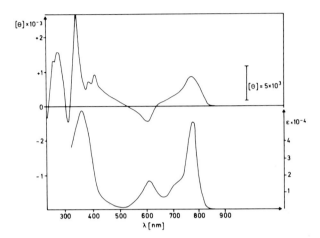

Fig. 3: Absorption (lower trace) and cd spectra of monomeric Bchl a̱gg
in methanol, from ref. 9.

Comparison of detergent and aqueous organic solvent induced complexes

As observed first by Komen (4), Bchl complexes with strongly red -
shifted absorptions are formed under two different conditions, e.g.
in mixtures of organic solvents (methanol, acetone, and others) containing ≳ 5o% water, and in solutions with detergents like sodium
dodecyl sulfate. Long wavelength absorbing forms of Bchl in mixtures
of organic solvents with water have also been obtained by others (5,6,7).
We have obtained rather similar absorption and cd spectra for complexes
of Bchl with the detergent, Triton X-1oo (figs. 1,2, see also ref.9), and
in the mixtures of acetone or methanol with water (figs. 4,5). In the
latter systems, the onset of complex formation with λmax ≳8oo nm is
at water concentrations of 45-5o%, and the absorption spectra are
rather similar up to the highest water concentrations studied, e.g. 95%.
These results reproduce the findings of Komen (4). It should be noted,
that the kinetics of the complex formation depend on the preparation procedure. The results shown in figs. 4,5 have been obtained by injection
of a Bchl stock solution in methanol or acetone into appropriate aqueous
organic mixture. The complex formation is slowed down (but the same

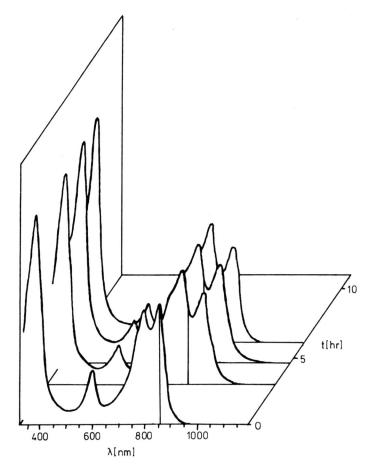

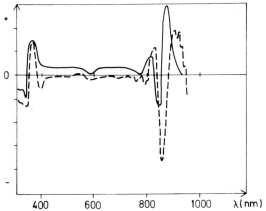

Fig. 4: Absorption spectra of Bchl $\underline{a}$gg after the dilution of an acetonic solution with water to a final concentration of 70% water.

Fig. 5: Circular dichroism spectra of the sample shown in Fig.4 after 0 (——) and 4.25 hrs (----).

spectrum is eventually obtained), if the stock solution is first
diluted with the respective pure solvent, and the resulting dilute
solution is then further diluted with water.

The absorption and cd spectra of Bchl in these solvent systems (figs.
4, 5), are very similar to the ones observed in detergent micelles
(fig. 6). Distinct differences are in the absorption ratios of the
two nir bands of the B830/939 complex (4:3 vs. 2;1 in Triton X-1oo),
and a more Gaussian appearance of the bands (less pronounced tails at
the red wings, fig. 6). In spite of this similarity, the size of the
aggregates seems to be rather different in the two systems. The
following observations suggest the presence of large aggregates in the
mixed aqueous/organic solvents. Solution of Bchl in the latter solvents
are optically clear, but loose there color upon prolonged standing (hrs),
due to the formation of a colored precipitate. Centrifugation in a small
laboratory centrifuge is already sufficient to pellet all the pigment
present, and the original spectrum is restored upon resuspension of the
precipitate. The pigments thus seems to be present in dense aggregates
of colloidal dimensions.

Fig. 6: Resolved spectra of
the two long - wavelength
absorbing forms, B86o and
B83o/93o of Bchl agg in
aqueous methanol.

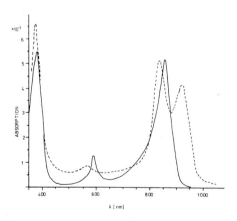

Katz et al.(8) and Scherz et al. (10) have earlier observed Bchl forms
with similar spectra but rather varying aggregation number. Long-wavelength absorbing species of Bchl are also formed in hydrated films of
the pigment (6-8). It then appears, that the spectrum depends mainly
on a specific next -neighbor relationship, which is rather similar
irrespective of the size of the aggregate.

Structure and complex formation of Bchl´s

The influence of the central Mg has been tested by a series of similar
experiments with Bchl agg and Bphe agg. Both form complexes in the mixed
aqueous-organic solvent system and with Triton X-100. The Bphe complexes have also strongly red-shifted absorptions (λmax $\approx$850 nm) and
show the same hyperchroism of the Qy band, with the Qx band being
almost reduced to zero (fig.7). There are noneless distinct spectroscopic differences among the complex of the two pigments. Bphe forms
only one type of complex, with an absorption maximum close to 850 nm
and a strong S-shaped cd signal in the nir (fig.8). This complex is
spectrosopically similar to the B860 complex of Bchl. The bandwidth
of the Bphe complexes are, however, considerably narrower than in
either form of Bchl complexes (fig.9), and there is essentially no
additional absorption in the nir correponding to either monomeric Bphe
(λmax $\approx$770 nm) or a B830/930 type complex. Whereas the absorption
spectra of the Bphe complexes are similar in different mixtures of
acetone with $\rangle$ 50% water, the cd spectra are rather different (fig.7,8)
and indicate a different arrangement in both forms.

Most of the results are similar to those obtained by Scherz and Parson
with Bphe in a solvent system containing LDAO in dilute acetic acid
(10, the same authors have recently also obtained rather pure B860
complexes of Bchl, private communication). We assume, therefore, a
similar arrangement as has been suggested from the theoretical analysis
of these authors (11), with only a slight overlap of the non-parallel
macrocycles. Krasnovskii et al. (5) have on the other hand suggested,

Bchl a$_p$ R =

Bchl a$_{gg}$ R =

Bphe = Bchl without central Mg

10-hydroxy-Bchl a   Bchl b

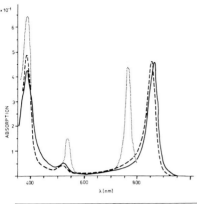

Fig. 7: Absorption spectra
of Bphe agg in aqueous
acetone (1:1, ———) and
(9:1, ----), and in pure
acetone (.....).

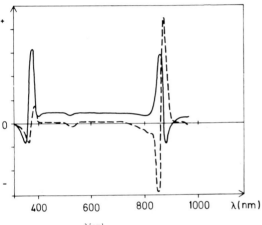

Fig. 8: Circular dichroism
spectra of Bphe agg in
aqueous methanol (1:1, ———)
and (9:1, ----).

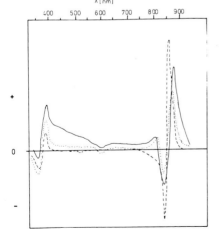

Fig. 9: Comparison of circular
dichroism of Bchl agg in tris
buffer containing Triton X-100
(0.1%, ———), and in aqueous
acetone (8:2, ...), and of
Bphe agg in aqueous (95:5,
-------).

that π-π interactions are predominant in the aggregation of Bphe,
which would rather favor a larger overlap of the two macrocycles.
Aggregates of this type are also found in vivo. Angerhofer et al.(16)
have identified Bphe species with rather similar spectra in whole
bacteria by fluorescence techniques. In particular do they exhibit only
a very small Qx band in the excitation spectra. The formation of complexes irrespective of the presence of the central Mg sets them apart
from the aggregates investigated in detail by Katz et al., in which the
Mg is essential for the coordination with other chlorophylls or bridging
ligands. A possible benefit of an aggregation which does not involve the
central Mg, is the fact that this would free the latter for coordination
e.g. with amino acid side chains in the protein (Scherz, private communication). This situation prevails in the excitation transfer protein
from Prostecochloris aestuarii (17).

To test the influence of the esterfying alcohol, a series of experiments
was carried with Bchl´s carrying different alcohols. Many former results
(4,9,10) have been obtained with Bchl isolated from Rs. rubrum, which is
esterfied with geranylgeranol (Bchl agg) (18),and no source for the
pigment given in the other studies (5 -8). When Bchl ap (esterified
with phytol, isolated from Rp. speroides) is used instead, the ability
to form complexes is strongly reduced. In mixtures of water with methanol
or acetone, the initial absorptions ≥ 800 nm are reduced by more than
50%, and they remain at a much lower level throughout the experiments,
which involved generally the observation over 24 hrs. In particular is the
formation of B830/930 strongly inhibited, and B860 somewhat stabilized.
The results are again essentially independent of the solvent (water/
methanol or water/acetone) and of the water concentration above a thres-
hold value of 50% up to 95%. The difference between Bchl agg and Bchl ap
is even more pronounced in the Triton X-100 micelles. With the pigment
esterfied with phytol, we have been unable to prepare any solutions
with significant absorptions - 800 nm by the procedure given above, and
only 50% complex is formed by the procedure of Rosenbauch and Scherz (15)
using aqueous formamide as the solvent.

The long chain terpenoid alcohols are a characteristic structural
feature of almost all chlorophylls, but little is known about their
function. To our knowledge, the complex formation with Triton X-100 is
the first example of a non-enzymatic (19) reaction which is profoundly
influence by these alcohols. It is noteworthy in this context, that the
high-resolution x-ray structure of a Bchl $\underline{a}$ -protein shows a rather
peculiar arrangement of the alcohol chains (which are phytol in this case),
which indicates that it may be involved in the binding of the pigment (17).
Scherz and Parson have recently observed a strong cooperativity in the
formation of the long-wavelength absorbing forms of Bchl, with a seed
formation preceeding the actual aggregation process (private communication). They suggested, that the alcohol could affect the first step of
this sequence.

There is as yet only marginal data on the influence of other structural
modifications on the complex formation. Bchl $\underline{b}$ esterfied with different
alcohols (20) gives only shoulders on the long-wavelength side of the
monomer absorption band. Since part of the pigment is oxidized during
the incubation, the significance of this finding is not clear. Prepa-
rations containing allomerized Bchl $\underline{a}$ (= 10-hydroxy-Bchl $\underline{a}$) do not
form long-wavelength absorbing complexes. This is also a tentative
explanation for the degradation of the Bchl complexes upon prolonged
standing (see figs. 1.4), because the pigment becomes gradually oxidized. The allomerization is, however, slowed down in the Triton X-100
complexes as compared to Bchl in solution, which can be used to
stabilize labile Bchl´s.

Taken together, the results indicate an important influence of the long
-chain terpenoid alcohol and the enolizable ß-ketoester system at the
isocyclic ring on the formation of long-wavelength absorbing complexes,
and only a lesser influence of the central Mg and the substituents at
ring 2. Further modifications are necessary to fully understand the
role of the different functional groups present in chlorophylls on this
type of aggregate formation.

CONCLUSIONS

Bchl agg and its demetalated derivative e.g. Bphe agg form strongly
red-shifted and Qy hypochromic complexes in micelles with the detergent,
Triton X-100 and in mixed organic-aqueous solvent systems containing
$\geqslant$ 50% water. Both types of complexes are aggregates of the pigments,
if judged from the exciton splittings in the cd spectra. According to
their similar absorption and cd spectra, they appear to have rather
similar next-neighbor relationships, but the aggregation number is
small (2-4) in the detergent micelles and probably much larger in the
aqueous organic solvents. The aggregation does not require the presence
of the central Mg, but is critically affected by the esterfying terpenoid
alcohol present in Bchl´s, and is inhibited by oxidation at C-10 of the
pigments. The complexes formed in detergent micelles may be a useful
model to analyze the influence of aggregation geometry on the spectra,
and a help towards understanding the spectra of Bchl-proteins.

ACKNOWLEDGEMENTS

This work was supported by the Deutsche Forschungsgemeinschaft, Bonn
(SFB 143). The expert technical assistence of C. Bubenzer is acknowledged. We thank A. Scherz and W. Parson for helpful discussions and
the communication of their results prior to publication.

REFERENCES

1.   Cogdell, R.J., Valentine, J.: Photochem. Photobiol. <u>38</u>, 769-772
     (1983)

2.   Okamura, M.Y., Feher, G.: in R.K. Clayton, W.R. Sistrom (eds.),
     The Photosynthetic Bacteria, Academic Press, New York, 1978

3.   Katz, J.J., Shipman, L.L., Cotton, T,M., Janson, T.R.: in
     D. Dolphin (ed.): The Porphyrins,Vol. V, chapter 9, Academic
     Press, New York, (1978); Krasnovskii, A.A., Bystrova, M.I.:
     Biosystems <u>12</u>, 181-194 (1980)

4.   Komen, J.G.: Biochim. Biophys. Acta 22, 9-15 (1956)

5.   Krasnovskii, A., Bistrova, M.I., Umrikhina A.V.: Dokl. Acad.
     Nauk. SSSR 235, 232-235 (1977)

6.   Ballschmiter, K., Katz, J.J.: Biochim. Biophys. Acta 256, 307-
     327 (1972)

7.   Cotton, T.M., VanDuyne, R.P.: J. Am. Chem.Soc. 103, 6020-6026
     (1981)

8.   Katz, J.J., Oettmeier, W., Norris, J.R.: Phil. Trans. R. Soc.
     Lond. B273 227-253 (1976)

9.   Gottstein, J., Scheer, H.: Proc. Natl. Acad. Sci. USA 80, 2231-
     2234 (1983)

10. Scherz, A., Parson, W.: Biochim. Biophys. Acta., in press

11. Scherz, A., Parson, W.: Biochim. Biophys. Acta., in press

12. Kramer, H.J.M., VanGrondelle, R., Hunter, C.N., Westerhuis, W.J.H.,
     Amesz, J.: BIochim. Biophys. Acta 765, 156-165 (1984)

13. Svec, W.A.: in D. Dolphin (ed.), The Porphyrins, Vol. V, 341-399,
     Academic Press, New York, 1978

14. Omata, T., Murata, N.: Plant Cell Physiol. 24, 1093-1100 (1983)

15. Rosenbauch, V., Scherz, A.: private communication, 1984

16. Angerhofer, T., Wolf, H.: private communication, 1983

17. Olson, J.: in R.K. Clayton, W.R. Sistrom, (eds), The Photosyn-
     thetic Bacteria, Academic Press, New York, 1978

18. Katz, J.J., Strain, H.H., Harkness, A.L., Studier, M.H., Svec,W.A.,
     Janson, T.R., Cope, B.T.: J. Am. Chem. Soc. 94, 7938-7939 (1972)

19. Schoch, S., Lempert, U., Rüdiger, W.: Z. Pflanzenphysiol. 83,
     427-436 (1977)

20. Steiner, R., Scheer, H.: Z. Naturforsch. 36c, 417-420 (1981)

Received July 27, 1984

Discussion

Woody: How many Bchl molecules are there in the reaction centers from
the purple photosynthetic bacteria?

Scheer: There are four Bchl and two Bphe. Two Bchl constitute the "special pair" primary donor, another one serves as the primary acceptor, and
the Bphe as the secondary acceptor. Only the spectrum of the special pair
and the active pheophytin is known in some detail from several difference
spectroscopy methods.

<u>Woody</u>:   What do you envision as the principal type of interaction stabilizing the Bchl dimers in micelles?

<u>Scheer</u>:   Since bacteriomethylpheophorbide does form the complexes, Mg-coordination and hydrophobic interactions do not appear essential. There remains π-π interactions, as has been suggested in films of Bchl and Bphe by Krasnovskii (1981). A possible modifying effect of the long-chain alcohol could be, that it affects the positioning of the tetrapyrrole heads in the Gouy-Chapman layer. The C-20 alcohol is rather large as compared to the dimensions of the hydrophobic core of the Triton X-100 micelles.

<u>McDonagh</u>:   When you mix derivatives that aggregate with derivatives that do not, do they coaggregate or do they behave independently?

<u>Scheer</u>:   The only mixture we tried was bacteriochlorophyll and bacteriopheophytin which did not produce significant changes.

Concluding Remarks

Martin Gouterman

Department of Chemistry, University of Washington, Seattle, USA

Tetrapyrroles - Linear and Cyclic

On being asked to sum up the relations between the porphyrins/chlorophylls and the bilins, I heard someone remark that it was like comparing benzene and hexatriene:

I think this is an instructive comparison. For the chemical properties of both are dominated by their π electrons. However, the following differences are important:

1) Benzene is structurally rigid. This stabilizes the excited states, and the molecule both fluoresces and phosphoresces. The monoanion and mono-cation radicals are easily obtained.

2) The excited states of the triene do not have strong emission. The molecule photoisomerizes. I do not know but doubt there is much stability to the radical cations and anions.

This same set of similarities and differences is obtained for the linear and cyclic tetrapyrroles, with much more complexity.

A principal "complexity" for the porphyrin is the ability to complex a metal. Thus the iron porphyrins become a cofactor in redox reactions. The Mg complex is used for photosynthesis, where excited state stability as

Optical Properties and Structure of Tetrapyrroles
© 1985 by Walter de Gruyter & Co., Berlin · New York

well as stability of the radical ions play a rôle. I tend to think this
system might have developed without Mg, using either the free base or
another closed shell metal. The iron porphyrins have much more flexibility
for electron transfer than would the porphyrin as a free base or with a
closed shell metal. The iron porphyrins can also complex to small ligands,
i.e. O_2. However, this can be done by other metals. Indeed, if Co were
more available, perhaps coboglobin might have been developed by nature to
transport oxygen. The inorganic chemical properties of the cyclic tetrapyrroles are not well developed in benzene - although dibenzene chromium
is known.

As with hexatriene, the photochemistry of the bilins is dominated by isomerization reactions. These are influenced by the mode of joining to a
protein. In the same way the redox properties of iron porphyrins are modified by the protein in which they are held. Thus, judging from the papers
of Zuber and Rapoport, the linkage of the bilin to the protein modulates
the rôle of the bilin in energy transport in a manner analogous to how the
linkage of heme to the protein modulates the electrochemical potential.
In many ways the chlorophylls are intermediate between hemes and bilins:
their biological variability is determined through side chains rather than
ligands.

With regard to techniques of study, we find similarity is greater than differences. Quantum theory has played a useful rôle for porphyrins and will
do so with the open tetrapyrroles. Quantum theory will surely be able to
predict spectral changes with isomerization just as it did much to help
understand porphyrin behaviour. X-ray study and structural study of the
protein-chromophore complexes will be necessary for a full understanding
just as with hemoglobin/myoglobin, the first proteins studied by X-ray
technique.

Thus, between the linear and the cyclic tetrapyrroles there are major differences but also major similarities. This conference has exposed them
both and has been very stimulating for that reason. Let us thank Professors
Blauer and Sund for organizing this conference and also the people in
Konstanz who were remarkable hospitable. And let us reconvene to explore
similarities and differences in 1994.

Index of Contributors

Adler, A.D. 3*

Andreoni, A. 121*

Beychok, S. 60,152,173,175*, 200,201,276

Blauer, G. 137*,152,153,155, 174,200,256,257,277,309, 328,422

Bode, W. 200,259,408,445*

Braslavsky, S.E. 367*

Buchler, J.W. 89,91*,104,105, 107*,119,153,173,225,277, 293,364,421,442,473,505

Cubeddu, R. 121*,133-136

Dörnemann, D. 43*,60,488,489*, 505

Dörr, F. 383*

Eaton, W.A. 227*

Eilfeld, P. 119,295,348,349*, 364-366

Falk, H. 40,258,281*,293-295, 308,347,348,364,366,423

Fuhrhop, J.-H. 16,19*,40,41, 105,119,174,277,294,309, 329,345,423,472

Geiselhart, P. 383*

Glazer, A.N. 411*

Gottstein, J. 507*

Gouterman, M. 3*,17,18,63*,89, 90,104,119,152,258,295,309, 327,347,380,381,408,442,449, 523*

Gregory, R.P.F. 153,475*,488, 505

Hefferle, P. 383*

Hoff, A.J. 17,134,380,442, 453*,472,473,505

Holten, D. 63*

Holzwarth, A.R. 135,174,295, 309,328,367*,380-382,409, 442,472

Huber, R. 408,421,442,445*

John, W. 383*

Ketterer, B. 136,261*,276-278,328

Kisslinger, J. 107*

Knoff, M. 91*

Köst, H.P. 422

Köst-Reyes, E. 397*

Lamola, A.A. 17,89,133,277, 294,308,311*,327-330,346, 347,365,381

Lehmann, T. 19*

Lightner, D.A. 297*

McDonagh, A.F. 134,155,297*, 308-310,329,364,422,522

Mindl, T. 383*

Myer, Y.P. 40,104,120,155, 173,201,203*,225,226,328, 346

Nagai, K. 157*,173,174

Otto, J. 397*

Paulke, B. 507*

*Symposium paper

Rapoport, H.   60,226,310,330,
    346,411*,421-423,505
Rüdiger, W.   349*
Schaffner, K.   367*
Scharnagl, Ch.   397*
Scheer, H.   40,60,89,154,257,
    294,308,327,347,364,365,
    383*,409,410,422,442,472,
    488,505,507*,521,522
Schejter, A.   227*
Schirmer, T.   445*,449
Schneider, S.   327,364,382,
    383*,397*,407-409,442
Senger,H.   43*,489*
Sidler, W.   445*

Song, P.-S.   17,41,89,136,
    152,257,295,327,331*,347-
    349,365,380-382,407,421
Sund, H.   17,155,309,409,442
Thümmler, F.   349*
Vig, I.   227*
Vogler, A.   17,89,107*,119,
    120,309
Woody, R.W.   17,41,152,153,
    200,226,239*,256-259,277,
    407,472,488,521,522
Zuber, H.   346,409,425*,442,
    443,445*,

Subject Index

Absorbance-detected magnetic resonance (ADMR) 453,464

Absorption difference spectra 73

  picosecond transition 64

Actinide porphyrins 12

Aetiobiliverdin-IV-γ 282

Aq complexes 85

Aggregates

  ferriheme 138-155
  quinidine 138-155

Albumin 262-264,303-305,310, 312-324

  bilirubin binding 312-324
  fluorescence lifetime of bilirubin-albumin complex 317-321
  tetrapyrrole binding 262-264

Allophycocyanin 390,427-429, 432,433

Anemia

  pernicious 7,9

Antimalarial drugs 138

Apohemoglobin 175,200

Bacteriochlorophyll 507-522

  aqueous organic solvent induced complexes 512-515
  circular dichroism 510,512, 517
  electronic spectra 507
  long-wavelength forms 507-522
  magnesium 507,515

Bacteriochlorophyll b 516

Bacteriopheophytin 508

Bilatrienes-abc 281-295
  light absorption 281-295

Bile pigments 5,286

Biliproteins 411-423

Bilirubin 7,9-11,261-278, 311-331

  binding to albumin 312-331
  fluorescence lifetime of bilirubin-albumin complex 317-321
  photoisomerization 297-310
  photolysis 299
  transport by proteins 261-278

Biliverdin 7

Biliverdindimethylester 282

Binding of phytochrome to membrane 343

Biosynthesis

  chlorophyll 60
  phycocyanobilin 60
  porphyrins 11
  tetrapyrrole 43-60

Blood 3,6,9

Bolaamphiphiles 29,35

Carotene-porphyrin-quinone triad 28,36

Catalase 8

Ce porphyrins 95

Cell lines 122

Chain assembly of hemoglobin 175

Charge transfer   63-90
   band   288
   bands of hemoprotcins   234,
      235

Chiroptical properties   281

Chlorin   5,17

Chlorina mutant of barley 477

20-Chloro-13^2-hydroxy-chloro-
   phyll a   503

20-Chloro-methylpheophorbide
   497

Chlorophyll   3,5,7-10,17,475-
   488
   biosynthesis   60
   -chlorophyll interactions
      239-259
   circular dichroism   239-259,
      475-488
   complex with nitroaromatics
      23
   dimer   24
   flavin complex   23
   globin complexes   250
   peptide   476
   quinone complex   23,36
   substituted globins   240

Chlorophyll RC I   489-505
   fluorescence spectra   495-
      502
   optical properties   489-505
   structure   489-505

Chloroplasts   475-488

Chromopeptides   397-410

Chromophore
   F$_{430}$ 56
   structure of phytochrome
      349-366

Circular dichroism   138-155,
   175-201,239-259,397-410,475-
   488,510,512,517

Circular dichroism of
   chlorophyll   475-488
   chlorophyll-chlorophyll in-
      teractions   239-259
   heme-heme interactions   239-
      259
   hemoglobin   177-179
   myoglobin   177,178
   phycocyanin   397-410

Clostripain   181,182

Coboglobin   12

Co complexes   74,76

Compartimentation   56

Conformation   338

Conformational change of phytochrome   339

Corrins   5

Cu complexes   77,89

10-Cyano-aetiobiliverdin-IV-γ
   284

Cyanobacteria   426

Cytochrome   8,9,11,175-201
   b$_2$ 189
   b5   189
   c   203-226,229-231,233,235,
         236,242
      heme configuration   203-
         226
   c$_3$ 239,242,244,248-250,
         253,257

Dark revision of Pfr   339

Deoxyhemoglobin   244

Diastereomerism   286

Dielectric constant   243

Digitonin-deoxycholate   480

Di-hematoporphyrin-ether   121

2,3-Dihydrobilatrienes-abc
  281-295

2,3-Dihydrodioxobilin   420

3,4-Dihydropyrromethenone
  420

4,5-Dioxovalerate   51

Dipyrromethenone   313
  chromophore   302,308

Doubledecker structure   95,96

Elasto-optic modulator   481

Electron transfer   116,117
  excited state   107-120

Electronic spectra   109,112,
  507

Emission spectra   66

Enantiomeric preponderance
  290

ESR   227-237

Eu porphyrins   95

Evolution of hemoglobin   180

Excited singlet   63-90

Excited state
  electron transfer   107-120
  kinetics   63-90

Excited triplet   63-90

Exciton
  interactions 26,432,508
  splitting   520
  theory   240,243,249,253,458

F_{430} (chromophore)   56

Fatty acid binding protein
  269,270

Fe complexes   83

Fe(II)-His stretching   160

Ferric heme, five-coordinate
  167,173

Ferriheme-quinidine complexes
  138-155

Five-coordinate ferric haem
  167,173

Flavins   23

Fluorescence   123,128
  decay of phycocyanin   386,
    387,391,393
  depolarization   383-395
  lifetime of bilirubin-
    albumin complex   317-321
  porphyrins   8
  phycocyanin   383-395
  spectra of phytochrome 368-
    371
  time-resolved   123

Folding pathway of hemoglobin
  176,194

ß-Globin intron/exon structure
  181

Globin chain genes   176

Globin conformation and heme
  binding   176

Glutamate-1-semialdehyde   51

Ground-state absorption
  spectra   66,67

Hammett correlation   12

Haem (see heme)

Heavy atom effect   113

Hematoporphyrin   121-136

Heme   3,4,7,8,261-278

   binding   175-201
   binding domain   176,180,196
   circular dichroism   239-253
   configuration in cytochrome
    c   203-226
   dicyanide   181-184
   ferriheme-quinidine complexes   138-155
   five-coordinate ferric heme
   167,173
   heme interactions   239-259
   proteins   4,11,203-226
   transport by proteins   261-
   278

Hemichromogen   186

Hemoglobin   8,10,11,157-174,
175-201,231,233,239,244,247,
253,258

   apohemoglobin   175,200
   calculated CD spectra   248
   α chain   162,164
   ß chain   164,165
   chain assembly   175,177
   circular dichroism   175-201
   conformation   175-201
   deoxyhemoglobin   244
   evolution   180
   experimental CD spectrum
   248
   five-coordinate ferric heme
   167,173
   folding pathway   176,194
   Hb M Boston   167,168,170
   Hb M Iwate   167,168,170
   heme binding   175-201
   heme iron   203-226
   kinetics of globin folding
   194
   lamprey   175
   optical properties   175-201
   oxy-hemoglobin   244
   quaternary structure   161,
   164

R→T transition   163
sickle cell   11
T structure   166,170
T→R transition   170

Hemopexin   265,266,277

   tetrapyrrole binding   265,
   266,277

Hemoprotein   11,227-237

   change transfer bands   234,
   235
   conformation and configuration   203-226
   ESR   227-237
   near-infrared spectra   227-
   237
   oxidation-reduction potential   227-237

Hexa-coordinated heme iron
209,220

High-spin

   ferric porphyrins   219,221
   heme c systems   217,220,222

Histidine in heme binding 181

History of porphyrins   3-18

Human hemoglobin   245,252

Hydrogen-deuterium exchange
in phytochrome 335

Hydrogen-tritium exchange in
phytochrome   338

Hydrophobicity of phytochrome
342

10-Hydroxy-bacteriochloro-
phyll a   516,519

Hyperbilirubinemia   312,321

Hyperchromism   508

Hypochromic complexes   520

Infrared spectra   102

Interchromophoric interactions
 in hemoglobin   251

Iron   (see also Fe)
  heme iron   203-226

Z,E Isomer   323,328,329

(Z→E) Isomerization   318,319

Isophorcabilin   352,353,358,
 360

Jaundice   6
  neonatal   297-310

Kinetics of globin folding
 194

Lactam-lactim tautomerism   285

Lactim ether   285

Lamprey hemoglobin   175

Lanthanoid
  bisphthalocyaninates   91
  bisporphyrinates   91,101
  porphyrins   12,91-105

Laser   123,124

Ligand dissociation   71

Ligand-field excited state
 114

Ligandin   266-269
  tetrapyrrole binding   266-
  269

Light absorption
  spectra   138-155
  of bilatrienes-abc   281-295
  of 2,3-dihydrobilatrienes-
   abc   281-295

Light-harvesting
  antennas   440,442
  complex   475
  pigment-protein complexes
   425-443,454
    domain structure   436
    structural organization
     425-443
    tetrapyrrole pigments
     425-443
  polypeptides   437

Linker peptides   384,427

Low-spin ferric
  heme   227,232
  porphyrins   207,209-211

Luminescence quenching   110

Lutetium bis(phthalocyaninate)   98

Lysine
  poly-α,L-Lysine   139,144,
   150

Membranes
  synthetic   21
  vesicle   19-41

Metal complexes
  of porphyrins   74-90
  of tetrapyrroles   91-105

Metalloporphyrin   9

Methene tautomerism   284

Methylpheophorbide RC I   497

Mg complexes   133

Mn complexes   85

MO calculations   288

Mo complexes   85

Model chromophore   242

Molecular model of phyto-
   chrome   331-348

Myoglobin   175-177,229,231,
   233,242,247,258,448

Myohematin   8

Near infrared spectra   102,
   227-237

Neonatal jaundice   297-310

Ni complexes   72

NMR   289,338,340,347
   spectra of phytochrome 338,
      340,347

NOESY   289

Non-covalent interactions   150

Octaethylporphyrin   94,97,99

Optical probes   331-348
   of phytochrome   331-348

Optical spectra of bilirubin
   312

Os complexes   80,107-120

Osmochromes   108

Oxidation-reduction potential
   of hemoproteins   227-237

Oxygen binding   176,186

Oxyhemoglobin   244

π cation radicals   12

Pd complexes   84,90

Pentacoordinate
   ferrous state   176,180
   heme iron   220
   iron   203-226

Pentapyrrin   283

Pernicious anemia   7,9

Peroxidase   7-9,11

Pfr, see Pr/Pfr

Pheophorbide
   20-Cl-methylpheophorbide
      497

Pheophytin
   a   492,494
   RC I   492,494,495

Phosphorescence   113,115
   quenching   115

Phosphorous porphyrin   13

Photoassociation   70

Photochemical centers   475

Photochemistry   71,107,114

Photoconversion of phyto-
   chrome   349-366

Photodissociation   70

Photo-excited bilirubin   312

Photofrin II   121,134

Photoisomerization
   of bilirubin   297-310,312
   of bilirubin-albumin com-
      plex   318,320,323,324,329

Photooxidation   111,115,117

Photophysical processes of
   bilirubin   311-331

Photoreceptor   331

Photosensitization   131

Photosubstitution   110,114

Photosynthesis   5-7,10,12,
108,453-473
    models   20,35

Photosystem I and II   475-488

Phototherapy
    of neonatal jaundice   297-310
    of tumors   121-136

Phototransformation of phytochrome   331

Phthalocyanine   6,9,10,12,13

Phycobiliprotein   411-423,
425-443
    chromopeptides   413
    cysteinylthioether bond   417
    molecular organization   425-443
    primary structure   414-417
    structure   425-443
    thioether bond   415

Phycobilisome   383,384,426,
428,431,434
    antenna   440,442

Phycocyanin   351,354,355,364,
383-395,397-410,427
    aggregation   383-395
    allophycocyanin   390,427-429,432,433
    chromopeptides   397-410
    circular dichroism   397-410
    fluorescence
        decay   386,387,391,393
        depolarization   383-395
    intra-chromophore interac-
    tions   400
    primary structure   399

protein-chromophore inter-
    actions   397
quantum mechanical model
    calculations   399

C-Phycocyanin   412,417-419,
421,429-431,433,434,445-449
    α chain conformation   447
    β chain conformation   448
    comparison to hemoglobin
    446
    crystal structure   445-449
    evolution   445-449
    homology to myoglobin   449
    monomer conformation   447
    trimer conformation   449
    X-ray analysis   445-449

Phycocyanobilin   411
    biosynthesis   60

Phycoerythrin   428
    B   412
    C   412
    R   412

Phycoerythrobilin   411

Phycoerythrocyanin   429,430

Phycourobilin   411

Phytochrome   281,331-348,
349-366,367-382
    binding to membrane   343
    chromophore   332
        structure   349-366
    comparison with rhodopsin
    359,360
    conformation   338,339
    conformational change   339
    fluorescence spectra   368-371
    hydrogen-deuterium exchange
    335
    hydrogen-tritium exchange
    338
    hydrophobicity   342
    molecular model   331-348
    NMR   338,340,347

oxidation   333,345,346
photochemistry   367-382
photophysics   367-382
phototransformation   332,
  339,342,343
Pr/Pfr
  dark revision   339
  photoconversion   349-366
  phototransformation   367-
  382
    calorimetry   367-382
Pr/Pfr phototransformation
  kinetics   367-382
  photophysical parameter
  367-382
quasielastic light scattering   338
spectra   368-371
structure   349-366
TNM oxidation   333,345,346
topography   332,342

Picosecond

  spectroscopy   65
  transient absorption   64,83

Polarizability   242

  theory   241,248,249,253

Poly-α,L-lysine   139,144,150

Porphin   10,12

Porphobilinogen   11

Porphyria   8-10

Porphyriazine   6,11

Porphyrin   3-6,11,12,17,261-
278

  absorption, picosecond tran-
    sition   64
  actinide   12
  Ag complexes   85
  assemblies   19-41
  biosynthesis   11
  carotene-porphyrin-quinone
    triad   28,36
  Ce porphyrin   95
  Co complexes   74,76

covalent redox pairs   19-41
Cu complexes   77,89
dimers   22
domains   35
Eu porphyrins   95
Fe complexes   83
fluorescence   8
ground-state absorption
  spectra   67
hematoporphyrin   121-136
history   3-18
lanthanoid complexes   12,
  91-105
Mg complexes   133
Mn complexes   85
Mo complexes   85
molecular complexes   19-41
octaacetic acid   30,40,41
octaethylporphyrin   94,97,
  99
Os complexes   80,107,120
phosphorous   13
Pd complexes   84,90
Pt complexes   84
π-radical   100,102
Ru complexes   80
Sc complexes   100
spectra   3-18,63-90
transition metal complexes
  63-90
transport by proteins   261-
  278
Zn complexes   133

PPP calculations   284

Pr/Pfr

  chromophores   332-338
  photoconversion   349-366
    low-temperature spectro-
    scopy   356-359
  phototransformation   332,
  339,342,343,367-382
    calorimetry   367-382
    kinetics   367-382
    photophysical parameter
    367-382

Pr porphyrins   95

Primary structure of cyanin
399

Protein
  -chromophore system 283
  -heme interaction 215

Protochlorophyllide dimer
  24

Pt complexes 84

Purple photosynthetic bacteria 426

Pyrrochlorophyllide derivative of hemoglobin 250-252

Pyrrole 8

Pyrromethenone chromophores
  298,306

Quantum yields of
  emission 111
  photooxidation 112

Quinidine complexes 138-155

Quinone-chlorophyll complex
  23,26

Radicals
  π-cation 12

Reaction centers (photosynthesis) 453-473

Redox
  chain 35
  pairs 19-41

Resonance Raman Spectroscopy
  157,203-226

Rhodopsin 350,359,360

Rotatory strength 144

Ru complexes 80

Sc porphyrin 100

Shemin pathway 44

Sickle cell hemoglobin 11

Single-photon timing 123

Siroheme 12

Sodium dodecyl sulphate 477

Solar energy 5

Spectra of porphyrins 3-18

Stercobilin 10

Surface-active agents 477

Tetranitromethane oxidation
  of phytochrome 333,345,346

Tetra(p-tolyl)porphyrin 94

Tetrapyrrole 17,261-278
  binding to albumin 262-
    264
  biosynthesis 43-60
  metal complexes 91-105
  transport by proteins 261-
    278

Thylakoid membrane 475

Time-resolved fluorescence
  122,130,131

TNM oxidation of phytochrome
  333,345,346

Transient absorption spectra
  63-90

Transition metal porphyrins
  63-90

Tripledecker structure 96

Triplet-minus-singlet absorbance difference spectra
  463-468

Triplet state   453,462

Tumor phototherapy   121-136

Ultracentrifugation   138,
  140,148

Urobilin   8

Vapour pressure osmometry
  288

Vectorial redox-chain   35

Vesicles

  domains in   19-41
  regions in   28

unsymmetric bilayers   30
unsymmetric monolayers 34

Vitamin B12   5,11,12

Wood-Fickett-Kirkwood treatment 290

X-ray analysis of C-phycocyanin   445-449

Z,E isomer   323,328,329

Z-lumirubin   302,308,310

Zn complexes   133

Walter de Gruyter
Berlin · New York

H. Sund
(Editor)

Pyridine Nucleotide-Dependent Dehydrogenases

Proceedings of the 2nd International Symposium on Pyridine Nucleotide-Dependent Dehydrogenases. Konstanz, West Germany, March 28-April 1, 1977. FEBS Symposium No. 49

1977. 17 cm x 24 cm. IV, 529 pages. Numerous illustrations and tables. Hardcover. DM 155,–; approx. US $ 64.60

G. Blauer
H. Sund
(Editors)

Transport by Proteins

**Proceedings of a Symposium on Transport by Proteins.
Konstanz, West Germany, July 9–15, 1978.
FEBS Symposium No. 58**

1978. 17 cm x 24 cm. XV, 420 pages. Numerous figures and tables. Hardcover. DM 145,–; approx. US $ 60.40

D. Pette
(Editor)

Plasticity of Muscle

**Proceedings of a Symposium on Plasticity of Muscle.
Konstanz, West Germany, September 23–28, 1979.**

1980. 17 cm x 24 cm. XXVI, 625 pages. Numerous figures. Hardcover. DM 160,–; approx. US $ 66.70

H. Sund
C. Veeger
(Editors)

Mobility and Recognition in Cell Biology

**Proceedings of a FEBS Lecture Course held at the University of Konstanz, West Germany, September 6–10, 1982
FEBS Lecture Course No. 82/09**

1983. 17 cm x 24 cm. XII, 586 pages. Numerous illustrations. Hardcover. DM 190,–; approx. US $ 79.20

Price are subject to change without notice

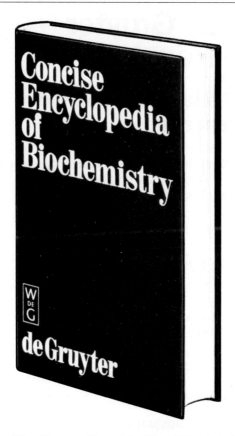